D0894633

THE COASTAL EVERGLADES

LONG-TERM ECOLOGICAL RESEARCH NETWORK SERIES

LTER Publications Committee

Grassland Dynamics: Long-Term Ecological Research in Tallgrass Prairie
Editors: Alan K. Knapp, John M. Briggs, David Hartnett, Scott L. Collins

Standard Soil Methods for Long-Term Ecological Research
Editors: G. Philip Robertson, David C. Coleman, Caroline S. Bledsoe, Phillip Sollins

Structure and Function of An Alpine Ecosystem: Niwot Ridge, Colorado
Editors: William D. Bowman, Timothy R. Seastedt

Climate Variability and Ecosystem Response at Long-Term Ecological Sites
Editors: David Greenland, Douglas G. Goodin, Raymond C. Smith

Biodiversity in Drylands: Toward A Unified Framework
Editors: Moshe Shachak, Steward T. A. Pickett, James Gosz, Avi Perevolotski

Long-Term Dynamics of Lakes in the Landscape: Long-Term Ecological Research on North Temperate Lakes
Editors: John J. Magnuson, Timothy K. Kratz, Barbara Benson

Alaska's Changing Boreal Forest
Editors: F. Stuart (Terry) Chapin, Mark W. Oswood, Keith Van Cleve, Leslie A. Viereck, Dave Verbyla

Structure and Function of A Chihuahuan Desert Ecosystem: The Jornada Basin Long-Term Ecological Research Site
Editors: Kris M. Havstad, Laura F. Huenneke, William H. Schlesinger

Principles and Standards For Measuring Net Primary Production in Long-Term Ecological Studies
Editors: Timothy J. Fahey and Alan K. Knapp

Agrarian Landscapes in Transition: Comparisons of Long-Term Ecological and Cultural Change
Editors: Charles Redman and David Foster

Ecology of the Shortgrass Steppe: A Long-Term Perspective
Editors: William K. Lauenroth and Ingrid C. Burke

A Caribbean Forest Tapestry: The Multidimensional Nature of Disturbance and Response
Editors: Nick Brokaw, T. Crowl, A. Lugo, W. McDowell, F. Scatena, R. Waide, M. Willig

Long-Term Response of A Forest Watershed Ecosystem: Clearcutting in the Southern Appalachians
Editors: Wayne Swank and Jackson Webster

Alaska's Changing Arctic: Ecological Consequences for Tundra, Streams, and Lakes
Editors: John Hobbie, Anne Giblin, Goerge Kling, Chris Luecke, Bruce Peterson, Gus Shaver

The Ecology of Agricultural Landscapes: Long-Term Research on the Path to Sustainability
Editors: Steven K. Hamilton, Julie E. Doll, Phil Robertson

The Coastal Everglades

THE DYNAMICS OF SOCIAL-ECOLOGICAL TRANSFORMATION IN THE SOUTH FLORIDA LANDSCAPE

Edited by Daniel L. Childers, Evelyn E. Gaiser, and Laura A. Ogden

Oxford University Press is a department of the University of Oxford. It furthers the University's objective of excellence in research, scholarship, and education by publishing worldwide. Oxford is a registered trade mark of Oxford University Press in the UK and certain other countries.

Published in the United States of America by Oxford University Press
198 Madison Avenue, New York, NY 10016, United States of America.

CIP data is on file at the Library of Congress
ISBN 978–0–19–086900–7

9 8 7 6 5 4 3 2 1

Printed by Integrated Books International, United States of America

The authors thank Stephen E. Davis III for use of his photograph for the cover of this book.

Contents

Preface

THIS BOOK PRESENTS a broad overview and synthesis of research on the coastal Everglades, a region that includes Everglades National Park, adjacent managed wetlands, and agricultural and urbanizing communities. Though the Everglades ecosystem spans nearly a third of the state of Florida, our focus on the coastal Everglades allows us to examine key questions in social-ecological science in the context of ongoing restoration initiatives. As this book demonstrates, our long-term research has facilitated a better understanding of the roles of sea level rise (SLR), water management practices, urban and agricultural development, and other disturbances, such as fires and storms, on the past and future dynamics of this coastal environment.

Contributors for this volume are all collaborators on the Florida Coastal Everglades Long Term Ecological Research Program (FCE LTER). The FCE LTER began in 2000 with a focus on understanding key ecosystem processes in a coastal wetland, while also developing a platform for and linkages to related work conducted by an active and diverse Everglades research community. The program is based at Florida International University in Miami but includes scientists and students from numerous other universities as well as staff scientists at key resource management agencies, including Everglades National Park and the South Florida Water Management District. In fact, the original proposal was conceived and written in 1998–99 with many of these agency scientists. Because of this collaborative approach, FCE LTER findings have had direct and often immediate impacts on resource management and restoration decision making. In this preface we present a brief history of the first 17 years of the FCE LTER and highlight key findings.

In the nine chapters of the book we present those findings using a broadly integrative approach. Notably, this book is not a review of everything FCE LTER researchers have learned in the last 17 years. Rather, it is an easily accessible synthesis of the impact of the FCE LTER on the environment and society of south Florida, on the understanding the ecology of coastal wetlands, and on advancing ecosystem science and theory.

Our primary research objective in the first 6-year phase of funding (FCE I; 2000–2006) was to determine how fresh water from oligotrophic marshes interacted with a marine source of the limiting nutrient, phosphorus (P), to control productivity in the oligohaline ecotone, the zone where fresh water and marine water mix in the coastal Everglades. *We hypothesized that ecosystem productivity would be greatest where fresh water, higher in nitrogen and dissolved organic matter (DOM) relative to coastal waters, meets marine waters, where P was more available.* A quasi-Lagrangian sampling design allowed us to track water flow and ecosystem properties along the main wetland drainages in Everglades National Park: Taylor Slough-Panhandle (TS/Ph) and Shark River Slough (SRS). Both transects extended from freshwater canal inputs to the Gulf of Mexico and the TS/Ph transect included Florida Bay (Fig. P.1). Our research documented a productivity peak in the oligohaline ecotone of the TS/Ph transect (Childers et al. 2006a) that was likely driven by brackish groundwater delivery of P to ecotone wetland and aquatic ecosystems (Price et al. 2006). In contrast, the SRS transect exhibited a wedge of

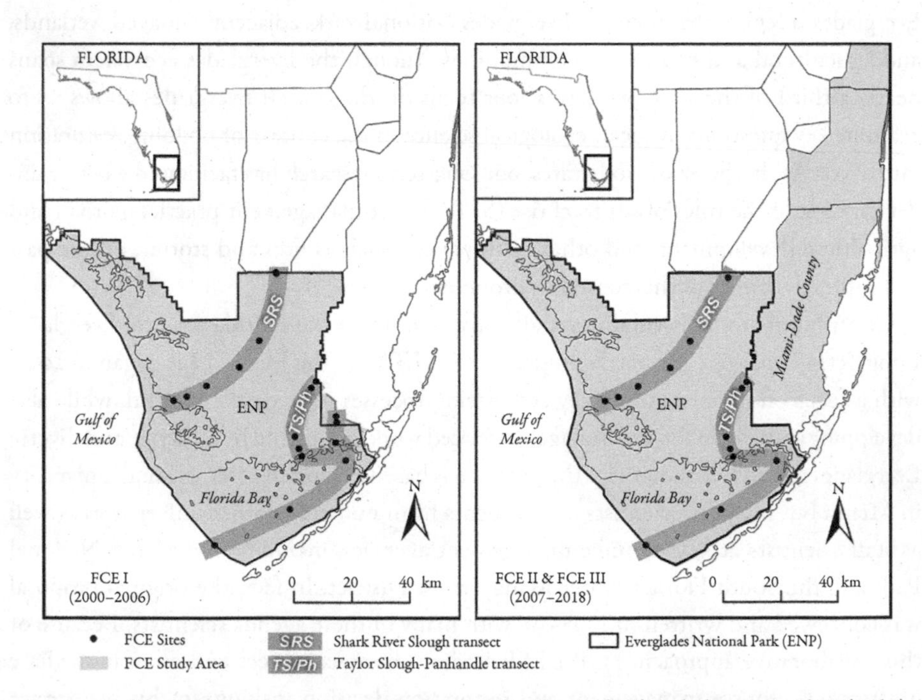

FIGURE P.1 The FCE LTER Program study domains for Phase I (left) and Phases II and III (right). Note that the light gray study area went from only including Everglades National Park (ENP) in Phase I to also including the urbanized coastal regions in Phases II and III.

increasing productivity toward the coast (Ewe et al. 2006), where tidal influences were strongest and the source of P was greatest (Childers et al. 2006b). This work demonstrated how Everglades estuaries are functionally "upside-down" relative to the classic estuary model, with seawater supplying limiting nutrients landward, rather than the other way around (Childers et al. 2006b; Trexler et al. 2006). These findings led us to expand our interdisciplinary research to examine how the restoration of freshwater flow would interact with climate variability to influence the shape of these estuarine productivity gradients.

Research conducted during our second 6-year funding cycle (FCE II; 2006–2012) was organized around hypotheses that could be tested through restoration projects aimed at enhancing freshwater inflow to the SRS transect, but not to the TS/Ph transect, during this period. This landscape-scale manipulation provided a "grand experiment" to test the hypothesis that *increasing inputs of freshwater will interact with climate and disturbance regimes to reduce nutrient supply, production, and detrital transport in nutrient-poor coastal systems.* Although this hypothesis remains central to FCE research, restoration plans did not proceed as expected. Anticipated restoration projects were stalled and reduced in scope, and new projects were considered. In lieu of a "grand experiment," we discovered that delays in freshwater restoration increased rates of saltwater encroachment into the freshwater Everglades, in both surface waters and groundwater. This intrusion of marine water increased the concentrations of P in the oligohaline ecotone and led to an increase in the density of salt-tolerant mangroves (Ross et al. 2000). Notably, long-term data from our eddy flux tower located in a mangrove forest near the Gulf of Mexico also showed that this forest sequestered more CO_2 from the atmosphere than most other ecosystems on the planet (Barr et al. 2011). Additionally, FCE research has shown that the organic material produced by these coastal plant communities is more prone to oxidation when freshwater flow is reduced and salinity exposure is enhanced (Chambers et al. 2013) These changes in the abundance of organic matter also influenced animal populations, including large predators that feed on detrital marine food webs and then transport this energy up-estuary when they migrate to freshwater marshes (Rosenblatt and Heithaus 2011).

During FCE II we also expanded our interdisciplinary reach considerably by incorporating social science into our core mission and by expanding the FCE boundaries beyond Everglades National Park to include southern Miami-Dade County (see Fig. P.1). This social-ecological growth in our research and perspective has been powerful and rewarding. For example, it has revealed how south Florida residents respond to changes in the Everglades. We found that those responses have sometimes led to conflicts among stakeholders over land and resource distribution, and that these conflicts ultimately delayed the restoration process (Ogden 2008). FCE research is also unique in its partnerships with government agency scientists who have "co-produced" FCE science with academic researchers since its inception. Through these valuable collaborations, we have begun to project future scenarios for the Everglades under a range of restoration alternatives, management plans, and climate change conditions. These scenario exercises have enabled us to inform restoration decisions that benefit both the Everglades and the people of south Florida who depend on the Everglades.

Integrated social-ecological conceptual models help define patterns and relationships in complex, dynamic systems while providing a context for formulating testable hypotheses at disciplinary levels to promote discoveries that reduce uncertainty (Carpenter et al. 2009). In our third phase of funding (FCE III; 2012–2018), we used such a conceptual framework that mechanistically linked the human and biophysical domains (*sensu* the press–pulse dynamics model of Collins et al. 2011) across spatial and temporal scales. Our conceptual model was developed with the recognition that humans and biophysical processes are intertwined at local to global scales, and the model showed them in the same "box" in order to emphasize joint vulnerabilities and relational dynamics. Spatial scales were connected through hydrodynamics (the relative supply of fresh and marine water to south Florida), which was the central focus of our FCE III research. We applied the model to three temporal contexts (past, present, future), allowing us to examine the effects of legacies on modern trends and future predictions (per Chapin et al. 2001). FCE III activities addressed four main goals: (1) to evaluate the source of sociopolitical conflicts over freshwater distribution, and how solutions that improve inflows to the Everglades mediate the effects of climate change (especially SLR) on freshwater sustainability in the coastal zone; (2) to determine how the balance of fresh and marine water supplies to the oligohaline ecotone will control the rates and pathways of C sequestration, storage, and export by influencing P availability, water residence time, and salinity; (3) to characterize spatial-temporal patterns in ecosystem sensitivity to, and legacies of, modifications of freshwater delivery to the Everglades that are driven by climate variability and land use change; and (4) to develop future scenarios of freshwater distribution and use that maximize the sustainability of the south Florida region in the face of global climate change, particularly SLR. Of particular note, SLR was a major focus of our FCE III research because south Florida is, literally, ground zero for its impacts.

Relative to these goals, FCE research continues to provide evidence of rapid shifts in ecosystem function in response to saltwater encroachment in the absence of large-scale restoration. We have shown that the chief impediment to restoring freshwater flows is the conundrum inherent in operating a water management system for multiple, often conflicting purposes, including flood control, water supply, and ecosystem health—a problem compounded by the technical challenges of massive-scale water control in southeast Florida. Ultimately, ecosystem restoration may be impossible without challenging flood protection entitlement or accounting for farmers' and residents' decision-making rationales and environmental attitudes (Schwartz 2014). Indeed, partnerships with regional farmers and policy makers have led to better management practices that have resulted in P reductions from the Everglades Agricultural Area. In addition, urban growth studies have shown that zoning has specific and quantifiable impacts that can improve forecasts of the effects of land and water use changes in the region and thereby improve decisions (Onsted and Roy Chowdhury 2014). These historical socioeconomic development patterns and constraints not only influence current decisions about restoration but will cause SLR adaptation to have strong social and environmental justice implications in South Florida (Craumer et al. 2014).

In the absence of freshwater restoration that is sufficient to improve freshwater delivery to Everglades estuaries, saltwater continues to intrude to the interior, particularly through groundwater, which is mobilizing P from the limestone bedrock into the soils and water of the ecotone (Koch et al. 2014; Zapata-Rios and Price 2012). This increase in availability of the limiting nutrient may stimulate further landward transgression of mangroves and account for increased accretion rates and reduced water depths there (Saunders et al. 2014; Smoak et al. 2013). However, there is increasing evidence that freshwater marsh soils exposed to new marine water sources may experience accelerated decomposition, leading to "peat collapse" (Chambers et al. 2013). Because peat collapse may be a positive feedback to saltwater intrusion, it is a key uncertainty in models of ecosystem response to SLR and has become a focus for FCE manipulative studies in both mesocosm and field experiments. While managed water releases into Everglades National Park do not appear to be offsetting intrusion, signs of improvement have been observed in the upper watershed of Taylor Slough, where some restoration of freshwater flows appears to be causing a convergence of trends in sawgrass (*Cladium*) biomass and productivity (Sullivan et al. 2014; Troxler et al. 2014), although these hydrologic improvements appear to also be associated with nutrient enrichment (Gaiser et al. 2014). Modeling efforts are focused on determining practical targets for freshwater flow restoration that maintain ambient nutrient concentrations and maximize soil stability in the ecotone under existing climate patterns (Briceño et al. 2013). Moving forward, specific restoration options will be combined with selected SLR trajectories in a scenario framework to explore possible futures for the Everglades. Thus far, analyses of variance in regional sea level data suggest that we will be on a statistically identifiable trajectory by 2020 to 2030 (Haigh et al. 2014), reinforcing the importance of long-term data to forecasting both change trajectories and their uncertainty.

The ability of the FCE program to advance theory on the dynamics of social-ecological systems while also informing and learning from the decision-making process has been largely cultivated through partnerships. Partnerships within the U.S. LTER Network were instrumental in developing a framework for long-term information collection and synthesis that embraces studies of the critical human dimension of ecosystem transformation (Collins et al. 2011). Network collaborations influenced the direction of FCE's programmatic evolution through the exchange of ideas and involvement of researchers with perspectives from the humanities to the physical sciences. In turn, the reliance of the FCE program on partnerships between academic and agency scientists to understand and influence transformation of this expansive but human-dominated ecosystem has enabled it to serve as a model of successful conveyance of research into policy making (i.e., Sullivan et al. 2014). Finally, FCE researchers have participated widely in international LTER

Network research and leadership, primarily because some of our most globally threatened and vulnerable coastlines are in the nearby and similarly structured tropical environments of the Caribbean and South America. By comparing properties of the Everglades with other subtropical and tropical wetlands, we have challenged ideas of novelty while revealing properties of ecosystems at the ends of gradients that are often ignored. We have also provided insights from and encouragement for long-term collaborative studies that inform resource management in similarly threatened coastal wetlands.

The FCE LTER Program has been funded by the National Science Foundation since its inception in 2000 (Grants No. DEB-1237517, DBI-0620409, and DEB-9910514). Additional funding for FCE-related research and activities has been generously provided by Everglades National Park, the South Florida Water Management District, the U.S. Geological Survey, the National Oceanic and Atmospheric Administration, the U.S. Department of Energy, Florida Seagrant, and others.

Daniel L. Childers
Evelyn E. Gaiser
Laura A. Ogden

References

Barr, J.G., V. Engel, T.J. Smith, and J.D. Fuentes. 2011. Hurricane disturbance and recovery of energy balance, CO_2 fluxes and canopy structure in a mangrove forest of the Florida Everglades. *Agricultural and Forest Meteorology* 153: 54–66.

Briceno, H.O., J.N. Boyer, J. Castro, and P. Harlem. 2013. Biogeochemical classification of south Florida's estuarine and coastal waters. *Marine Pollution Bulletin* 75: 187–204.

Carpenter, S. R., H.A. Mooney, J. Agard, D. Capistrano, R.S. DeFries, S. Diaz, T. Dietz, A.K. Duraiappah, A. Oteng-Yeboah, H.M. Pereira, C. Perrings, W.V. Reid, J. Sarukhan, R.J. Scholes, and A. Whyte. 2009. Science for managing ecosystem services: beyond the Millennium Ecosystem Assessment. *Proceedings of the National Academy of Sciences USA* 106: 1305–1312.

Chambers, R.M., R.L. Hatch, and T. Russell. 2014. Effect of water management on interannual variation in bulk soil properties from the eastern coastal Everglades. *Wetlands* 34(Suppl 1): 47–54.

Chapin, F.S., III, O. Sala, and E. Huber-Sannwald. 2001. *Global Biodiversity in a Changing Environment: Scenarios for the 21st Century*. New York: Springer-Verlag.

Childers, D.L., J.N. Boyer, S.E. Davis III, C.J. Madden, D.T. Rudnick, and F.H. Sklar. 2006a. Relating precipitation and water management to nutrient concentration patterns in the oligotrophic "upside down" estuaries of the Florida Everglades. *Limnology and Oceanography* 51(1): 602–616.

Childers, D.L. 2006b. A synthesis of long-term research by the Florida Coastal Everglades LTER Program. *Hydrobiologia* 569(1): 531–544.

Collins, S.L., S.R. Carpenter, D.L. Childers, T.L. Gragson, N.B. Grimm, J.M. Grove, S.L. Harlan, A.K. Knapp, G.P. Kofinas, J.J. Magnuson, W.H. McDowell, J.M. Melack, L.A. Ogden, D. Ornstein, G.P. Robertson, M.D. Smith, S.R. Swinton, and A. Whitmer. 2011. An integrated conceptual framework for socio-ecological research. *Frontiers in Ecology & the Environment* 9(6): 351–357.

Craumer, P.R., H. Gladwin, and S. Mic. 2014. The sea in the 'hood: spatio-temporal modeling of water level and settlement in Miami-Dade County, Florida. Annual Meeting of the Association of American Geographers, Tampa, Florida, April 11, 2014.

Ewe, S.M.L., E.E. Gaiser, D.L. Childers, V.H. Rivera-Monroy, D. Iwaniec, J. Fourqurean, and R.R. Twilley. 2006. Spatial and temporal patterns of aboveground net primary productivity (ANPP) along two freshwater-estuarine transects in the Florida Coastal Everglades. *Hydrobiologia* 569(1): 459–474.

Gaiser, E.E., P. Sullivan, F. Tobias, A. Bramburger, and J.C. Trexler. 2014. Boundary effects on benthic microbial phosphorus concentrations and diatom beta diversity in a hydrologically-modified, nutrient-limited wetland. *Wetlands* 34(Suppl 1): 55–64.

Haigh, I.D., T. Wahl, E.J. Rohling, R.M. Price, C.B. Pattiaratchi, F.M. Calafat, and S. Dangendorf. 2014. Timescales for detecting a significant acceleration in sea level rise. *Nature Communications* 5: 3635.

Koch, G.R., S. Hagerthy, E. Gaiser, and D.L. Childers. 2014. Examining Everglades floc transport using a sediment tracing technique. *Wetlands* 34: S123–S133.

Ogden, L.A. 2008. The Everglades ecosystem and the politics of nature. *American Anthropologist* 100: 21–32.

Onsted, J., and R. Roy Chowdhury. 2014. Does zoning matter? A comparative analysis of landscape change in Redland, Florida using cellular automata. *Landscape and Urban Planning* 121: 1–18.

Price, R.M., P.K. Swart, and J.W. Fourqurean. 2006. Coastal groundwater discharge—an additional source of phosphorus for the oligotrophic wetlands of the Everglades. *Hydrobiologia* 569: 23–36.

Rosenblatt, A.E., and M.R. Heithaus. 2011. Does variation in movement tactics and trophic interactions among American alligators create habitat linkages? *Journal of Animal Ecology* 80: 786–798.

Ross, M.S., J.F. Meeder, J.P. Sah, L.P. Ruiz, and G.J. Telesnicki. 2000. The Southeast saline Everglades revisited: 50 years of coastal vegetation change. *Journal of Vegetation Science* 11: 101–112.

Saunders, C.J., M. Gao, and R. Jaffé. 2015. Environmental assessment of vegetation and hydrological conditions in Everglades freshwater marshes using multiple geochemical proxies. *Aquatic Sciences* 77(2): 271–291.

Schwartz, K. The anti-politics of biopolitical disaster on Florida's coasts. Western Political Science Association, Seattle, Washington, April 2–4, 2014.

Smoak, J.M., J.L. Breithaupt, T.J. Smith, and C.J. Sanders. 2013. Sediment accretion and organic carbon burial relative to sea-level rise and storm events in two mangrove forests in Everglades National Park. *Catena* 104: 58–66.

Sullivan, P., E.E. Gaiser, D.D. Surratt, D.T. Rudnick, S.E. Davis, and F.H. Sklar. 2014. Wetland ecosystem response to hydrologic restoration and management: the Everglades and its urban-agricultural boundary (FL, USA). *Wetlands* 34: 1–8.

Trexler, J.C., E.E. Gaiser, and D.L. Childers. 2006. Interaction of hydrology and nutrients in controlling ecosystem function in oligotrophic coastal environments of South Florida. *Hydrobiologia* 569(Special Issue): 1–2.

Troxler, T., D.L. Childers, and C.J. Madden. 2014. Drivers of decadal-scale change in southern Everglades wetland macrophyte communities of the coastal ecotone. Wetlands 34(Suppl 1): 81–90.

Zapata-Rios, X., and R.M. Price. 2012. Estimates of groundwater discharge to a coastal wetland using multiple techniques: Taylor Slough, Everglades National Park, USA. *Hydrogeology Journal* 20: 1651–1668.

Contributors

Bill Anderson
College of Arts, Sciences & Education and Department of Earth and Environment
Florida International University
Miami, Florida 33199
andersow@fiu.edu

Jordan Barr
Elder Research
Charlottesville, Virginia 22903
jordangbarr777@gmail.com

James Beerens
U.S. Geological Survey
Wetland and Aquatic Research Center
Fort Lauderdale, Florida 33314
jbeerens@usgs.gov

Ross Boucek
Bonefish & Tarpon Trust
Marathon, Florida 33050
ross@bonefishtarpontrust.org

Joseph N. Boyer
Center for Research & Innovation
Plymouth State University
Plymouth, New Hampshire 03264
jnboyer@plymouth.edu

Laura Brandt
U.S. Fish and Wildlife Service
Davie, Florida 33314
laura_brandt@fws.gov

Josh Breithaupt
Department of Biology
University of Central Florida
Orlando, Florida 32816
Joshua.Breithaupt@ucf.edu

Henry Briceño
Southeast Environmental Research Center
Florida International University
Miami, Florida 33199
bricenoh@fiu.edu

Mike Bush
Audubon Dakota
Fargo, North Dakota 58102
mikerbush@gmail.com

Edward Castañeda-Moya
Southeast Environmental
Research Center
Florida International University
Miami, Florida 33199
edward.castaneda@fiu.edu

Jessica Cattelino
Department of Anthropology
University of California, Los Angeles
Los Angeles, California 90095
jesscatt@anthro.ucla.edu

Randolph Chambers
Keck Environmental Lab
College of William & Mary
Williamsburg, Virginia 23188
rmcham@wm.edu

Sean Charles
Department of Biological Sciences
Florida International University
Miami, Florida 33199
scharo56@fiu.edu

Daniel L. Childers
School of Sustainability
Arizona State University
Tempe, Arizona 85287
dan.childers@asu.edu

Rinku Roy Chowdhury
Graduate School of Geography
Clark University
Worcester, Massachusetts 01610
RRoyChowdhury@clarku.edu

Ligia Collado-Vides
Department of Biological Sciences
Florida International University
Miami, Florida 33199
colladol@fiu.edu

Mark Cook
South Florida Water Management
District
West Palm Beach, Florida 33406
mcook@sfwmd.gov

Carlos Coronado
South Florida Water Management
District
West Palm Beach, Florida 33406
ccoron@sfwmd.gov

Stephen E. Davis
III
Everglades Foundation
Palmetto Bay, Florida 33157
sdavis@evergladesfoundation.org

Tom Dreschel
John F. Kennedy Space Center
Kennedy Space Center, Florida 32899
thomas.w.dreschel@nasa.gov

Carl Fitz
EcoLandMod, Inc.
Fort Pierce, Florida 34946
carlfitz3@gmail.com

James Fourqurean
Center for Coastal Oceans Research and
Department of Biological
Sciences
Florida International University
Miami, Florida 33199
fourqure@fiu.edu

Tom Frankovich
Southeast Environmental Research
Center
Florida International University

Miami, Florida 33199
Thomas.Frankovich@fiu.edu

Jose D. Fuentes
Department of Meteorology and Atmospheric Science
Pennsylvania State University
University Park, Pennsylvania 16802
jdfuentes@psu.edu

Evelyn Gaiser
Department of Biological Sciences and Southeast Environmental Research Center
Florida International University
Miami, Florida 33199
gaisere@fiu.edu

Michael Heithaus
College of Arts, Sciences & Education and Department of Biological Sciences
Florida International University
Miami, Florida 33199
heithaus@fiu.edu

Charles Hopkinson
Department of Marine Sciences
University of Georgia
Athens, Georgia 30602
chopkins@uga.edu

Rudolf Jaffé
Department of Chemistry and Biochemistry and Southeast Environmental Research Center
Florida International University
Miami, Florida 33199
jaffer@fiu.edu

Gregory R. Koch
Zoo Miami Foundation
Miami, Florida 33177
gregrkoch@gmail.com

John Kominoski
Department of Biological Sciences and Southeast Environmental Research Center
Florida International University
Miami, Florida 33199
jkominos@fiu.edu

Laurel Larsen
Department of Geography
University of California, Berkeley
Berkeley, California 94720
laurel@berkeley.edu

Christopher Madden
South Florida Water Management District
West Palm Beach, Florida 33406
cmadden@sfwmd.gov

Sparkle Malone
Department of Biological Sciences
Florida International University
Miami, Florida 33199
sparkle.malone@fiu.edu

Philip Matich
Texas Research Institute for Environmental Studies
Sam Houston State University
Huntsville, Texas 77341
matich.philip@shsu.edu

Agnes McLean
Everglades National Park
National Park Service
Homestead, Florida 33030
Agnes_McLean@nps.gov

Christopher McVoy
South Florida Engineering and Consulting
Lake Worth, Florida 33460
cmcvoy@gmail.com

Gregory B. Noe
U.S. Geological Survey
Hydrological-Ecological Interactions Branch
Reston, Virginia 20192
gnoe@usgs.gov

Steve Oberbauer
Department of Biological Sciences
Florida International University
Miami, Florida 33199
Steven.Oberbauer@fiu.edu

Laura A. Ogden
Department of Anthropology
Dartmouth College
Hanover, New Hampshire 03755
Laura.A.Ogden@dartmouth.edu

Jeff Onsted
California Department of Conservation
Sacramento, California 95814
jeffrey.onsted@conservation.ca.gov

René Price
Department of Earth and Environment
Florida International University
Miami, Florida 33199
pricer@fiu.edu

Jennifer Rehage
Department of Earth and Environment
Florida International University
Miami, Florida 33199
rehagej@fiu.edu

Jennifer Richards
Department of Biological Sciences
Florida International University
Miami, Florida 33199
richards@fiu.edu

Victor Rivera-Monroy
Department of Oceanography and Coastal Sciences
Louisiana State University
Baton Rouge, Louisiana 70803
vhrivera@lsu.edu

Adam Rosenblatt
Biology Department
University of North Florida
Jacksonville, Florida 32224
adam.rosenblatt@unf.edu

Dave Rudnick
Everglades National Park
National Park Service
Homestead, Florida 33030
david_rudnick@nps.gov

Jay Sah
Southeast Environmental Research Center
Florida International University
Miami, Florida 33199
jay.sah@fiu.edu

Amartya Saha
MacArthur Agro-ecology Research Center
Archbold Biological Station
Lake Placid, Florida 33852
riparianbuffer@gmail.com

Nick Schulte
Environmental Studies Program
University of Colorado Boulder
Boulder, Colorado 80303
Nicholas.Schulte@Colorado.edu

Katrina Schwartz
Environmental Social Scientist
Collaborator, Florida Coastal Everglades LTER Program
Chapel Hill, North Carolina 27517
Katrina.Z.S.Schwartz@gmail.com

Len Scinto
Department of Earth and Environment and Southeast Environmental Research Center
Florida International University
Miami, Florida 33199
scintol@fiu.edu

Fred Sklar
South Florida Water Management District
West Palm Beach, Florida 33406
fsklar@sfwmd.gov

Joseph Donny Smoak
Environmental Science
University of South Florida
St. Petersburg, Florida 33701
smoak@mail.usf.edu

Greg Starr
Department of Biological Sciences
University of Alabama
Tuscaloosa, Alabama 35401
gstarr@ua.edu

Joel Trexler
Department of Biological Sciences
Florida International University
Miami, Florida 33199
Joel.Trexler@fiu.edu

Tiffany Troxler
Sea Level Solutions Center
Florida International University
Miami, Florida 33199
troxlert@fiu.edu

Kevin Whelan
South Florida/Caribbean Inventory & Monitoring Network
National Park Service
Palmetto Bay, Florida 33157
kevin_r_whelan@nps.gov

Walter Wilcox
South Florida Water Management District
West Palm Beach, Florida 33406
wwilcox@sfwmd.gov

Jeffrey R. Wozniak
Department of Biological Sciences
Sam Houston State University
Huntsville, Texas 77341
wozniak@shsu.edu

1

Introduction

Daniel L. Childers, Evelyn Gaiser, and Laura A. Ogden

MODERN-DAY SOUTH FLORIDA is iconic in many ways: the glitz and glamour of South Beach; the snaggle-tooth skyline of condo towers that line its shores; the lush tropical vegetation, azure waters, and single-season warm climate; the hustle, bustle, and variety of richly diverse human cultures; the endless beaches that invite relaxation; and, of course, the Everglades. All of the anthropocentric icons of the region, as well as all of the people who live there, are intrinsically linked to and dependent on the Everglades. Yet it is an interesting, if somewhat unsettling, contradiction of south Florida that so few people there are aware of this connection and codependence. In this chapter we summarize the history of the Everglades landscape, of human habitation and use of that landscape, and of the more recent modifications that have created a highly designed, engineered, and managed Everglades that is scarcely half the size it was 150 years ago. In 1947 Marjory Stoneman Douglas declared in her book *The Everglades: River of Grass* that "there are no other Everglades in the world." In many ways her quote is still true, because the manmade landscape that is the modern Everglades is also unique. In this volume we present a better understanding of how the nexus of scientific knowledge, societal demands, and human decisions has transformed and will continue to transform the Everglades landscape and the south Florida region.

The Everglades landscape is only about 6,000 years old. Up until this time, as post–Wisconsin Glaciation sea levels were rising rapidly and south Florida's land mass was shrinking equally rapidly, the land south of Lake Okeechobee was drier and was mainly oak savannah. A hint of what this landscape might have looked like at that time can be seen today in some parts of central Florida. But as the Holocene climate began to stabilize and sea level rise slowed about 6,000 years ago, south Florida began to receive more rain. Slowly the landscape transformed into a vast wetland, and Lake Okeechobee became much more connected to the Everglades proper as it more routinely overflowed to the south. And the Everglades was born.

Early indigenous peoples were in south Florida for this transformation, though we are only now beginning to understand the ways in which they adapted to and transformed this dynamic landscape. Their use of the landscape almost certainly changed as the climate and the land became wetter, but the Everglades has always known some degree of human habitation and manipulation. Up until several centuries ago, human uses and impacts were minimal relative to the scale of the landscape, but it is safe to say that the Everglades developed and evolved with input from *Homo sapiens*.

When the Spanish arrived in Florida in 1513, there were over 20,000 native people living in the Greater Everglades region, the majority and most powerful of whom were Calusa, who lived along the southwest coast (Griffin 1996). Much of the former Calusa territory is now within Everglades National Park (ENP). The Calusa, part of what archaeologists call the "Glades Culture," were prosperous and socially complex. Hernando de Escalante Fontaneda, who was a captive of the Calusa for 17 years, estimated that there were about 50 indigenous towns loosely networked into a tribute-paying political hierarchy (Griffin 1996).

The Calusa (as well as the people who lived for a few thousand years before them) built dramatic shell mounds and earthworks throughout the southwestern coast of Florida. These earthworks were built in varied configurations that served multiple purposes—including horseshoe-shaped protective jetties, inland transportation canals, and elevated plateaus used as village sites. Chokoloskee Island, a coastal fishing village, exemplifies the later type of earthwork. Chokoloskee is built on a 24.3-ha shell mound that is 6.1 m high in places, the largest such mound in the southeastern United States. Less noticeably but equally significant, evidence of several thousand years of human use can be found on almost all of the tree islands within the Everglades' freshwater sloughs and marshes.

Spanish contact with Florida's indigenous populations resulted in their near total annihilation throughout the state (Milanich 1999). In the Everglades, the Spanish colonial period was an era of rapid native depopulation and cultural change. The Calusa, and other indigenous groups, engaged in trade, obtaining metal goods and tools from the Spanish; many became Christian and learned to speak Spanish; and they famously captured, enslaved, and often sacrificed the survivors of Spanish shipwrecks. With these interactions came the spread of diseases to which Everglades native people had little immunity, such as measles and smallpox. By the midpoint of the first Spanish colonial period, these once powerful and prosperous Everglades tribes were living in scattered groups with scant material evidence to indicate their prior rich spiritual and economic life. Within another hundred years almost all of the Glades-era people were gone from the Everglades, with most of the remaining families reportedly boarding Spanish ships for Havana in 1763 (Griffin 1996, 201).

For the next several decades the Everglades was largely deserted. Some native people who survived Florida's colonial period may have held on in isolated camps, and there are reports of Cuban fishermen and "Spanish Indians" along the Everglades coast during this time. Certainly there are considerable gaps in our understanding of the Everglades' human history during Florida's Spanish, English, and French colonial periods. But it seems very likely that Everglades indigenous groups who survived Florida's colonial and early American periods (Calusa, Tequesta, and others) may have united with indigenous groups who were being pushed south into the Florida peninsula from Georgia, Alabama, and northern Florida. Still, for the most part and for a brief period in time, the Everglades was largely depopulated and abandoned.

The contemporary Seminole in south Florida trace their ancestry through 12,000 years of indigenous history in the southeastern United States. Originally living in Creek Confederacy towns in the southeast, indigenous people now known as the Seminole (and later Seminole and Miccosukee) moved into Spanish northern Florida in the early 18th century (Sturtevant and Cattelino 2004). Living in northern Florida, this amalgam of southeastern Indigenous peoples, as well as descendants of African American slaves, practiced traditional town-based agriculture and raised livestock (Griffin 1996; Porter 1996). The expanding European and later American presence in north and central Florida forcibly dislocated the Seminole multiple times—including the deportation of a majority of the population to Indian Territory in Oklahoma during the mid-1800s (Covington 1993).

The few hundred Seminole who remained moved south into the Everglades and the Big Cypress Swamp. Here they lived in dispersed extended-family matrilineal "camps" and practiced small-field horticulture and hunting. As the archaeologist Frank Griffin notes, the camp settlement pattern grew out of a need for mobility and practices of guerilla warfare, yet proved to be a remarkable adaptation to the Everglades environment (Griffin 1996). From early on, Seminole hunters participated in the commercial hide market, selling alligator hides and otter pelts at trading posts in Fort Lauderdale, Miami, and other locations, as well as engaging in subsistence hunting for deer and other animals.

The first Americans to explore the interior Everglades were probably soldiers participating in the Second Seminole War (1835–1842), one of a series of bloody conflicts Andrew Jackson initiated to forcibly relocate the Seminole from Florida. When these conflicts ended, the population of Seminole living in south Florida had dwindled to about 200 (Griffin 1996). These survivors tended to avoid contact with settlers, instead staying in their village camps. Here, they grew sugarcane, corn, sweet potatoes, and other vegetables; fished; and hunted for birds, deer, alligator, and other animals. Despite the 19th century's wars of removal, as well as the continued pressure by white settlers on land and game during the early part of the 20th century, the Florida Seminole maintained their economic and political sovereignty.

Charles B. Vignoles, an Irish-born civil engineer, appears to be the first to refer to the swamps of south Florida as "Ever-glades," reflecting the landscape's vastness (Vignoles 1824). Whether as soldiers, explorers, or surveyors, early explorers and settlers encountering the Everglades during the mid- to late 1800s were simply overwhelmed by its spatial extent. Estimates suggest that the predrainage Everglades spanned roughly 11,000 km^2 and was dominated by an alternation of gray-green sawgrass ridges and deeper water sloughs interspersed with higher tree islands.

American settlement in the Everglades was bound up in the nation's larger political and economic history and the state's territorial claims. When Florida gained statehood in 1845, much of the southern portion of the state was underwater for periods of the year and, according to federal law, these wetlands remained under the control of the U.S. government. At that time, wetlands were considered impediments to the developmental interests of the nation, with agricultural production considered a national security issue (Vileisis 1997).

The federal "Swamp and Overflowed Land Act" of 1850 provided a mechanism for transferring wetlands to states, though this transfer was contingent on the lands being "reclaimed" for agricultural production. In south Florida, drainage schemes sputtered along until 1881, at which time Florida governor William Bloxham sold Philadelphia millionaire Hamilton Disston 16,187 km^2 of Everglades land. As part of this deal, and subsequent deals with other investors, Disston was granted ownership of half of any lands his company drained. Disston's efforts were quite successful; in little over a decade he drained over 200 km^2 of land. He dug a key drainage canal connecting Lake Okeechobee to the Caloosahatchee River and by so doing hydrologically disconnected the Kissimmee watershed from the Everglades (the Lake Okeechobee Phase of drainage; see Box 1.1). Other entrepreneurs quickly followed his lead (Light and Dineen 1994; McCally 1999).

Canal dredging and marsh drainage began in earnest in 1905, though, when Napoleon Bonaparte Broward was elected governor. His campaign platform had one plank: if elected, he would drain the Everglades. And he gave it his best shot! By the 1920s, the combination of shunting Kissimmee Valley water to the east and west coasts, early levees on Lake Okeechobee, and four large northwest-to-southeast canals so effectively drained the Everglades that widespread burning of the dried peat soils became a concern (the Muck Canal Phase of drainage; see Box 1.1). In the 1920s the Tamiami Trail was built from Miami to Naples. The culverts installed under the roadbed may have been adequate to handle the remaining trickle of sheet flow, but later, as Everglades water management shifted toward conservation, the Tamiami Trail became the southern dam for the two largest impoundments—Water Conservation Areas (WCA) 3A and 3B. Until the recent construction of a 1.6-km bridge, the only way for water to flow south into ENP was through four water control structures (S-12A, S-12B, S12C, and S12D), built in the late 1960s. It is important to note that ENP, established in 1947, lies wholly south of the Tamiami Trail.

In 1926 and 1928 two strong hurricanes struck south Florida. In both cases Lake Okeechobee overflowed and the subsequent flooding caused considerable loss of life and property. The response to these two disasters was the Herbert Hoover Dike, built to permanently isolate Lake Okeechobee. Major hurricanes again in 1947 and 1948 spurred congressional action to tame Everglades hydrology, and the U.S. Army Corps of

BOX 1.1
THE THREE PHASES OF EVERGLADES DRAINAGE

Christopher McVoy and Thomas W. Dreschel

The earliest observations of the Everglades were made by explorers of Lake Okeechobee. Along its south shore, they reported custard apple swamp and a long shoreline of sawgrass marsh. These wetland communities, the shoreline soils, and direct observations all indicated that water passed directly out of the lake and into the Everglades during much of most years. Lake Okeechobee provided a conduit to the Everglades for the water draining from the Kissimmee Valley, effectively doubling the Everglades catchment area.

Drainage engineers seeking to make the rich soils of the Everglades available for agriculture quickly realized that they could do so only if water levels in Lake Okeechobee were first lowered substantially. The connection they dredged from the lake to the headwaters of the Caloosahatchee River in the 1880s lowered lake levels by as much as 1.5 m, cutting off (for a period of time) the outflows into the Everglades and initiating a century of drastic changes to Everglades hydrology. Due to its importance, this period (1880–1896) has been referred to as the *Lake Okeechobee Phase* of Everglades drainage (McVoy et al. 2011).

Hamilton Disston's drainage companies appear to have declined after his death in 1896 (Knetsch 2006), and as a result the dredged connection to the Caloosahatchee River became less effective. Governor Napoleon Bonaparte Broward's election in 1904 renewed interest in draining the Everglades, initiating the *Muck Canal Phase* of Everglades drainage (1906–1930). Under Governor Broward, Lake Okeechobee water levels were lowered still further, a muck dike on the southern shore of the lake was improved, and four major muck canals were dredged from the lake through the Everglades. These canals, running south and southeast through the Everglades to the Atlantic coast, had a dramatic effect on Everglades hydrology (McVoy et al. 2011).

The Lake Okeechobee and Muck Canal Phases together separated the lake from the Everglades and eliminated inflows to the Everglades in all but the most extreme years. These phases lowered water depths over large areas, and agriculture and small settlements began to move into the region just to the south of the lake, in the former Sawgrass Plains. In extreme years, the muck dike proved inadequate, and hurricanes in 1926 and 1928 devastated these communities. This prompted construction of the Herbert Hoover Dike between 1932 and 1938. The muck canals and the dike were so effective at draining the Everglades that soil scientists became alarmed at the rapid loss of the exposed organic soils, both by microbial oxidation and by outright peat fires.

By the late 1940s, it became apparent that it was unlikely that all of the Everglades would be converted to agriculture and that the remaining portions needed protection from desiccation. This, in combination with an exceptional flood in 1947, led to the *Impoundment Phase* of Everglades drainage—a joint state and federal effort. This

Central and Southern Florida Flood Control Project created new canals and improved existing ones. However, the main focus was on construction of many miles of levees to create impoundments called WCAs. These WCAs allowed for higher water levels in the Everglades, retaining some of the water that was formerly lost to the ocean. This was beneficial but carried ecological costs, including the elimination of sheet flow, increased fluctuations in water depth, and unnatural variability in flows to ENP.

Engineers' Comprehensive Plan of 1948 was born (the Impoundment Phase of drainage; see Box 1.1). Under this plan the major construction of new canals, levees, and water control structures in the early 1950s included a large levee and associated canal that ran from West Palm Beach to southern Miami-Dade County. This tie-back levee was designed to protect the rapidly urbanizing cities along the Atlantic coast from Everglades flooding. With one exception—the 8.5 Square Mile Area—this tie-back levee remains the westernmost boundary of urban development; it also delineates the eastern edge of the remaining Everglades.

The Comprehensive Plan included re-dredging the upper portions of the Hillsboro, North New River, and Miami Canals in the late 1950s, as well as major pumps and new levees that formally defined the nearly 300,000-ha Everglades Agricultural Area (EAA). The EAA played an important role in south Florida's cultural and political history because its creation coincided with the fall of the Batista government to the Castro regime in Cuba. The migration of Cubans into south Florida that followed this political transition included the economic power of Cuba's large sugar industry, and the EAA quickly became dominated by sugarcane. Drainage of the deep Everglades peat soils in the EAA has led to dramatic losses of these rich organic soils—in some cases as much as two-thirds of the original peat. As a result, the EAA is now considerably lower than Lake Okeechobee and the remaining Everglades to the south. It is kept dry by the nearly constant pumping of rainwater out of the basin. For many years, this nutrient-rich agricultural runoff was directed into the downstream Everglades, creating massive eutrophication-related impacts in these naturally nutrient-poor (oligotrophic) marshes. This created huge management challenges that led to political conflicts, court battles, and ultimately a shift in decision-making toward Everglades restoration.

The last major effort to tame and bound the remaining Everglades included the construction of more canals and levees in the early 1960s that compartmentalized the landscape into five massive basins, known as WCAs. A dizzying array of water control structures were built to manage the movement of water into and between the WCAs, and this large network of structures is still used by the South Florida Water Management District to control the hydrologic regime of south Florida.

The Florida Coastal Everglades Long Term Ecological Research Program (FCE LTER) focuses mostly on ENP. Despite very strong legal protection and being relatively free of drainage canals and levees, ENP is hydrologically (and hence ecologically) the most severely altered portion of the remaining Everglades. Physics trumps legal protection here: water drains unimpeded out of ENP and to the Gulf of Mexico, so any reductions in inflows translate directly into drier conditions. These ongoing reductions, begun in the 1880s (see Box 1.1), have diminished water depths, shortened hydroperiods, and reduced water flow velocities (McVoy et al. 2011). These drier conditions have had major ecological impacts on ENP. For example, widespread oxidative losses of peat have fundamentally altered the area by (1) altering the topographic and hence hydrologic relationships between Shark River Slough and its flanking marshes, (2) reducing the elevation difference between ridges and sloughs, and (3) altering water chemistry.

Peat soils of various depths once covered essentially all of what is now ENP (Fig. 1.1, upper panel). Exposed limestone "pinnacle rock" was restricted primarily to the upland pinelands of the Miami Rock Ridge. In ENP marshes, most of this rock was covered by peat. This prevalence of peat would have meant softer water conditions, in strong contrast to the calcium-rich alkaline water that now typifies ENP marshes. By 1955, the postdrainage loss of this peat had significantly altered the transverse cross-section of much of ENP (Fig. 1.1, lower panel). Significantly, the greatest losses of peat thickness and elevation occurred where the peat was thickest—in the center of Shark River Slough. This depressed and deepened the central slough relative to the marginal marshes that

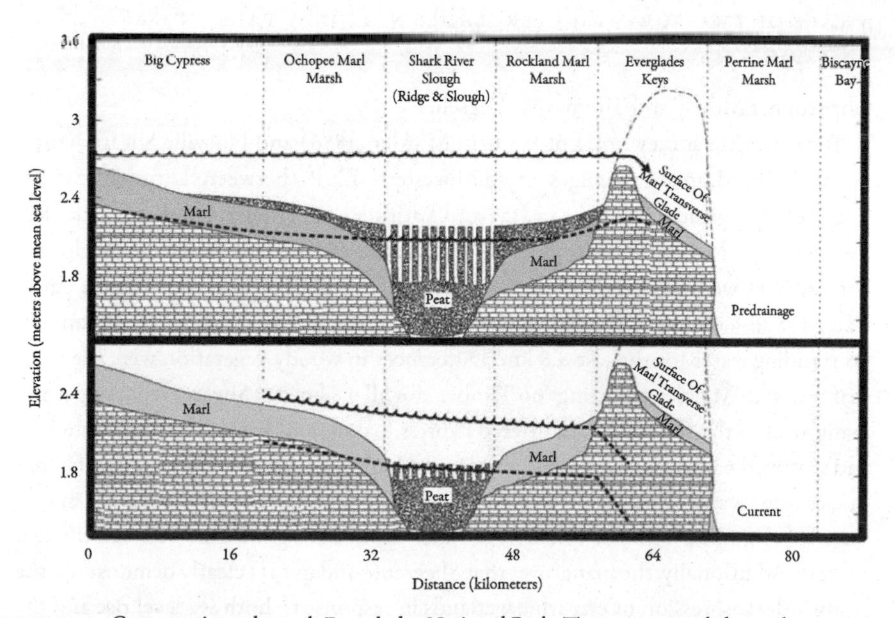

FIGURE 1.1 Cross sections through Everglades National Park. The upper panel shows the predrainage peat and water levels to be compared with the lower panel showing the current lowered peat surfaces and reduced water levels. See McVoy *et al.* (2011) for details.

are now known as the "marl prairies" (Fig. 1.1, lower panel). Surveyor William Mickler's descriptions of this area in 1886 are consistent with the flatter topography shown in Figure 1.1 (Box 1.2). Drainage has effectively created two adjacent environments—the central slough peat marshes and the fringing marl prairies—from what in 1886 was a single, wide peat marsh.

In summary, the FCE LTER study area in ENP is arguably the most hydrologically altered portion of the remaining Everglades landscape. The area is considerably drier than predrainage conditions, with shallower water depths and shortened hydroperiods. Overall flow volumes and flow velocities have been reduced. Peat oxidation simultaneously has *increased* the topographic difference between Shark Slough and the flanking marl marshes—a phenomenon called "dishing"—and has *decreased* ("flattened") the differences between ridges and sloughs in the corrugated ridge and slough marshes of the central slough (see Fig. 1.1). The loss of peat soils covering limestone has also shifted the area from soft water to hard water conditions. Finally, topographic flattening and increased stem densities within sloughs have blurred the distinction between the original sawgrass ridges and sloughs.

Contemporary water management in the Everglades and south Florida is a remarkably effective, if complex, system. The result is one of the most designed, engineered, and managed landscapes in the country—perhaps in the world. The focus today, and for the

BOX 1.2

A TALE OF TWO SURVEYS: EVERGLADES NATIONAL PARK IN 1886 AND 1955

Christopher McVoy and Thomas W. Dreschel

The township survey notes of William Mickler (1886) and Norville Shearer (1955) illustrate the dramatic changes in southwestern ENP (between Latitude: 25° 21′ 50.84807, Longitude: 80° 52′ 23.34523 and Latitude: 25° 21′ 51.71601, Longitude: 80° 46′ 38.64634; McVoy et al. 2011). Pronounced differences were observed even though both surveys were performed in January. Mickler reported at least 45.7 to 76.2 cm of water for all 9.6 km, whereas Shearer reported only 30.5 cm for the first 4.8 km and no standing water for the last 4.8 km. Differences in woody vegetation were the most striking, with Mickler reporting "no Timber" for all 9.6 km but Shearer reporting dense mangroves in the first 1.6 km; scattered cypress, palms, and bay in the second 1.6 km; and scattered palms, palmetto, and pines in the higher 6.4 km. The observed shift from thick sawgrass and "Everglades" in 1886 to low sawgrass, saw palmetto, and pines in 1955 suggests a significant decrease in average water depth during the time between the two surveys. Additionally, the mangroves that Shearer found in 1955 clearly demonstrate the landward transgression of estuarine wetlands in response to both sea level rise and the dramatic reductions in freshwater flows.

last several decades, is on Everglades restoration. The socio-politics of this effort will ultimately determine the degree to which restoration will undo some of the last 150 years of canal and levee construction versus engineering the system even more, but with a different goal. In other words, the question remains: Will the restoration effort free the Everglades to function in a somewhat more natural fashion, or will it continue to control the Everglades but with a more restoration-focused management approach?

We would be remiss if we did not point out two final and critical points about this unique landscape. The first is related to the fact that south Florida is low-lying and very flat. The highest point on the Miami Rock Ridge—a limestone outcrop along the western shore of Biscayne Bay—has an elevation of scarcely 7 m and the highest point in ENP is scarcely 1 m above sea level. Because of this, the south Florida landscape represents ground zero for sea level rise, which is probably the most certain uncertainty of climate change. Sea levels are rising at about 0.5 cm per year in this region, and all projections by the Intergovernmental Panel on Climate Change (IPCC) have this rate accelerating in the near future. Notably, none of those projections account for uncertainties about the increasingly unstable West Antarctic Ice Sheet and the rapidly melting Greenland Ice Sheet. If the water frozen in either of those landbound ice masses makes it to the ocean, sea levels may rise 6 to 7 m—and very quickly. Even today, parts of Miami Beach flood during many full and new moon tides (i.e., spring tides) as the tide literally rises through the storm drains and into streets and buildings. In response, the city is installing an array of flow flaps and pumps in the storm drains, at considerable expense, but this is a temporary measure at best and can scarcely be thought of as a sustainable adaptation to the problem. There is no effective way to wall off the ocean, as has been done in the Netherlands and as has been attempted in New Orleans, because the sea will simply flow under levees through the highly porous limestone bedrock. As sea levels rise, mangroves will continue their slow, steady transgression landward across ENP and south Florida's human population will be increasingly vulnerable to flooding, from storms and "sunny day flooding," and to saltwater intrusion into the Biscayne Aquifer, which is the source of south Florida's drinking water.

And this brings us to the second critical point: the drinking water for nearly all of south Florida's population and visitors comes from the shallow Biscayne aquifer. South Florida's bedrock is very porous and conductive limestone, and it is this karst that supports the Biscayne aquifer. Early European explorers used to fill their ship's water barrels at freshwater artesian well "boils" found throughout Biscayne Bay. Today, a series of wellfields draw drinking water from the Biscayne aquifer, from depths as shallow as 15 to 20 m. These wellfields have had to be moved west a number of times because of saltwater intrusion, and today they are located along the western margin of south Florida's urban areas. Water from the Everglades continues to recharge this aquifer, almost in real time, although no longer at a rate sufficient to power underwater springs in Biscayne Bay. This recharge water first percolates through the oligotrophic peat soils of Everglades marshes, which is likely part of the reason that south Florida's drinking water is among the cleanest in the United States. Thus, the sustainability of south Florida's population and economy is inextricably linked to, and dependent on, the Everglades. This fact, which is little known beyond scientists and water managers, is why Everglades restoration is so critical. And this is why it is so vitally important that the restoration process be informed by the best scientific knowledge available. And that is the fundamental goal of the synthesis we present in this book.

The book is organized into nine broad-brush, integrative chapters, including this introduction. In the second chapter we discuss how views of the Everglades as an icon have shaped its modern condition and plans for its future. Water is obviously central to everything Everglades, and we focus on this critical aspect in Chapter 3. Our modifications of the Everglades over the last 150 years have fragmented the Everglades landscape with levees while increasing hydrologic connectivity with thousands of kilometers of canals. We explore these seemingly contradictory characteristics in Chapter 4 while also making important connections between south Florida and a range of more global phenomena. Just as water is central to the Everglades, phosphorus (P) is an important driver of both ecological and political processes because Everglades wetlands are so oligotrophic and P is the limiting nutrient. We synthesize a huge body of research on the biogeochemical importance of P in Chapter 5 while also exploring the "social life of P" in south Florida. There is a seeming contradiction in the fact that Everglades wetlands are P-limited and nutrient-deplete, yet they remain remarkably productive. We explore this incongruity in Chapter 6 with our synthesis of carbon cycling. The Everglades landscape is well adapted to, and is perhaps somewhat dependent on, natural disturbances such as fires, freezes, and tropical storms. The system must also tolerate anthropogenic disturbances, from invasive species to our centuries-long construction endeavor. In Chapter 7 we synthesize what we have learned about disturbance as a driver of Everglades dynamics. No synthesis of past findings would be complete without an application of that synthesis to the future—in both decision-making and scenarios. To this end, we go "back to the future" in Chapter 8. And we conclude our synthesis with Chapter 9, where we demonstrate how the field of ecology, as well as its education and application, has been positively influenced by our Everglades research findings and contributions.

As the managing editors of this Everglades research synthesis volume, we hope you will find it to be enlightening and informative, a valuable reference to inform our future, and an enjoyable read.

References

Cattelino, J., and W. Sturtevant. 2004. Florida Seminole and Miccosukee. In *Handbook of North American Indians*, vol. 14, *Southeast*, R.D. Fogelson, volume editor, W.C. Sturtevant, general editor, 429–449. Washington, DC: Smithsonian Institution.

Covington, J.W. 1993. *The Seminoles of Florida*. Gainesville: University Press of Florida.

Griffin, J.W. 1996. *Fifty Years of Southeastern Archaeology: Selected Works of John W. Griffin*, edited by P.C. Griffin. Gainesville: University of Florida Press.

Light, S.S., and J.W. Dineen. 1994. Water Control in the Everglades: A Historical Perspective. In *Everglades: The Ecosystem and Its Restoration*, edited by S.M. Davis and J.C. Ogden, 47–84. Delray Beach, FL: St. Lucie Press.

McCally, D. 1999. *The Everglades: An Environmental History*. Gainesville: University Press of Florida.

McVoy, C.W., W. Park Said, J. Obeysekera, J. VanArman, and T.W. Dreschel. 2011. *Landscapes and Hydrology of the Predrainage Everglades*. Gainesville: University Press of Florida.

Milanich, J.T. 1999. *Laboring in the Fields of the Lord: Spanish Missions and Southeastern Indians*. Washington, DC: Smithsonian Institution Press.

Porter, K.W. 1996. *The Black Seminoles: History of a Freedom-Seeking People*. Gainesville: University Presses of Florida.

Vignoles, C. 1824. *The History of the Floridas, From the Discovery by Cabot, in 1497, to the Cession of the Same to the United States, in 1821. With Observations on the Climate, Soil and Productions*. Brooklyn, NY.

Vileisis, A. 1997. *Discovering the Unknown Landscape: A History of American's Wetlands*. Washington, DC: Island Press.

2

The Everglades as Icon

Laura A. Ogden and Joel Trexler with Daniel L. Childers, Evelyn Gaiser, and Katrina Schwartz

In a Nutshell

- The Everglades has captured the imagination as a "frontier," a "river of grass," and a "unique ecosystem" that is home to alligators and large rookeries of wading birds.
- Humans have occupied the Everglades for thousands of years and they have modified it throughout that history, but modern engineering modifications have altered its size, ecological structure, and function.
- Several features of the Everglades stand out: its large size and peninsular location; its karstic geology and oligotrophy; its "upside-down" estuaries that receive the limiting nutrient from the ocean rather than the watershed; its geologic history; and its latitude (25°N) and subtropical climate with a distinct wet and dry season. The Everglades is the largest extant example of a Caribbean karst wetland.
- The success of efforts to restore the Everglades depends on the resiliency of historical ecological processes that shaped the ecosystem. Though some areas have been irrecoverably changed, large portions of the ecosystem retain the potential for historical functions. Restoration of a historical flow of water that is low in phosphorus is a critical element for maintenance, conservation, and recovery of the iconic ecological features of the Everglades.

Introduction: Understanding Icons

An icon is a person or thing that has symbolic worth, often representing something of cultural value. Yet, like all signs and symbols, the values associated with icons are complex and sometimes contradictory. Consider, for example, Ansel Adams's iconic images of Yosemite National Park. For many, his dramatic photographs have come to define the ways we know and appreciate the American West—as an epic wilderness, outside the boundaries of civilization. Yet, his portrayal of Yosemite as emblematic of sublime Nature is a selective account that strategically ignores the landscape's social life and history. As William Deverell (2006) notes, "Adams erased people simply by aiming his camera above and beyond them."

In this chapter, we explore the ways the Everglades has become "iconic," or valued as a location, a habitat, and an idea, while paying close attention to the ways icons bring only certain truths into view. The ecological, cultural, and economic significance of this landscape is hard to overstate. The Everglades is central to our understanding of the challenges of settlement in the southeastern United States, as well as to the current challenges to the sustainability of a region that is home to millions of people. It is home, too, to threatened and endangered species, and remains the largest representative of Caribbean karstic wetlands, a globally distinctive ecological system. Thousands of international visitors come to Everglades National Park (ENP) hoping to experience "primeval" nature, while archeologists continue to find a deeper history of human life in Everglades areas once considered impenetrable by European explorers and settlers of the region. The thesis of this chapter is that the Everglades is appropriately considered an icon in multiple ways and to broad audiences, but like all icons, the Everglades is much more complex than is suggested by most popular representations of the landscape.

The Everglades is both a U.S. national park and a UNESCO World Heritage site, reflecting its iconic status to a broad and international community. As an example, the National Parks and Conservation Association (NPCA), which is the preeminent advocacy organization for the U.S. National Park System, has identified the Everglades as one of 13 iconic landscapes in the United States. In addition to the Everglades, the NPCA list includes the Colorado River, the California Desert, and Bristol Bay, Alaska. Though the NPCA did not articulate criteria for membership in its iconic landscape shortlist, it seems likely that the ability of a landscape to represent some diverse aspect of the American wilderness experience and related conservation values was a key reason for inclusion.

Landscapes become icons when they are recognized for their ability to represent socially significant environmental values. Sometimes these environmental values speak to scientific and/or conservation concerns (e.g., biodiversity or rare biota). At other times, they remind us of our cultural history (e.g., significant battlefields or agrarian landscapes), while some places become icons because they reflect broader ideas about nature itself ("wilderness," for example). That said, there is never a consensus or singular vision that accounts for a landscape's iconic status. Instead, landscapes from Bristol Bay to the Maine Woods, to borrow a couple of the NPCA's sites, are iconic in multiple, and sometimes competing, ways. Bristol Bay can be iconic to a commercial fisherman *and* to a marine ecologist, though sometimes in ways that reflect conflicting values.

We might do better to speak of a landscape's iconography—a lexicon of images that reflect its diverse symbolic, economic, and political values. It is important to understand that landscape iconography, as we are using the term, includes the arena of environmental politics. Consider the ways in which competing values associated with the Pacific Northwest's forests became distilled into an iconography of spotted owls, "tree huggers," and working-class livelihoods. The Everglades, like the forests of the Pacific Northwest, has a long history of resource conflict and associated litigation, with conflicts in the Everglades mainly stemming from the use of, and competing demands for, water. As we show in this chapter, environmental politics also shape Everglades iconography, since ways of understanding the Everglades are bound up in the politics of Everglades restoration and, more broadly, in conflicting values concerning the region's economic, social, and environmental future.

The iconography of the Everglades, as we will examine, includes several divergent yet powerful images that continue to shape the politics of Everglades use, protection, management, and restoration. These images include the Everglades as a "frontier," a "river of grass," and a "unique ecosystem," all associated with Marjory Stoneman Douglas's classic book, *The Everglades: River of Grass*, originally published in 1947. In this chapter, we use our research and insights to closely interrogate these images, not to discount them, as they all speak to different truths about the landscape. Our hope is that this engagement allows us to better understand how this iconography works and the role it plays in the politics of the Everglades.

The Frontier: Worthless, Empty of People and History

As we described in Chapter 1, the Everglades has been transformed by long-term climatic changes and by multiple cycles and types of land use change, including contemporary practices of drainage, development, and engineered water management. For the most part, the changes to the Everglades landscape over the past century were enabled by familiar American practices associated with the "frontier." Frederick J. Turner offered a famous account of the frontier's place in the making of American identity and history. In his book initially published in 1920, Turner argued that America's early history entailed a shifting, westward transformation of lands and people that lay outside the "pale of civilization." In other words, becoming America was predicated on a settlement approach where land was valued for its development potential while being simultaneously viewed as empty of people (i.e., of value) and of history. For Euro-Americans first settling in south Florida, the Everglades was a frontier lacking intrinsic worth, people, and history. Within the iconography of this frontier, the value of the Everglades lay solely in its potential to be reclaimed (i.e., drained) and cultivated. Without reclamation, the

Everglades was considered miasmic, dangerous, worthless, and empty of human history and potential.

As a frontier landscape, ideas about the Everglades were bound up in the nation's larger political and economic history and in territorial claims by the state. When Florida gained statehood in 1845, much of the southern portion of the state was under water for periods of the year and, by federal law, these wetlands remained under control of the U.S. government. At that time, wetlands were considered impediments to the development interests of the nation, with agricultural production considered a national security interest (Vileisis 1997). The federal Swamp and Overflowed Land Act of 1850 provided a mechanism for transferring wetlands to states with the contingency that these lands be "reclaimed" for agricultural production. Drainage schemes sputtered along until 1881, when Philadelphia millionaire Hamilton Disston acquired 16,187 km^2 of Everglades land. Disston's efforts were fairly successful; in little over a decade he drained over 200 km^2 of land and created the first major drainage canals in the region (see Chapter 1 for more on this history of drainage).

Countering popular ideas of an unpeopled landscape, evidence from Everglades tree islands has shown that south Florida has a long history of human occupation. When the Spanish arrived in Florida in 1513, there were over 20,000 Native American people living in the Greater Everglades region, the majority and most powerful of whom were the Calusa, who lived along the southwest coast (Griffin 1996). The Calusa (as well as the people who lived for a few thousand years before the Calusa) built dramatic shell mounds and earthworks throughout south Florida. Chokoloskee Island, a coastal fishing village, is actually a human-built 24.3 ha shell mound that is 6 m high in places; it is the largest such mound in the southeastern United States (Marquardt 2010). Less noticeable but equally significant, evidence of several thousand years of human use can be found on most of the tree islands that remain in the Everglades landscape (Ardren et al. 2016; Carr 2002; Coultas et al. 2008). We are only now beginning to consider how the legacy of these intentional transformations of the landscape has affected the ecology and evolution of the contemporary Everglades (Graf et al. 2008). Within the vast landscape mosaic that is the Everglades, the greatest concentration of prehistoric human productivity occurred on tree islands. In a sea of watery grass, tree islands offered high, dry ground and shelter against storms while they also provided dry habitat for huntable mammals, reptiles, and birds. For approximately 5,000 years, tree islands have been the nexus points for an array of cultural practices and significance in the Everglades (Fradkin 2016; Schwadron 2006).

In addition to being centers of prehistoric human activity, tree islands are also "ecological hotspots" in an otherwise nutrient-poor Everglades wetland landscape. Hotspots occur where the flow paths of nutrients converge, whether these vectors of nutrient transport are waterborne, windborne, animal, or human (McClain et al. 2003). Wetzel et al. (2005, 2009) defined tree island ecological hotspots as areas with high nutrient concentrations and productivity, and more rapid rates of nutrient cycling, relative to the surrounding marsh landscape. Phosphorus (P) accumulates in tree islands, and tree island soils have up to 100 times more P compared with the surrounding marsh (Orem et al. 2002; Troxler-Gann and Childers 2006; Wetzel et al. 2005, 2009). Everglades tree islands leach some of their productivity and excess nutrients to downstream marshes, enhancing productivity that cascades up Everglades food webs (Trexler et al. 2002; Turner et al. 1999). It is likely that fishing and foraging have always been more productive on and immediately downstream of tree islands compared with open marsh away from tree islands.

Plant productivity and structural complexity are also markedly greater on tree islands relative to the surrounding marsh (Sklar and van der Valk 2002). Research on tree island soils suggests that prehistoric contributions to these soils are much higher than previously thought (Coultas et al. 2008), suggesting that humans had a key role in the actual formation and accretion of many tree islands. In one example, it appears that Archaic people began visiting tree islands as long as 5,000 to 6,000 years ago, when their soils were quite shallow (about 10 cm thick over the underlying limestone bedrock). The island under study now has more than 2 m of soil over the limestone. This soil contains numerous bones and a few artifacts, but also sand, which is not found in the interior Everglades, strongly suggesting that these soils were largely built by human activities. These inland activities, along with the Calusa shell middens along the coast, dramatically increased tree island heights. This was happening as both the sea level was rising and inland marsh water levels were increasing, as south Florida became steadily wetter in the late Holocene. These land-building activities created important topographic diversity across the Everglades landscape. Craighead (1971) suggested that prehistoric people may also have introduced plant species to the hammocks and tree islands that they were creating and maintaining in the Everglades. For example, plants important to the Calusa such as soapberry, southern hackberry, persimmon, and mulberry can be found throughout Paradise Key, a large tree island that became the cornerstone of efforts to protect the Everglades during the early 1900s.

Robert Kohler's (2006) description of the "inner frontier" aptly characterizes the development patterns of south Florida in the early decades of the 20th century, where "the boundaries between wild and settled were unusually extensive and permeable." While south Florida's commercial center was once Key West, over a hundred miles south of the mainland, struggling settlements along the coastal ridge, such as Miami and Coconut Grove, became increasingly important tourist destinations and cities in their own right. Ironically, the proximity of these new and growing settlements to the Everglades produced a growing awareness of the threats to its unique species and habitats. For instance, the Ingraham Highway, which was completed in 1916, linked Miami to Paradise Key, enabling visiting naturalists and wealthy patrons to access, study, and enjoy one of the Everglades' most significant hammocks without much difficulty. While popular ideas about the Everglades continue to evoke a frontier, appreciation of the intimate reciprocal link of human history in the region to the ecology of the Everglades continues to grow.

A River of Grass

Douglas opens *The Everglades: River of Grass* with a passage that highlights two characteristics of the Everglades landscape that have gelled in the public's mind: that it is "unique" and is a "river of grass." The passage begins with the simple statement: "There are no other Everglades in the world." Douglas then richly evokes the subtle beauty of this "river of grass," which she characterizes as "the central fact of the Everglades of Florida." Douglas was one of the Everglades' most vocal advocates, and it would be difficult to overestimate the significance of this passage to the ways we understand the contemporary Everglades. For example:

This passage is displayed on the entrance wall of the main ENP visitor center.
President Barack Obama quoted the passage in his 2015 Earth Day speech highlighting the importance of the Everglades for regional climate change mitigation.
This passage is quoted or cited by nearly every author who writes about the Everglades (including many of the authors of this book).

Douglas's romantic attitude toward the Everglades was not shared by the soldiers, explorers, or surveyors who first encountered it during the mid- to late 1800s. Instead, they were simply overwhelmed by its spatial extent and the challenges of moving through its uncharted and seemingly impenetrable sawgrass marsh. For example, Major Archie Williams led the first expedition to cross the Everglades from north to south in 1883 on a trip funded by the New Orleans *Times-Democrat* newspaper. Williams's first impressions of the great Everglades prairies south of Lake Okeechobee portrayed a landscape of unrelenting, unlimited sameness. While those who encountered the Everglades in the 1800s likely did not appreciate the ecological diversity of the landscape, their emphasis on its vastness was not exaggerated. Estimates suggest that the predrainage Everglades spanned roughly 28,000 km^2 and was dominated, as it still is, by vast prairies of gray-green sawgrass.

Yet Douglas's characterization of the Everglades as a river, rather than a prairie or swamp, has proven to be a lasting and powerful image. Is this a technically accurate description of the Everglades? In some ways, our appreciation of the Everglades' river-like qualities corresponded with a growing awareness of the grave impacts of diking and draining on the region's hydroecological characteristics. The flowing nature of the Everglades was not of primary concern to the engineers who compartmentalized and diverted its waters in the 20th century. However, by the 1980s it became clear that the loss of the topographic relief in hydrologically isolated areas was a major impact of replumbing (McVoy et al. 2011; SCT 2003). Historically, the Everglades topography was a directional "ridge and slough," "corrugated" and "patterned" relief similar in appearance to braided streams, floodplains, and river deltas around the world (Palmer and Poff 1997). This relief, parallel to the direction of flow and perpendicular to direction of marsh drying, allows for a sequential formation of drying pools from the system's edge to center as the dry season progresses (Fig. 2.1; Larsen et al. 2007). And this dynamic is critical in creating a moving banquet of high concentrations of small fish and crustaceans for apex predators such as wading birds and alligators during the spring nesting season. The loss of flow, and subsequent loss of corrugated relief, diminished the presence of these high concentrations of prey necessary to sustain apex predators. Recent research, including experimental re-creation of relatively high-velocity pulses of water, has further delineated the impacts of flow on biophysical and biogeochemical functions necessary to maintain key ecosystem functions (CISRERP 2014). Sustained flows of 3 to 5 cm sec^{1} are required to suspend and displace flocculent benthic organic matter (Larsen et al. 2011). However, maximum wet-season flows in today's Everglades rarely approach this velocity.

A river is a conveyance for excess water to reach base level, and movement can be on the surface or underground (Leopold 1994). The flow of rivers is a mechanism for the transport of particles, leading to erosion, channel formation, and deposition. The Everglades remains a conveyance of surface and ground water from the north to the south, providing important contributions of freshwater to Florida Bay and the southern Everglades estuarine river systems (Shark River, Harney River, Rodgers River, Lostmans River, and others). By definition, it is appropriately considered a river. However, the suspension and displacement of sediment, another characteristic often considered typical of rivers, is now a function limited to pulsed events in very wet years in the vast majority of the system. The very slow flows typical of the contemporary, compartmentalized Everglades challenge its characterization as a river.

Larsen et al. (2007) describe the Everglades as an anabranching river, or a river with multiple divergent channels. Anabranching rivers are often found in conditions of seasonal flow, with parallel channels acting to enhance discharge and separated by vegetated semipermanent islands (Nanson and Knighton 1996). These differ from braided rivers in the stability of the islands or ridges that separate channels; the topographic features of braided rivers are ephemeral, while in anabranching rivers they are stable and long-lived. Maintenance of the branching nature of the Everglades is believed to require flowing water capable of suspending and displacing the organic flocculent material that is continuously produced in Everglades marshes (Larsen et al. 2011, but see Kaplan et al. 2012). Larsen et al. (2011) reported that a flow of 3 to 5 cm sec^{-1} is necessary to flush this benthic organic flocculent material from sloughs and keep them from filling in. It is believed that flows in excess of 4 cm sec^{-1} were common in the historical Everglades and were responsible for creation and maintenance of the ridge and slough "corrugated" topography.

Historically, the Everglades was a river, but is the modified landscape still a river? If restoration is able to restore historical flow velocities, recent experimental work suggests that organic matter displacement will return, potentially regaining functions that have been on hiatus since channelization and diversion of historical flows. In this sense, restoration initiatives have the potential to restore flow and the processes that maintain ridge

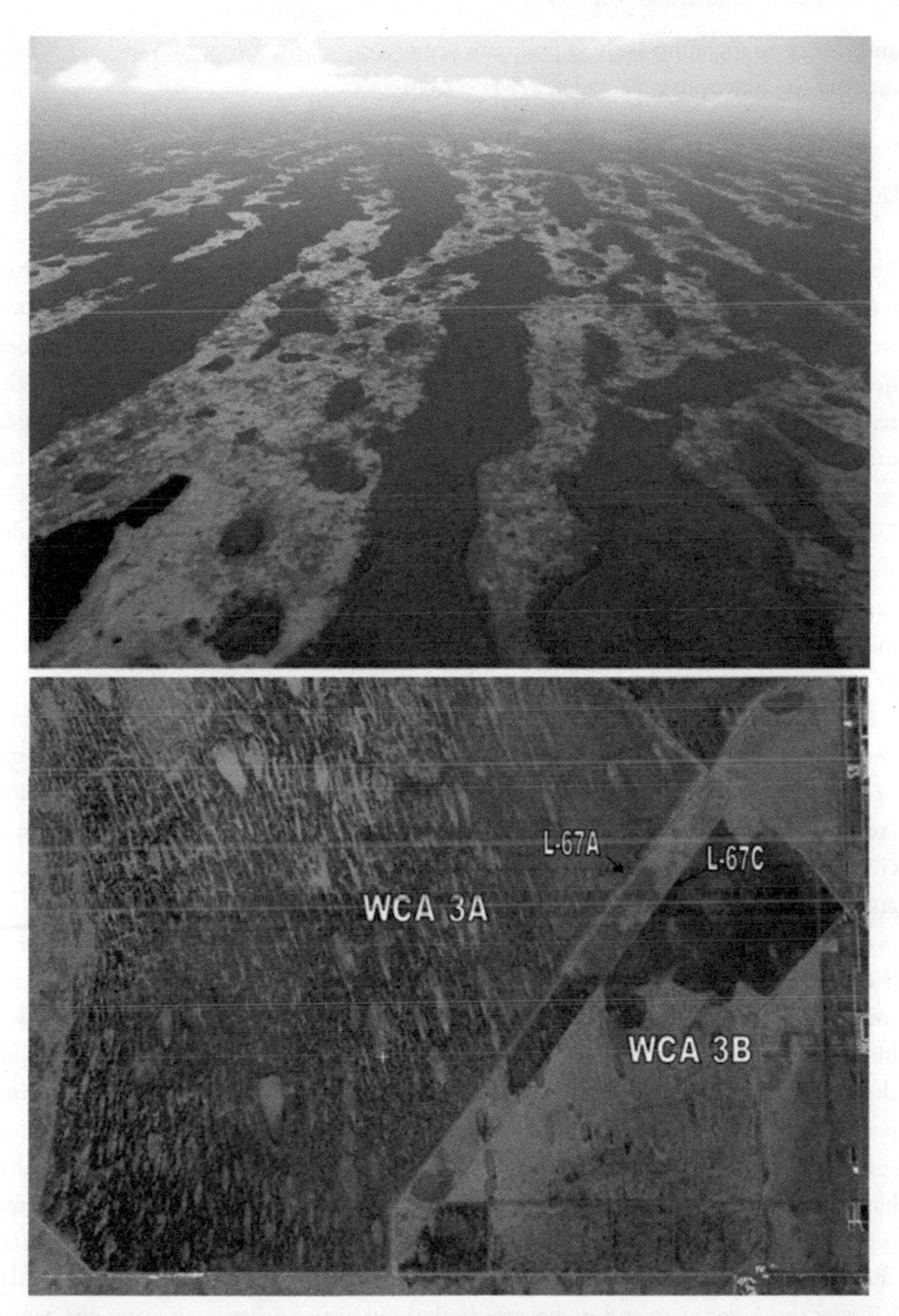

FIGURE 2.1 Top panel: An example of the "corrugated" ridge and slough considered typical of the historical Everglades and remaining in only a fraction of its historical range. Bottom panel: Google Earth image illustrating a region with ridge-and-slough relief intact (WCA 3A) and degraded (WCA 3B). The L67A and L67C canals and levees were constructed in the late 1960's and block surface flow from entering WCA 3B, which only receives water from rain and percolation from underground flow. Historically, the main Everglades flow path passed through WCA 3B.

and slough topography. If these programs are successful, the Everglades will again have many of the functions expected of a river ecosystem.

Uniqueness

The opening lines of Douglas's *The Everglades: River of Grass* have become a cherished reminder of the landscape's value and mystery: "There are no other Everglades in the world. They are, they have always been, one of the unique regions of the earth, remote, never wholly known." The public has taken up this idea of "uniqueness" as a fact of the region. Today, it is common to hear government leaders and elementary school children call the Everglades unique. But is this technically correct? Are no other regions of the world similar to the Everglades? To answer this, we must begin by defining features of the Everglades in order to compare it to other ecosystems. Several features of the Everglades stand out: its large size and peninsular location, its karstic geology and oligotrophy, its "upside down" estuaries, its geologic history, and its latitude (25°N) and "tropical" climate arising from its close proximity to the Caribbean Sea and the warming influences of the Florida Current.

The *peninsula effect* describes a common pattern of biogeography characterized by decreasing biodiversity, from the top to the tip of a peninsula, for taxa that have little or no ability to disperse across salt water. Florida displays such a pattern in its flora and fauna. For example, native amphibians and reptiles and freshwater fish decrease in species numbers from north to south, with the lowest richness in the Everglades (Means and Simberloff 1987; Trexler 1995). Dispersal by terrestrial animals to the tips of peninsulas is limited to one direction, which for the Everglades is from the north. Since the Everglades' extensive freshwater wetlands are less than 6,000 years old, the time for freshwater wetland species to colonize has been relatively short. Certainly, Everglades-like habitats existed in deep lakes north of modern-day Lake Okeechobee throughout the late Pleistocene and Holocene, and these environments could have seeded contemporary Everglades communities (Quillen et al. 2013). However, the diversity of habitats south of Lake Okeechobee is also limited by the karstic geology and hard water that has a neutral or slightly alkaline pH. In contrast, aquatic habitats with high tannin and soft, low-pH water characterize central and north Florida. Species adapted to such water chemistry have no home in south Florida in general, or in the Everglades in particular (Trexler 1995). For plant and animal groups that can disperse across salt water, the peninsula effect in the Everglades is moderated by the presence of tropical taxa that have colonized from the south. Thus, some components of botanical and microbial biodiversity are enriched by these additions, as are coastal salt-tolerant and marine fishes (LaHée and Gaiser 2012; Loftus and Kushlan 1987).

While the Everglades is not an area of high species richness, it is home to some rare and endangered taxa, at least as defined by U.S. law. For example, the Florida population of the snail kite (*Rostrhamus sociabilis*) is a federally designated endangered species whose only U.S. population is found in, and north of, the Everglades and whose population declined from approximately 3,200 to 1,500 individuals from the mid-1990s to the early 2000s (Martin et al. 2007). The subspecies of kite found in Florida is indistinguishable from kites found in Cuba, but genetic analysis has separated them from kites found in Central and South America (Haas et al. 2009). The species as a whole is not considered in extreme peril, and the conservation status of the Cuban population is stable (Raffaele et al. 1998). Thus, the snail kite is a relatively widespread species with a fragile, isolated U.S. population that is limited to the wetlands of south Florida that include the Everglades.

As we noted, the Everglades is home to a relatively small number of endemic taxa. One such species is the Cape Sable seaside sparrow (*Ammodramus maritimus mirabilis*), which evolved as a geographic race in the relative isolation of the southern Everglades (Walters et al. 2000). Such geographic races, sometimes considered subspecies and sometimes species, are common globally and result from the history of the environment where they live and isolation of the taxa. The Florida panther (*Puma concolor coryi*) is another example of a geographically isolated endemic population that has been considered genetically and morphologically distinct from a larger distribution of cougars (Johnson et al. 2010). The Liguus Tree Snail (*Liguus fasciatus*) is yet another example of a tropical species found in south Florida and Cuba, with locally distinct populations (Hillis et al. 1991; Jones et al. 1981). Thus, the geography and geologic history of the Everglades have combined to yield a flora and fauna that is a mix of temperate and tropical taxa, yet is overall depauperate and lacking in endemism.

The seasonal hydrology of the Everglades permits this oligotrophic system to support large populations of apex predators (Gaiser et al. 2012). During the wet season, prey species are spread across a large landscape at relatively low density, making foraging by apex predators relatively inefficient. Water depths in sloughs are too great for long-legged wading birds to feed—leg length determines the depth at which these birds can catch their fish and crustacean prey (Frederick et al. 2009; Gawlik 2002). As water depths drop in the dry season, high-quality foraging patches are created along the drying edges of the landscape (Beerens et al. 2011; Russell et al. 2002). Birds and alligators take advantage of these patches to stuff themselves. As a result, the early dry season coincides with nesting for these species. Wading bird success is particularly dependent on a productive wet season, followed by a gradual recession and formation of pools in the ridge and slough topography that have high prey concentrations. The large landscape size, microtopography created by flowing water, and seasonal rainfall combine to create protracted periods of ample food supply in enough years to sustain long-lived predators. Similarly, during the dry season, river and creek channels of Everglades estuaries fill with freshwater species escaping the drying marshes, and marine and freshwater predators benefit from these seasonal pulses of abundant prey (Boucek and Rehage 2013; Matich and Heithaus 2014; Rosenblatt and Heithaus 2011).

The karst template of the Everglades affects water chemistry in fundamental ways that support distinctive, but not unique, microbiologic communities in the form of extensive and highly productive periphyton mats (Gaiser et al. 2012). These mats are essential to the Everglades food web. They are the base of the food web and the bulk of organic matter that sustains animal life, and they provide structural predation refuges for small animals (Trexler et al. 2015). The result is an unusual Eltonian food web with a large biomass of primary producers sustaining a relatively small biomass of aquatic invertebrates and fishes, most of which have life cycles of less than a year and small adult body sizes (Gaiser et al. 2015; Turner et al. 1999). It is curious that so much primary production appears to go unconsumed in an aquatic ecosystem, and one reason is the high amounts of calcium carbonate in the water. Everglades periphyton mats are dominated by cyanobacteria that secrete voluminous quantities of extracellular polysaccharides embedded with calcium carbonate crystals that precipitate during photosynthesis (Hagerthey et al. 2011). These cyanobacteria also produce toxic compounds believed to act as chemical deterrents to grazers. Edible species of algae gain protection from these inedible partners in the mats that are held together by the extracellular polysaccharides. Additionally, animals small enough to live inside the mats largely escape predation by small fish patrolling on the outside and are free to forage on edible components of the mats, such as diatoms and nonfilamentous green algae (Trexler et al. 2015).

Many of these features are found in other karstic wetlands around the Caribbean (Gaiser et al. 2012). For example, extensive calcareous periphyton mats are characteristic of karstic wetlands in the Yucatan, southern Belize, and Jamaica (La Hée and Gaiser 2012). In Belize and the Yucatan, these karstic wetlands support Eltonian pyramids of biomass that are similar to the Everglades (Gaiser et al. 2012). For example, aquatic snails, widely found as dominant herbivores in freshwater systems, are in low biomass and numbers in Caribbean karstic wetlands, including the Everglades (Ruehl and Trexler 2011). While these other Caribbean systems may support nesting wading bird populations and crocodilians, they do not support the large populations that characterized the Everglades historically. This difference may be related to the large size of the Everglades—the landscape provides much larger areas to produce and concentrate nesting-season prey (Lake 2011).

The estuaries of the Everglades are also functionally distinct from classically defined estuaries. Estuaries are aquatic systems where fresh water mixes with ocean water. In most estuaries, the fresh water is supplied by rivers that are the dominant source of both nutrients and organic matter, and this upstream subsidy is largely responsible for the high productivity that characterizes these ecosystems. By contrast, the estuaries of the Everglades receive freshwater inflows from oligotrophic marshes, not from nutrient-rich rivers. In fact, the Gulf of Mexico supplies P—the limiting nutrient—to Everglades estuaries. This unique hydrologic situation creates gradients of increasing ecosystem productivity as one progresses down-estuary from the freshwater end member to the coast. We refer to Everglades estuaries as "upside down" because the nutrient supply and productivity patterns are quite literally opposite of most other estuaries (Childers et al. 2006).

In summary, the Everglades has many aspects of its ecology in common with other karstic wetlands of the Caribbean basin, but these characteristics are unusual when compared with most other aquatic systems. Thus, the Everglades has a distinctive ecology that could be considered unique because of the combination of its Caribbean karst geology and large aerial extent. For Douglas, the Everglades was a uniquely mysterious landscape, "never wholly known." The more we learn about this landscape, though, the less unknown and perhaps less unique it is, but the mystery and adventure it evokes continue to draw visitors from around the world and to inspire support for its conservation.

Future of Iconic Status in Doubt?

By the mid-20th century, concerns over changes to the landscape and the related endangerment and loss of characteristic plant and animal species brought increased national and international attention to the Everglades. Scientists working for state and federal resource agencies focused their attention on understanding and cataloging these losses. During the 1980s, environmental organizations and activists were able to use this research to demand a shift in the way natural resource agencies approached their stewardship responsibilities, ultimately leading to more holistic and collaborative approaches to Everglades science and management (Ogden 2008). This revitalized scientific and political attention produced a broad understanding of the catastrophic impacts that drainage and development have had on the Everglades (Davis and Ogden 1994). Importantly, this shift in attention has altered the iconic status of the Everglades. Instead of being an exemplar of unique wilderness, the ecosystem has instead become famous for changes to the landscape and related efforts at restoration. For example, since 1993 the United Nations Educational, Scientific and Cultural Organization (UNESCO) has included ENP on its list of World Heritage Sites in Danger, reflecting continuing concerns over the future of both the park and the Greater Everglades Ecosystem.

Like many rivers and wetlands in the world today, the Everglades landscape has been dramatically engineered and managed to meet social and economic goals. Building on earlier efforts at drainage and flood control (see Chapter 1), the U.S. Army Corps of Engineers (USACE) and its state partners have transformed the river of grass into the "world's largest plumbing works" through construction of the Central & Southern Florida Flood Control (C&SF) Project, which began in the early 1950s. South Florida is consequently a radically human-altered landscape. But it is far from the only one. Consider the countless river valleys that have been inundated and the river deltas that have been desiccated by the planet's 50,000 mega-dams. Or consider Central Asia's once-massive Aral Sea, reduced by Soviet-era irrigation diversions to a tenth of its original size and surrounded by barren plains laden with salt and toxic chemicals.

What sets the C&SF Project apart from these dams and river diversions, however, is the scope and complexity of its engineered components and the management interventions and energy required to operate the system. As noted in Chapter 1, USACE engineers compartmentalized the "River of Grass" with levees and canals into the 1,215-km^2 Everglades Agricultural Area (EAA) and the 3,445-km^2 Water Conservation Areas (WCAs), built a 160-km "tie-back levee" along its eastern border to protect developed areas on the Atlantic coastal ridge, extended the Herbert Hoover Dike almost entirely around Lake Okeechobee, channelized the meandering Kissimmee River, and installed roughly a thousand pumps and water control structures. The entire system is jointly managed by the USACE, which controls water levels in Lake Okeechobee and the WCAs, and the South Florida Water Management District, which moves water through the rest of the system from a control room in its West Palm Beach headquarters. This degree of structural and operational complexity is necessary because south Florida is one of the most challenging places in the world to manage water—thanks to both its geomorphology (low elevation, flat topography, and porous geology) and its intense urban and agricultural development.

Reflecting these changes, some ecologists have applied the concept of the "novel ecosystem" to the Everglades (Zweig and Kitchens 2010). A novel ecosystem is one that "differs in composition and /or function from present and past systems as a consequence of changing species distributions, environmental alternation through climate and land use change, and shifting values about nature and ecosystems" (Hobbs et al. 2013a, 2013b). This concept has been criticized for its lack of precise definitions and reliance on system dynamics with thresholds that cannot be retraced (Murcia et al. 2014). While such threshold dynamics, called alternative stable states, have been recognized for some time (Fig. 2.2), not all human-impacted systems have crossed such thresholds, and it may be difficult to determine if they have. Thus, there is no clear delineation for what is and is not a novel ecosystem. A major motivation for the novel ecosystem discussion is to force ecologists to acknowledge that ecosystems change and humans have major impacts on those trajectories (Hobbs et al. 2013a).

Kitchens and Zweig (2010) suggested that the Everglades has been so transformed by human actions that it is no longer the Everglades, and therefore it cannot be restored to recapture historical ecological function or species composition. The challenges created by human alterations of the Everglades are outlined in detail throughout this book, including nutrient enrichment, compartmentalization, loss of area, and introduction of species. Certainly, areas of the Everglades are fundamentally changed from their historical state, with limited hope of restoring some functions or returning to past states (Ewel 2013). Zweig and Kitchens (2010) focused on the loss of ecosystem size (50% of its historical area), peat loss, and degradation of topography. Some of these impacts cover large areas, but others are more localized in sections of the Everglades. Notably, what remains of the Everglades is still the largest undeveloped tract of land in the United States east of the Mississippi River. Given these changes to its size and

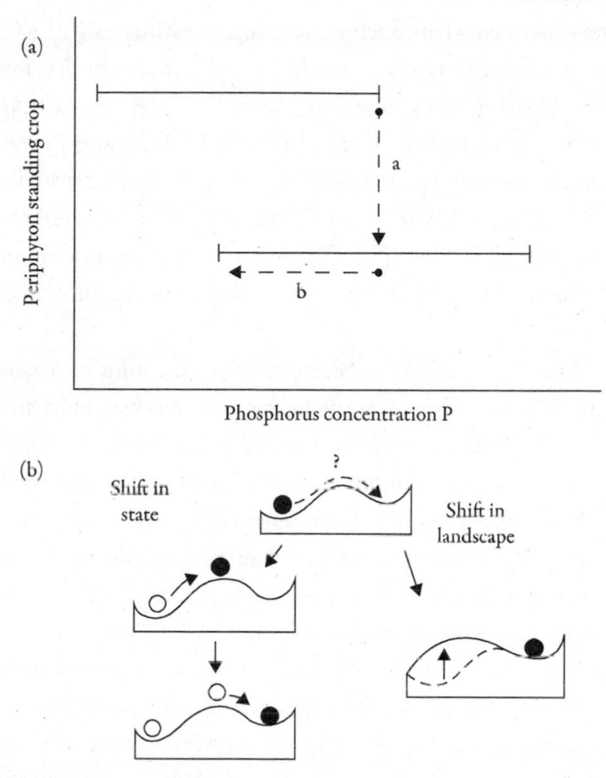

FIGURE 2.2 Alternative stable states in the Everglades. A) Phosphorus (P) enrichment in the Everglades leads to loss of the periphyton mat (indicated by 'a'). Once P is introduced to the environment, addition of low P water will not restore mats in a phenomenon called hysteresis (indicated by 'b'). Re-drawn from Dong et al. (2002). B) Phosphorus addition → loss of periphyton mats → cascading implications for aquatic food web function (Trexler et al. 2015) producing an alternative stable system state (left sequence). If a system can return to its former state, it is characterized as resilient. However, if enough P is added, the landscape of possible states has changed so that the system may not return (right sequence). In the Everglades, this occurs when P enrichment causes a state change to a cattail monoculture (redrawn from Beisner et al. 2003).

spatial extent, does the iconic Everglades still exist? If not, can it be rehabilitated, if not restored?

The Comprehensive Everglades Restoration Plan (CERP) was founded on optimism that some historical ecological structures and functions remain and can be conserved, and that some of those that have been lost can be recovered (Sklar et al. 2005). Debates about novel ecosystems are largely a critique of environmental management that assumes ecosystem stasis. Since no ecosystems are truly static, forcing systems into some image of a past state, however noble this goal may be, is a fundamentally flawed approach. Even in the absence of human intervention, many ecosystems are unlikely to be the same today as they were a century ago. Recognizing this reality, CERP legislation and supporting activities are very practical in their goals. A period of development of goals for CERP called "the Re-Study" produced a hydrologic model that envisioned less water in the ecosystem than during predrainage estimates (USACE 1999). Thus, CERP was never intended to re-create the historical system; this has long been known to be impossible (USACE 1999: Summary, pp. IX–X). A recent update to the restoration goals of CERP, called the Central Everglades Planning Process (CEPP), has further narrowed the hydrologic targets for restoration as an intermediate step toward the original goals of CERP (USACE 2014). Thus, CERP is expected to produce a very practical and streamlined version of the historical ecosystem, not an unattainable historical icon. Our question is: Will this narrow vision for the future of the ecosystem be the Everglades?

Several areas of the Everglades have been altered to the point of passing into an alternative stable state. The most obvious are areas that have received nutrient enrichment via canal inflows, which leads to a eutrophic state defined by a loss of periphyton mats, succession to a cattail monoculture, highly P-enriched soils, and periodic anoxic conditions (Childers et al. 2006; Gaiser et al. 2006; Hagerthey et al. 2008; Ogden et al. 2005; Osborne et al. 2011). There is little likelihood that these marshes will return to the vegetative structure or trophic function that characterizes karstic Caribbean wetlands (Gaiser et al. 2012), though aggressive management interventions may return some botanical and ecological features (Hagerthey et al. 2014). Other areas of the Everglades have been drained, farmed, then later overtaken by non-native plants when farming was abandoned or forced out (Ewel 2013; Smith et al. 2011). Attempts to restore some of these areas are under way, though recovery of their historical vegetation is not expected because the remediation process of removing the P-enriched soil has changed their elevational profile (Dalrymple et al. 2003). Much peat soil has been lost from Everglades marshes because of over-drying, oxidation, and intense fires, particularly in the northern end of the ecosystem (McVoy et al. 2011), while in the southern end of the ecosystem these soils are further threatened by saltwater encroachment associated with sea level rise (Chambers et al. 2014). Finally, non-native species (plants, reptiles, fish, aquatic snails, algae, and even birds) have invaded the ecosystem, some of which are now found throughout the entire landscape. A fraction of these invasives (e.g., old world climbing fern, melaleuca, Brazilian pepper, and Burmese python) have had demonstrable, widespread impacts beyond their presence, and there is little hope of extirpating most of these taxa (CISRERP 2014). In summary, our actions to control and manage water in the Everglades since the start of the 20th century have substantially modified ecosystem structure and function from its historical condition.

However, elements of the iconic ecosystem remain. While we have lost aspects of the historical Everglades, large areas of the Everglades remain in conditions probably indistinguishable from the historical ecosystem, particularly at the scale experienced by an observer. If one travels in an airboat into much of the Shark River Slough in ENP, or WCA 3A south of the Alligator Alley, and stands on the seat to scan the horizon, it is likely that the view is indistinguishable from that of the pre-Columbian human inhabitants or subsequent explorers standing in their boats. The expanse of such areas remains huge compared to other preserved habitats in the eastern United States.

In many respects these vast expanses are fundamentally resilient, hardly representing a new ecological trajectory unprecedented and unpredictable from past conditions and current ecological science. Labeling the Everglades as a novel ecosystem shifts the paradigm away from scientific approaches that seek to understand the system's resiliency, as well as management approaches directed at supporting system resilience even under changing social-ecological conditions. Hobbs et al. (2013a) challenged ecologists to envision the future of ecosystems rather than looking to the past. Accelerating sea level rise and uncertain future climates may put the future of the Everglades in doubt (Aumen et al. 2015). Scheffer et al. (2015) discussed the importance of maintaining a "safe operating space" for iconic ecosystems facing uncertain futures because of climate change and global change. This is an apt characterization of ongoing efforts to manage the Everglades. Increasing freshwater flow recaptures fundamental processes that maintain ecosystem function while buying time against saltwater intrusion resulting from rising seas (Catano et al. 2015; Pearlstine et al. 2010). Though the long-term future of the ecosystem is challenged by sea level rise, the immediate future should benefit from the plans of CERP and CEPP. Whether the Everglades still exists as an icon or a novel ecosystem, the benefits of planned restoration activities are clear. Whether the public will support investment in a novel ecosystem instead of the iconic Everglades is less clear. Since the designation of novel ecosystem is largely semantic, we believe that there is no benefit in doing so.

Conclusion

Douglas's classic *The Everglades: River of Grass* brought widespread attention to the Everglades, and her iconic book became a catalyst for public support for Everglades protection. Her words continue to shape the ways in which we view and value the landscape. In this chapter, we have explored the ways in which contemporary research both confirms and complicates her vision of the Everglades as a river of grass and as a unique landscape that lies outside the boundaries of civilization. In doing so, we hope to draw attention to the landscape's social and ecological characteristics that motivate our continued intellectual curiosity and commitments to the future of the Everglades. Like all complex social-ecological systems, the Everglades is dynamic. Rather than simply reclassifying it as a novel ecosystem, we have drawn on a rich array of archaeological, historical, and ecological research to understand its resilience in the face of change. Restoration efforts of CERP are justified in optimism that recapturing core ecological features of the Everglades (river-like flow, distinctive oligotrophic food webs, and lengthened hydroperiods with climatically driven seasonal fluctuation) will return and/or maintain the system within its historical state as an iconic Caribbean karstic wetland.

Acknowledgments

We wish to thank Laurel Larsen, University of California, Berkeley; William Loftus, ret. U.S. Geological Survey; and Jeff Kline, Everglades National Park, for providing suggestions on technical issues addressed in this chapter.

References

Ardren, T., J.P. Lowry, M. Memory, K. Flanagan, and A. Busot. 2016. Prehistoric human impact on tree island lifecycles in the Florida Everglades. *The Holocene* 26(5): 772–780.

Aumen, N.G., K.E. Havens, G.R. Best, and L. Berry. 2015. Predicting ecological responses of the Florida Everglades to possible future climate scenarios: introduction. *Environmental Management* 55: 741–748.

Beerens, J.M., D.E. Gawlik, G. Herring, and M.I. Cook. 2011. Dynamic habitat selection by two wading bird species with divergent foraging strategies in a seasonally fluctuating wetland. *Auk* 128: 651–662.

Beisner, B.E., D.T. Haydon, and K. Cuddington. 2003. Alternative stable states in ecology. *Frontiers in Ecology and the Environment* 1: 376–382.

Boucek, R.E., and J.S. Rehage. 2013. No free lunch: displaced marsh consumers regulate a prey subsidy to an estuarine consumer. *Oikos* 122: 1453–1464.

Carr, R.S. 2002. The archaeology of tree islands. In *Tree Islands of the Everglades*, edited by F.H. Sklar and A. van der Valk, 187–206. Dordrecht, the Netherlands: Kluwer Academic Publishers.

Catano, C.P., J.M. Beerens, L. Brandt, K.M. Hart, F.J. Mazzotti, S. Romanach, L. Pearlstine, and J.C. Trexler. 2015. Using scenario planning to evaluate the impacts of climate change on wildlife populations and communities in the Florida Everglades. *Environmental Management* 55: 807–823.

Chambers, L.G., S.E. Davis, T. Troxler, J.N. Boyer, A. Downey-Wall, and L.J. Scinto. 2014. Biogeochemical effects of simulated sea level rise on carbon loss in an Everglades mangrove peat soil. *Hydrobiologia* 726(1): 195–211.

Childers, D.L., J.N. Boyer, S.E. Davis III, C.J. Madden, D.T. Rudnick, and F.H. Sklar. 2006. Relating precipitation and water management to nutrient concentration patterns in the oligotrophic "upside down" estuaries of the Florida Everglades. *Limnology & Oceanography* 51(1): 602–616.

CISRERP (Committee of Independent Scientific Review of Everglades Restoration Progress). 2014. *Progress Toward Restoring the Everglades. The Fifth Biennial Review—2014*. Washington, DC: National Academies Press.

Coultas, C.L., M. Schwadron, and J.M. Galbraith. 2008. Petrocalcic horizon formation and prehistoric people's effect on Everglades tree island soils, Florida. *Soil Survey Horizons* 49: 16–21.

Craighead, F.C. 1971. *Trees of South Florida*. Coral Gables, FL: University of Miami Press.

Dalrymple, G.H., R.F. Doren, N.K. O'Hare, M.R. Norland, and T.V. Armentano. 2003. Plant colonization after complete and partial removal of disturbed soils for wetland restoration of former agricultural fields in Everglades National Park. *Wetlands* 23: 1015–1029.

Davis, S.M., and J.C. Ogden (eds.). 1994. *Everglades: The Ecosystem and Its Restoration*. Boca Raton, FL: St. Lucie Press.

Deverell, W. 2006. "Niagara magnified": finding Emerson, Muir, and Adams in Yosemite. In *Yosemite: Art of an American Icon*, edited by A. Scott, 9–22. Berkeley: University of California Press.

Dong, Q., P.V. McCormick, F.H. Sklar, and D.L. DeAngelis. 2002. Structural instability, multiple stable states, and hysteresis in periphyton driven by phosphorus enrichment in the Everglades. *Theoretical Population Biology* 61: 1–13.

Ewel, J.J. 2013. Case study: hole-in-the-donut, Everglades. In *Novel Ecosystems: Intervening in the New Ecological World Order*, edited by R.J. Hobbs, E.S. Higgs, and C.M. Hall, 11–15. New York: Wiley-Blackwell.

Fradkin, A. 2016. Early human settlement and natural formation of the Florida Everglades, USA: the icthyoarchaeological evidence. *Journal of Archaeological Science: Reports* 8: 463–469.

Frederick, P., D.E. Gawlik, J.C. Ogden, M.I. Cook, and M. Lusk. 2009. The White Ibis and Wood Stork as indicators for restoration of the Everglades ecosystem. *Ecological Indicators* 9: S83–S95.

Gaiser, E.E., E.P. Anderson, E. Castañeda-Moya, L. Collado-Vides, J.W. Fourqurean, M.R. Heithaus, R. Jaffe, D. Lagomasino, N. Oehm, R.M. Price, V.H. Rivera-Monroy, R. Roy Chowdhury, and T. Troxler. 2015. New perspectives on an iconic landscape from comparative international long-term ecological research. *Ecosphere* 6(10): 1–18.

Gaiser, E.E., D.L. Childers, R.D. Jones, J.H. Richards, L.J. Scinto, and J.C. Trexler. 2006. Periphyton responses to eutrophication in the Florida Everglades: cross-system patterns of structural and compositional change. Pt. 2, *Limnology and Oceanography* 51(1): 617–630.

Gaiser, E.E., J.C. Trexler, and P.R. Wetzel. 2012. The Florida Everglades. In *Wetland Habitats of North America: Ecology and Conservation Concerns*, edited by D.P. Batzer and A.H. Baldwin, 231–252. Berkeley: University of California Press.

Gawlik, D.E. 2002. The effects of prey availability on the numerical response of wading birds. *Ecological Monographs* 72: 329–346.

Graf, M., M. Schwadron, P.A. Stone, M. Ross, and G.L. Chmura. 2008. An enigmatic carbonate layer in Everglades tree island peats. *Eos* 89(12): 117–118.

Griffin, J.W. 1996. *Fifty Years of Southeastern Archaeology*, edited by P.C. Griffin. Gainesville: University Press of Florida.

Grunwald, M. 2006. *The Swamp: The Everglades, Florida and the Politics of Paradise*. New York: Simon & Schuster.

Haas, S.E., R.T. Kimball, J. Martin, and W.M. Kitchens. 2009. Genetic divergence among Snail Kite subspecies: implications for the conservation of the endangered Florida Snail Kite *Rostrhamus sociabilis*. *IBIS* 151: 181–185.

Hagerthey, S., B. Bellinger, K. Wheeler, M. Gantar, and E. Gaiser. 2011. Everglades periphyton: a biogeochemical perspective. *Critical Reviews in Environmental Science and Technology* 41(S1): 309–343.

Hagerthey, S.E., M.I. Cook, R.M. Kobza, S. Newman, and B.J. Bellinger. 2014. Aquatic faunal responses to an induced regime shift in the phosphorus-impacted Everglades. *Freshwater Biology* 59: 1389–1405.

Hagerthey, S.E., S. Newman, K. Rutchey, E.P. Smith, and J. Godin. 2008. Multiple regime shifts in a subtropical peatland: community-specific thresholds to eutrophication. *Ecological Monographs* 78: 547–565.

Hillis, D.M., M.T. Dixon, and A.L. Jones. 1991. Minimal genetic variation in a morphologically diverse species (Florida tree snail, *Liguus fasciatus*). *Journal of Heredity* 82: 282–286.

Hobbs, R.J., E.S. Higgs, and C.M. Hall. 2013a. Introduction: why novel ecosystems. In *Novel Ecosystems: Intervening in the New Ecological World Order*, edited by R.J. Hobbs, E.S. Higgs, and C.M. Hall, 3–8. New York: Wiley-Blackwell.

Hobbs, R.J., E.S. Higgs, and C.M. Hall. 2013b. Defining novel ecosystems. In *Novel Ecosystems: Intervening in the New Ecological World Order,* edited by R.J. Hobbs, E.S. Higgs, and C.M. Hall, 58–60. New York: Wiley-Blackwell.

Jenkins, D.G., and D. Rinne. 2008. Red herring or low illumination? The peninsula effect revisited. *Journal of Biogeography* 35: 2128–2137.

Johnson, W.E., D.P. Onorato, M.E. Roelke, E.D. Land, M. Cunningham, R.C. Belden, R. McBride, D. Jansen, M. Lotz, D. Shindle, J. Howard, D.E. Wildt, L.M. Penfold, J.A. Hostetler, M.K. Oli, and S.J. O'Brien. 2010. Genetic restoration of the Florida Panther. *Science* 329: 1641–1645.

Jones, A.L., E.C. Winte, and O.L. Bass, Jr. 1981. *The Status of Florida Tree Snails (Liguus fasciatus), Introduced to Everglades National Park*. South Florida Research Center Report T-622.

Kaplan, D.A., R. Paudel, M.J. Cohen, and J.W. Jawitz. 2012. Orientation matters: patch anisotropy controls discharge. *Geophysical Research Letters* 39: L17401.

Kohler, R. 2006. *All Creatures: Naturalists, Collectors, and Biodiversity, 1850–1950*. Princeton, NJ: Princeton University Press.

La Hée, J., and E.E. Gaiser. 2012. Benthic diatom assemblages as indicators of water quality in the Everglades and three tropical karstic wetlands. *Freshwater Science* 31: 205–221.

Lake, P.S. 2011. *Drought and Aquatic Ecosystems: Effects and Responses.* Chichester, UK: Wiley-Blackwell.

Larsen, L., N. Aumen, C. Bernhardt, V. Engel, T. Givnish, S. Hagerthey, J. Harvey, L. Leonard, P. McCormick, C. McVoy, G. Noe, M. Nungesser, K. Rutchey, F. Sklar, T. Troxler, J. Volin, and D. Willard. 2011. Recent and historic drivers of landscape change in the Everglades ridge, slough, and tree island mosaic, *Critical Reviews in Environmental Science and Technology* 41(S1): 344–381.

Larsen, L., J.W. Harvey, and J.P. Crimaldi. 2007. A delicate balance: ecohydrological feedbacks governing landscape morphology in a lotic peatland. *Ecological Monographs* 77: 591–614.

Leopold, L.B. 1994. *A View of the River*. Cambridge, MA: Harvard University Press.

Loftus, W.F., and J.A. Kushlan. 1987. Freshwater fishes of southern Florida. *Bulletin of the Florida State Museum, Biological Sciences* 31: 147–344.

Marquardt, W.H. 2010. Shell mounds in the southeast: middens, monuments, temple mounds, rings, or works? *American Antiquity* 75(3): 551–570.

Martin, J., W.M. Kitchens, and J.E. Hines. 2007. Importance of well-designed monitoring programs for the conservation of endangered species: case study of the snail kite. *Conservation Biology* 21: 472–481.

Matich, P., and M.R. Heithaus. 2014. Multi-tissue stable isotope analysis and acoustic telemetry reveal seasonal variability in the trophic interactions of juvenile bull sharks in a coastal estuary. *Journal of Animal Ecology* 83: 199–213.

McClain, M.E., E.W. Boyer, C.L. Dent, S.E. Gergel, N.B. Grimm, P.M. Groffman, S.C. Hart, J.W. Harvey, C.A. Johnston, E. Mayorga, W.H. McDowell, and G. Pinay. 2003. Biogeochemical hot spots and hot moments at the interface of terrestrial and aquatic ecosystems. *Ecosystems* 6: 301–312.

McVoy, C.W., W.P. Said, J. Obeysekera, J. VanArman, and T. Dreschel. 2011. *Landscapes and Hydrology of the Pre-drainage Everglades*. Gainesville: University Press of Florida.

Means, D.B., and D. Simberloff. 1987. The peninsula effect: habitat-correlated species decline in Florida's herpetofauna. *Journal of Biogeography* 14: 551–568.

Murcia, C., J. Aronson, G.H. Katton, D. Moreno-Mateos, K. Dixon, and D. Simberloff. 2014. A critique of the "novel ecosystem" concept. *Trends in Ecology & Evolution* 29(10): 548–553.

Nanson, G.C., and A.D. Knighton. 1996. Anabranching rivers: their cause, character and classification. *Earth Surface Processes and Landforms* 21(3): 217–239.

National Parks and Conservation Association (NPCA). 2013. America's Iconic Landscapes, map. http://www.archive-org-2013.com/open-archive/1275060/2013-01-31/3d1fd3785ffe942a8a75c09cbdf8985e.

Ogden, J.C., S.M. Davis, T.K. Barnes, K.J. Jacobs, and J.H. Gentile. 2005. Total system conceptual ecological model. *Wetlands* 25: 955–979.

Ogden, L. 2008. The Everglades ecosystem and the politics of nature. *American Anthropologist* 101(1): 21–32.

Orem, W.H., D.A. Willard, H.E. Lerch, A.L. Bates, A. Boylan, and M. Comm. 2002. Nutrient geochemistry of sediments from two tree islands in Water Conservation Area 3B, the Everglades, Florida. In *Tree Islands of the Everglades*, edited by F.H. Sklar and A.G. van der Valk, 153–186. Dordrecht, the Netherlands: Kluwer Academic Publishers.

Osborne, T.Z., S. Newman, P. Kalla, D.J. Scheidt, G.L. Bruland, M.J. Cohen, L.J. Scinto, and L.R. Ellis. 2011. Landscape patterns of significant soil nutrients and contaminants in the Greater Everglades Ecosystem: past, present, and future. *Critical Reviews in Environmental Science and Technology* 41(S1): 121–148.

Palmer, M.A., and N.L. Poff. 1997. The influence of environmental heterogeneity on patterns and processes in streams. *Journal of the North American Benthological Society* 16: 169–173.

Pearlstine, L.G., E.V. Pearlstine, and N.G. Aumen. 2010. A review of the ecological consequences and management implications of climate change for the Everglades. *Journal of the North American Benthological Society* 29: 1510–1526.

Quillen, A., E. Gaiser, and E. Grimm. 2013. Diatom-based paleolimnological reconstruction of regional climate and local land-use change from a protected sinkhole lake in southern Florida, U.S.A. *Journal of Paleolimnology* 49: 15–30.

Raffaele, H., J. Wiley, O. Garrido, A. Keith, and J. Raffaele. 1998. *A Guide to the Birds of the West Indies*. Princeton, NJ: Princeton University Press.

Rosenblatt, A.E., and M.R. Heithaus. 2011. Does variation in movement tactics and trophic interactions among American alligators create habitat linkages? *Journal of Animal Ecology* 80: 786–798.

Ruehl, C.B., and J.C. Trexler. 2011. Comparisons of snail density, standing stock, and body size among freshwater ecosystems: a review. *Hydrobiologia* 665: 1–13.

Russell, G.J., O.L. Bass, and S.L. Pimm. 2002. The effect of hydrological patterns and breeding-season flooding on the numbers and distribution of wading birds in Everglades National Park. *Animal Conservation* 5: 185–199.

Scheffer, M., S. Barrett, S.R. Carpenter, C. Folke, A.J. Green, M. Holmgren, T.P. Hughes, S. Kosten, I.A. van de Leemput, D.C. Nepstad, E.H. van Nes, E.T.H.M. Peeters, and B. Walker. 2015. Creating a safe operating space for iconic ecosystems. *Science* 347: 1317–1319.

Schwadron, M. 2006. Everglades tree islands prehistory: archaeological evidence for regional Holocene variability and early human settlement. *Antiquity* 80(310): Project Gallery. http://antiquity.ac.uk/ProjGall/schwadron/index.html.

Science Coordination Team (SCT). 2003. The role of flow in the Everglades ridge and slough landscape. South Florida Ecosystem Restoration Working Group, U.S. Geological Survey. http://sofia.usgs.gov/publications/papers/sct_flows/i.

Scott, J.C. 1998. *Seeing Like a State: How Certain Schemes to Improve the Human Condition Have Failed*. New Haven, CT: Yale University Press.

Sklar, F.H., M.J. Chimney, S. Newman, P. McCormick, D. Gawlik, S.L. Miao, C. McVoy, W. Said, J. Newman, C. Coronado, G. Crozier, M. Korvela, and K. Rutchey. 2005. The ecological-societal underpinnings of Everglades restoration. *Frontiers in Ecology and the Environment* 3: 161–169.

Sklar, F.H., and A.G. van der Valk. 2002. Tree islands of the Everglades: an overview. In *Tree Islands of the Everglades*, edited by F.H. Sklar and A.G. van der Valk, 1–18. Dordrecht, the Netherlands: Kluwer Academic Publishers.

Smith, C.S., L. Serra, Y.C. Li, P. Inglett, and K. Inglett. 2011. Restoration of disturbed lands: the hole-in-the-donut restoration in the Everglades. *Critical Reviews in Environmental Science and Technology* 41(S1): 723–739.

Trexler, J.C. 1995. Restoration of the Kissimmee River: a conceptual model of past and present fish communities and its consequences for evaluating restoration success. *Restoration Ecology* 3: 195–210.

Trexler, J.C., E.E. Gaiser, J.S. Kominoski, and J. Sanchez. 2015. The role of periphyton mats in consumer community structure and function in calcareous wetlands: lessons from the Everglades. In *Microbiology of the Everglades Ecosystem*, edited by J.A. Entry, A.D. Gottlieb, K. Jayachandrahan, and A. Ogram, 155–179. Boca Raton, FL: CRC Press.

Trexler, J.C., W.F. Loftus, C.F. Jordan, J. Chick, K.L. Kandl, T.C. McElroy, and O.L. Bass. 2002. Ecological scale and its implications for freshwater fishes in the Florida Everglades. In *The Everglades, Florida Bay, and Coral Reefs of the Florida Keys: An Ecosystem Sourcebook*, edited by J.W. Porter and K.G. Porter, 153–181. Boca Raton, FL: CRC Press.

Troxler-Gann, T., and D.L. Childers. 2006. Relationships between hydrology and soils describe vegetation patterns in three seasonally flooded tree islands of the southern Everglades, Florida. *Plant & Soil* 279: 271–286.

Turner, A.M., J.C. Trexler, F. Jordan, S.J. Slack, P. Geddes, and W. Loftus. 1999. Targeting ecosystem features for conservation: standing crops in the Florida Everglades. Conservation Biology 13: 898–911.

U.S. Army Corps of Engineers (USACE). 1999. Central and Southern Florida Comprehensive Review Study. Final Integrated Feasibility Report and Programmatic Environmental Impact Statement. http://141.232.10.32/pub/restudy_eis.aspx

U.S. Army Corps of Engineers (USACE). 2014. Central and Southern Florida Project Comprehensive Everglades Restoration Plan. Central Everglades Planning Project Final Integrated Project Implementation Report and Environmental Impact Statement. http://www.saj.usace.army.mil/Portals/44/docs/Environmental/CEPP/01_CEPP%20Final%20PIR-EIS%20Main%20Report.pdf

Vileisis, A. 1997. *Discovering the Unknown Landscape: A History of American's Wetlands*. Washington, DC: Island Press.

Walters, J.R., S.R. Beissinger, J.W. Fitzpatrick, R. Greenberg, J.D. Nichols, H.R. Pulliam, and D.W. Winkler. 2000. The AOU Conservation Committee review of the biology, status, and management of Cape Sable Seaside Sparrows: final report. *Auk* 117: 1093–1115.

Wetzel, P.R., A.G. van der Valk, S. Newman, C.A. Coronado, T.G. Troxler-Gann, D.L. Childers, W.H. Orem, and F.H. Sklar. 2009. Heterogeneity of phosphorus distribution in a patterned landscape, the Florida Everglades. *Plant Ecology* 200: 83–90.

Wetzel, P.R., A.G. van der Valk, S. Newman, D.E. Gawlik, T. Troxler-Gann, C.A. Coronado-Molina, D.L. Childers, and F.H. Sklar. 2005. Nutrient redistribution key to maintaining tree islands in the Florida Everglades *Frontiers in Ecology* 3: 370–376.

Zweig, C.L., and W.M. Kitchens. 2010. The semiglades: the collision of restoration, social values, and the ecosystem concept. *Restoration Ecology* 18(2): 138–142.

3

Water, Sustainability, and Survival

René Price and Katrina Schwartz with Bill Anderson, Ross Boucek, Henry Briceño, Mark Cook, Carl Fitz, Michael Heithaus, Jeff Onsted, Jennifer Rehage, Victor Rivera-Monroy, Rinku Roy Chowdhury, and Amartya Saha

In a Nutshell

- Large-scale alterations in water flow in south Florida have provided some relief from catastrophic flooding but at the expense of the Everglades and of groundwater supplies.
- Too much fresh water is discharged to the coast, particularly to estuaries east and west of Lake Okeechobee.
- This loss of water "to tide" means there is not enough water to rehydrate the Everglades and to reduce saltwater intrusion into the Biscayne Aquifer.
- Climate change will bring altered rainfall patterns, increased evapotranspirative losses, and accelerating sea level rise, all of which will transform the coastal Everglades via peat collapse, salinity intrusion, and mangrove transgression.
- Climate change will also further expose the limitations of the south Florida water management system to provide both fresh water and flood protection.

Introduction

Florida is located at the desert latitudes of the world, but its peninsular oceanic setting between tropical maritime and continental weather systems makes it the wettest place in the eastern United States. Despite its high rainfall, Florida is aptly nicknamed the Sunshine State, and most of the rainfall is lost through evapotranspiration (ET; Abtew 1996, 2004). Rainfall and ET thus constitute the most important components of the water cycle in south Florida (McPherson and Halley 1997), and together they heavily influence the availability of freshwater resources—both surface water and groundwater.

The Greater Everglades watershed encompasses 28,000 km^2 from the headwaters of the Kissimmee River, just south of Orlando, to Florida Bay and Biscayne Bay (Fig. 3.1; Light and Dineen 1994). Until it was replumbed in the 20th century for flood control and agricultural drainage, the entire area was hydrologically connected (see Chapter 1; McVoy et al. 2011). Surface water flowed south without interruption, driven by a gradual elevation decline from the Kissimmee to Lake Okeechobee and then across the entire southern peninsula. Before discharging into the Gulf of Mexico, the slow-moving sheet flow percolated into the underlying Biscayne Aquifer. The Biscayne is one of the planet's

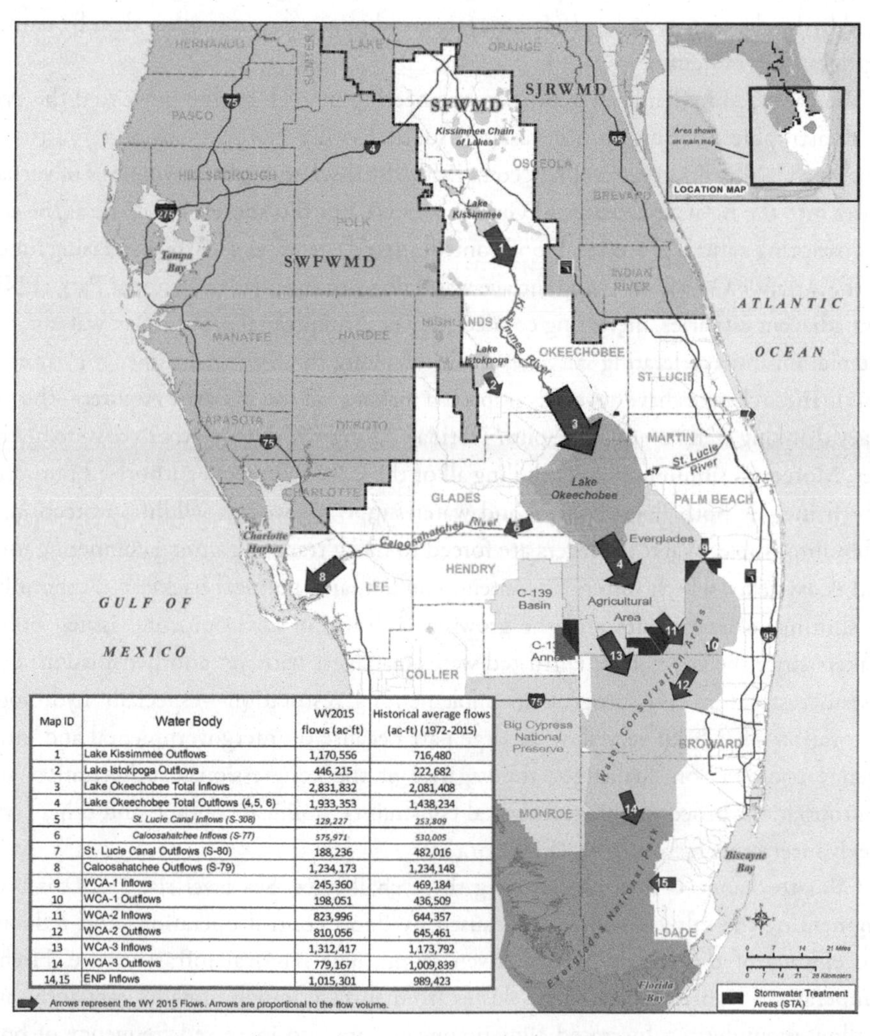

Map ID	Water Body	WY2015 flows (ac-ft)	Historical average flows (ac-ft) (1972-2015)
1	Lake Kissimmee Outflows	1,170,556	716,480
2	Lake Istokpoga Outflows	446,215	222,682
3	Lake Okeechobee Total Inflows	2,831,832	2,081,408
4	Lake Okeechobee Total Outflows (4,5, 6)	1,933,353	1,438,234
5	*St. Lucie Canal Inflows (S-308)*	*129,227*	*253,809*
6	*Caloosahatchee Inflows (S-77)*	*575,971*	*530,005*
7	St. Lucie Canal Outflows (S-80)	188,236	482,016
8	Caloosahatchee Outflows (S-79)	1,234,173	1,234,148
9	WCA-1 Inflows	245,360	469,184
10	WCA-1 Outflows	198,051	436,509
11	WCA-2 Inflows	823,996	644,357
12	WCA-2 Outflows	810,056	645,461
13	WCA-3 Inflows	1,312,417	1,173,792
14	WCA-3 Outflows	779,167	1,009,839
14,15	ENP Inflows	1,015,301	989,423

FIGURE 3.1 Water year 2015 (WY2015) flows in the Greater Everglades watershed, compared to historical average flows (table insert). Source: Abtew and Ciuca (2016).

most productive aquifers and boasts some of the purest water, as it is recharged through the nutrient-poor Everglades peat soils and filtered further by the limestone bedrock. When the aquifer recharges during the wet season, groundwater levels rise above the land surface, contributing to the surface water flow to which the wetland ecosystem is adapted. Rising water levels are a nuisance for farmers, who need to keep the roots of their crops above the groundwater table, and urban residents, who demand that flooding be minimized. As we saw in Chapters 1 and 2, hydrologic modification of the Everglades culminated in the 1950s and 1960s with the U.S. Army Corps of Engineers' (USACE) massive Central & Southern Florida Flood Control (C&SF) Project, which today controls virtually all surface water movement, apart from rainfall, throughout the region. Water flows still link the major components of the Kissimmee–Okeechobee–Everglades watershed, but the upstream–downstream connection is governed by a managed hydraulic system of canals, levees, gates, and pump stations rather than by natural topographic gradients.

The managed hydraulic system is an engineering marvel, having prevented the type of catastrophic flooding that devastated south Florida in the 1920s and 1940s (see Chapter 1). This flood control is accomplished by discharging huge volumes of surface water into the Atlantic Ocean and Gulf of Mexico. But this success has come at the cost of disrupting natural processes. Flood control altered the water cycle by increasing runoff to the Atlantic Ocean, reducing surface water flows to Everglades National Park (ENP) and adjacent estuaries, increasing connectivity of groundwater and surface water across the region, and accelerating saltwater intrusion along the entire coastline (McVoy et al. 2011). These changes have adversely impacted recharge of the Biscayne Aquifer—the primary drinking water source for south Florida—along with many other ecosystem services. Moreover, simultaneously fulfilling all of the C&SF Project's authorized purposes, which include both flood control and water supply, as well as wildlife protection, is often impossible. Water managers are forced to make tradeoffs among competing users and ecosystems, which fosters discontent, conflict, and political backlash. Responding to shifting societal values and the growing visibility of environmental issues, policy makers since the 1980s have entrusted water managers with yet another mission: ecosystem restoration. But progress in implementing restoration—especially hydrologic restoration—has been very slow, in large part because of intergovernmental and interagency coordination challenges, the vagaries of the congressional authorization and appropriations processes, biogeophysical constraints, political conflict, and other complexly interacting factors (CISRERP 2014).

Climate change is now exacerbating these challenges. Sea level rise (SLR) is both augmenting the saltwater intrusion caused by flood control operations and reducing the efficacy of the largely gravity-driven stormwater removal infrastructure. Higher temperatures lead to increased ET losses from surface impoundments, wetlands, and shallow groundwater. Increased climatic uncertainty and increased frequency of both wet and dry extremes compound the operational dilemmas of multipurpose water resource management. Thus, south Florida exemplifies the escalating tension between climate change impacts and growing human demand for an increasingly limited freshwater supply—a tension that poses challenges for coastal ecosystems and communities worldwide, and that will shape the future of the Everglades as a coupled social-ecological system.

We begin this chapter by describing the components and drivers of the south Florida water cycle, including human water use. Next, we discuss how climate change is expected to impact the water cycle and how the cycle shapes ecological patterns across the landscape, from animal movements to vegetation growth. We conclude with an exploration of the complex sociopolitical and environmental challenges water managers confront in operating the flood control system while attempting to rehabilitate ecosystem functions. As noted in Chapters 1 and 2, humans have a long history in south Florida, and their influence on the water cycle of the region, particularly since the 1950s, cannot be overstated.

The South Florida Water Cycle

Water is almost perpetually in motion—flowing in a stream, evaporating into the sky, falling from clouds, crashing on a beach, percolating underground. On the Earth's surface, water is temporarily stored in ponds, lakes, wetlands, rivers, and oceans. Below ground, it is stored as groundwater in aquifers. Water is contained within vegetation and as water vapor in the atmosphere that blankets our planet. The residence time, or length of time water remains in any of these reservoirs, varies from a few hours in a rainwater puddle to millions of years for fossil groundwater in an enclosed aquifer. These and other processes are collectively conceptualized as the *water cycle*. Thanks to water's dynamism, managers and scientific investigators alike must develop methods for quantifying the fluxes (rate of flow) and stocks (reservoirs) of water over time. Such quantification for a specific region over a specific period of time is known as a *water budget* or *water balance*.

This section describes the major components of south Florida's water cycle and water budget across natural, agricultural, and urban areas. Rainfall (the main input) and ET (the main output) are the most influential components of the water cycle in the region, and the relationship between the two affects other components, such as runoff into the sea and groundwater recharge. For example, Figure 3.2 shows monthly and seasonal variation in the water budget for Shark River Slough, ENP's major drainage. Rainfall creates a seasonal freshwater pulse that flows across the Everglades into the sea. This pulse decreases in late winter and spring (February–May), when ET is also very high, causing seawater to intrude seasonally into creeks and aquifers (Saha et al. 2012).

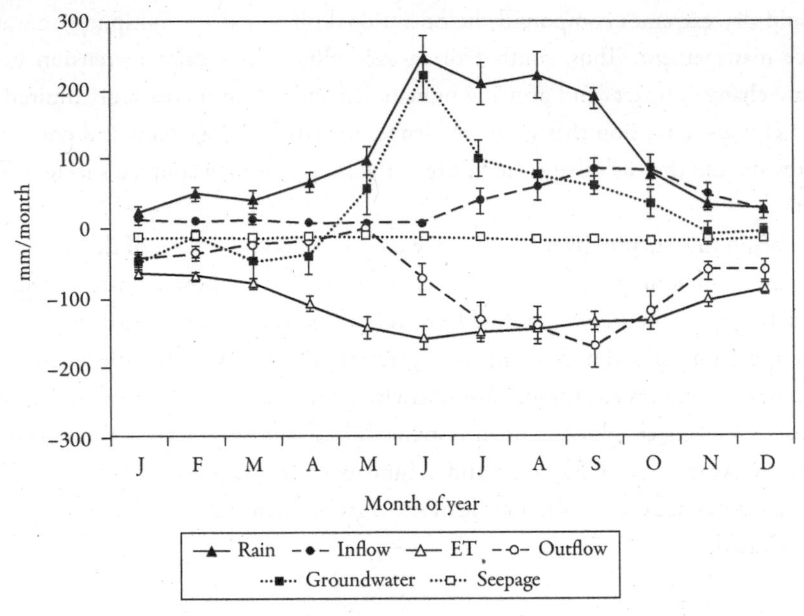

FIGURE 3.2 Average monthly water budget components (2002-2008), with error bars, for the Shark River Slough, Everglades National Park (Source: Saha et al. 2012). Negative amounts on the y-axis indicate water leaving the system (or drainage basin), such as outflows or evapotranspiration (ET).

PRECIPITATION

Rainfall in south Florida averages about 1,200 mm annually and is characterized by pronounced wet and dry seasons. More than 70% falls during the wet season (June–October), with peaks in June and in September and October, at the high point of tropical cyclone activity (Fig. 3.3; Abtew 1996, 2004; Price et al. 2008). Most wet-season rainfall originates as evaporated Atlantic Ocean seawater, but around 7% to 12% is evaporated freshwater from the Everglades or Lake Okeechobee (Abtew 2001; Price et al. 2006). Dry-season rainfall is associated with cold fronts moving from north to south across the peninsula roughly once a week. As is discussed further in Chapter 4, seasonal variability in rainfall is highly correlated with the El Niño-Southern Oscillation (ENSO) and the Atlantic Multi-decadal Oscillation (AMO; Moses et al. 2013). The combined influences of these teleconnections manifest in wet–dry cycles with roughly six years of drought typically following six years of above-average rainfall (Childers et al. 2006). Despite the relatively flat topography of south Florida, rainfall varies spatially with a general increase from southwest to northeast—in other words, from the Florida Keys (drier) to the Palm Beach coastline (wetter; Fig. 3.4; Moses et al. 2013). The urbanized southeast coastal region, comprising Miami-Dade, Broward, and Palm Beach counties, generally receives the most rain (see Fig. 3.4; Table 3.1).

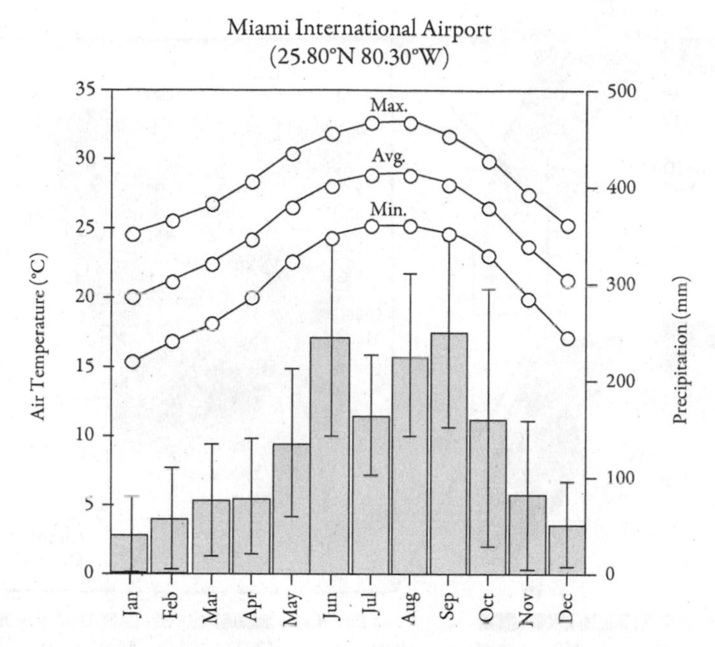

FIGURE 3.3 Thirty-year (1981–2010) average (avg.), maximum (max.) and minimum (min.) monthly air temperature (lines) and precipitation (bars) with standard error at Miami International Airport (data source: Arguez et al. 2010).

EVAPOTRANSPIRATION

The biggest loss of water from south Florida is through ET (see Fig. 3.2), which includes both evaporation from open and standing water in bays, wetlands, canals, roads, and fields, and transpiration by plants. Annual ET values in natural areas of south Florida range from 1,220 mm in the north to 1,370 mm in the south, with lower values in urban and agricultural areas due to lower water availability (Abtew et al. 2003). Solar radiation is the principal driver of ET, but air temperature, relative humidity, and wind speed also play a role. Another factor is variation in daily and longer-term water uptake and water-use efficiency by individual plant species and plant communities. For instance, broadleaf plants and trees transpire more than grasses, and ET is higher in mangroves than sawgrass, except during the dry season, when higher salinity in creeks depresses mangrove transpiration (Barr et al. 2014; Lagomasino 2014; Schedlbauer et al. 2011). Transpiration also tends to increase during the wet season, when photosynthetic rates are highest.

Some of the variability in ET estimates may be attributed to differences in assessment methods (Douglas et al. 2009; Wu and Shukla 2013). Because there is no method for directly measuring ET at a landscape scale (especially for woody vegetation or in areas with a broad mix of plant communities), researchers have employed several approaches to estimate differences in ET among vegetative communities in the Everglades. These include meteorological data-based vapor transport models (Abtew 1996, 2004; German

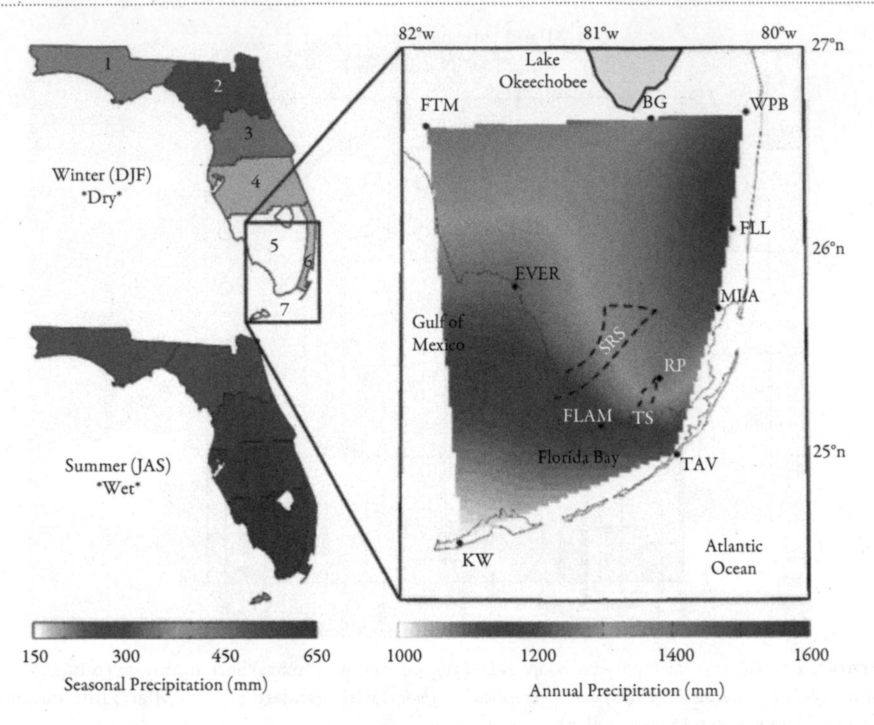

FIGURE 3.4 Left: State-wide seasonal patterns of rainfall; right: Mean annual rainfall in south Florida (source: Moses et al. 2013). See page 12 in color plates.

2000; Saha et al. 2011, 2012; Zapata-Rios and Price 2012), eddy flux measurements (Barr et al. 2014; Schedlbauer et al. 2011), sap flow and diurnal change in the water-table level (Sullivan 2012; Villalobos 2010), and remote sensing (Lagomasino 2014).

A significant portion of the south Florida landscape is vegetated, including natural areas (ENP, Big Cypress National Preserve, Loxahatchee National Wildlife Refuge) as well as farmland and the Water Conservation Areas (WCAs). These landscapes support diverse plant communities that vary in their phenology, flood tolerance, water uptake, and, consequently, rates of ET. Like straws of various sizes sucking water out of the ground, these communities differentially affect local hydrologic conditions. In the marsh, for instance, diurnal transpiration-driven groundwater drawdown is greater beneath the more upland and drier (High Head) plant community of tree islands as compared to the surrounding lower-elevation communities, and this causes groundwater inflow from the latter to the former (Ross et al. 2000 Saha et al. 2010; Sullivan et al. 2012, 2013; Troxler et al. 2014; Wang et al. 2011). Studies of sap flow and diurnal groundwater levels in a region with hardwood hammocks bordered by monospecific stands of the invasive exotic *Schinus terebenthifolius* showed that the largest daytime water-table depression occurred beneath hardwood hammocks, due to high water uptake and ET by hammock plants (Villalobos 2010). Tree islands have higher transpiration per unit area than the

TABLE 3.1.

South Florida Water Management Model (SFWMM) Average Annual Water Budgets for Selected Hydrologic Basins over a 36-Year Record for Current Water Management Practices

Basin	Area (km^2)	Rain (hm^3)	ET (hm^3)	Struct_ out (hm^3)	Struct_ out (hm^3)	Pumpage (hm^3)	SW_in (hm^3)	SW_out (hm^3)	Lseep_ in (hm^3)	Lseep_ out (hm^3)	GW_in (hm^3)	GW_ out (hm^3)	Irrig_in (hm^3)
Lake Okeechobee	1,886	2,056	2,615	2,540	2,012	0	0	0	0	0	0	0	0
EAA+	2,538	3,250	2,378	1,206	2,135	0	0	0	0	0	83	22	0
WCAs	3,377	4,258	4,033	2,266	1,679	0	281	79	2	616	2	434	0
Big Cypress	3,067	4,157	3,091	259	323	0	68	1,184	0	0	38	58	0
ENP	2,559	3,424	3,102	1,669	857	0	440	15	49	365	58	308	0
LEC	5,284	7,801	3,834	648	4,140	1,977	141	165	931	2	630	353	315
Florida Bay	1,999	2,103	3,316	0	0	0	180	0	0	0	0	0	0
System-wide	20,710	27,049	22,367	8,589	11,144	1,979	1,233	2,430	982	982	835	1,176	315

From the SFWMM v6.0, 2050 Base run.

ET, evapotranspiration; Struct, water control structure; SW, surface water flow; Lseep, seepage through a levee; GW, groundwater flow; Irrig, irrigation. All flows in millions of cubic meters or cubic hectometres.

The EAA+ basin includes Holey Land, Rotenberger Tract, STAs, and the 298 basin. WCAs include all five WCAs.

The ENP basin in the SFWMM does not include any of the mangrove regions, with the exception of a small region associated with Taylor Slough.

The LEC basin is the sum of the developed areas of eastern Palm Beach, Broward, and Miami-Dade counties.

IrrIn is landscape/golf course irrigation inputs from the combination of public water supply wells and treated wastewater.

FL Bay rainfall and ET are the average from 1970–2002 (source: Price et al. 2007); and SW_in is an average from 1970–1995 (source: Nuttle et al. 2000).

surrounding wetlands, as inferred from stable isotope studies indicating that tree islands take in water from the surrounding marsh (Saha et al. 2010; Wang et al. 2011).

ET in natural areas peaks during summer, but because irrigated crops are grown year-round in south Florida, wintertime rates of ET are higher in agricultural areas. Methods for computing ET for the latter depend on crop type and seasonality and are far simpler than for heterogeneous natural areas (Kisekka et al. 2010). For example, a study using lysimeters on rice fields in the Everglades Agricultural Area (EAA) measured daily ET losses ranging from 3.6 to 10.9 mm d^{-1} for spring crops (March–July), 3.3 to 11.7 mm d^{-1} for summer crops (May–September), and 1.8 to 7.6 mm d^{-1} for fall crops (September–November; Shih et al. 1982). Mean total losses were 800 mm, 740 mm, and 450 mm for spring, summer, and fall, respectively. Rice is a water-demanding crop with a partly flooded stage of growth, and other crops have lower water demands. Kisekka et al. (2012) produced an ET-based tool for north, central, and south Florida that allows farmers to calculate their irrigation requirements based on local rainfall, ET estimates, soil type, and crop water requirements.

The mix of pervious and impervious surfaces in urban areas poses a challenge in estimating ET rates, leading researchers to adopt methods such as measuring diurnal variation in soil moisture (Trout and Ross 2006). Extensive irrigation of parks and golf courses throughout the dry season can increase ET during those months, and the generally shallow water table is accessible to most tree species in urban south Florida, as indicated by their ongoing photosynthesis in the non-rainy months and by the fact that they do not require watering. While there are no published data on the water sources of these mostly exotic subtropical and tropical tree species across various neighborhoods, it is likely that flood control operations allow these flood-intolerant trees to survive in the urbanized coastal zone. Higher temperatures associated with urban heat island effects can also increase ET rates. Analyzing data from 57 weather stations throughout Florida for the period 1950–2007, Winsberg and Simmons (2009) identified this effect as the main cause of an observed increase in the length of the hot season.

SURFACE WATER

Surface water is the most visible portion of the water cycle. A total of 6.4 billion m^3 (1.69 trillion gallons) of surface water is contained in south Florida's lakes, ponds, impoundments, and wetlands, and a complex built infrastructure—comprising 6,600 km of canals and levees, 71 pump stations, and some 600 additional water control structures—moves this water throughout the greater Everglades watershed. At just over 1,900 km^2, Lake Okeechobee is the largest surface water body in Florida, holding 3.79 billion m^3 (1 trillion gallons). Okeechobee means "big water" in the Seminole language, but despite its size, Lake Okeechobee is remarkably shallow, with an average depth of 3 m and maximum depth of only 4 m. Lake Okeechobee is almost entirely enclosed by the Herbert Hoover Dike, a 9-m-high earthen levee built in response to devastating hurricanes in 1928 and 1947–48 that caused the lake to overflow its southern shore. Managers store surface water in Lake Okeechobee during the wet season and discharge it as needed to the EAA for irrigation, to the WCAs for aquifer recharge and wetland restoration, to constructed treatment marshes for nutrient removal (see Chapter 5), and to the Atlantic Ocean (via the St. Lucie Canal and River) and Gulf of Mexico (via the Caloosahatchee River) for flood control during extreme wet periods.

GROUNDWATER

Most of south Florida's water supply comes from shallow surficial aquifers that are recharged by rainfall, from surface water bodies (including canals), and sometimes from deeper aquifers. Groundwater connects directly with surface water, and the two act as a single hydraulic system. As the aquifer is recharged by rain or water impoundment, the groundwater table rises until it reaches the soil surface and contributes to surface water levels and flow. Because canals in south Florida are dug directly into the limestone bedrock that holds this aquifer, they augment this connectivity (see Chapter 4).

The region's largest surficial aquifer, and one of the planet's most productive, is the Biscayne. Extending over 10,000 km^2 and holding up to 34 billion m^3 (9 trillion gallons) of water (nine times more than Lake Okeechobee), the Biscayne Aquifer serves as the primary drinking water source for the densely populated urban tri-county area of Miami-Dade, Broward, and Palm Beach counties plus the Florida Keys. Composed primarily of limestone riddled with solution holes (sinkholes) and sponge-like conduits, it is highly permeable and porous (over 50% porosity in places) and thus extremely susceptible to contamination (Cunningham 2004; Difrenna et al. 2008; Manda and Gross 2006). Water quality in the Biscayne Aquifer is nevertheless very good, aside from high calcium concentrations and areas closer to the coast where salt water has intruded. Intrusion into the Biscayne Aquifer has increased in the last century due to lowered water levels resulting from canal dredging and groundwater pumping, increased demand, and SLR (Blanco et al. 2013). Water-control structures located seaward of water supply wells currently aid in reducing the landward extent of intrusion, but accelerating SLR threatens the capacity of some of these structures.

The larger Florida Aquifer System (FAS) underlays the entire Florida peninsula (Miller 1997). Because it is geologically older (Tertiary period) than the Biscayne (Pleistocene epoch), its carbonate rocks have weathered more extensively, forming larger sinkholes and caves. In central and north Florida, the FAS is located near the land surface and is the dominant water source for residents. In south Florida, the FAS is separated from the Biscayne Aquifer by the 180-m sequence of clays, siltstones, and sandy limestones of the Hawthorn Group, which confine it at a depth of about 300 m. Despite containing some salts (about 10,000 mg L^{-1} total dissolved solids), water in the upper Floridan Aquifer is used in small quantities to augment water supply, by mixing it with Biscayne Aquifer water, water from deep aquifer storage and recovery, and, most recently, reverse-osmosis desalination.

REGIONAL WATER BUDGETS

Every watershed or water body receives water (inputs) from rain, surface waters, and groundwater, and discharges water (outputs) through ET, surface water outflows, groundwater outputs, and extraction for human use. A water budget represents the sum of inputs minus the sum of outputs (see Table 3.1 for estimates based on 36-year averages [1964–2000] from the South Florida Water Management Model). Lake Okeechobee has a net precipitation deficit—in a typical year, it loses more water via ET than it receives from rainfall—but annual surface water inputs to the lake typically exceed outputs. Of the water discharged from the lake, 54% is sent to the Caloosahatchee and St. Lucie rivers and the remainder to the EAA (see Fig. 3.1). The dynamics of groundwater exchange with Lake Okeechobee are unknown. The region's other land-based basins—the EAA, WCAs, ENP, Big Cypress, and the urban areas (lower east coast [LEC])—have a net precipitation surplus. Groundwater exchange plays an important role in each of these basins, with inputs (including from pumping) having the most impact in the urban areas. The WCAs tend to have the highest losses to both groundwater and seepage. Florida Bay has a net precipitation deficit, but unlike Lake Okeechobee, its freshwater inputs (primarily runoff from Taylor Slough and the C-111 basin area of ENP) are small.

WATER RESOURCE USE AND DEMAND

In Florida, fresh water belongs to the state and is allocated to public utilities, power plants, farms, and commercial enterprises by five water management districts via consumptive use permits. South Florida's 8.1 million residents, plus farms and industries, use over 11.36 billion liters per day (South Florida Water Management District [SFWMD] 2014). The SFWMD is responsible for a 46,600-km^2 area that is divided for planning purposes into five service areas: Upper and Lower Kissimmee, Upper and Lower East Coast, and Lower West Coast (Fig. 3.5). Water supply plans incorporate a 20-year planning horizon and are updated every 5 years. The area of greatest relevance to coastal Everglades restoration is the 15,800-km^2 LEC, which includes the urban tri-county area, most of Monroe County, and the rural eastern portions of Hendry and Collier counties. It was home to 30% of the state's population in 2010. Public water supply is the LEC's largest water use category (52%), followed by agriculture (34%). The remaining 14% comprises residential self-supply (private wells), recreation and landscaping, industrial use, and electricity generation. The surficial aquifer system, including the Biscayne, is the primary water source, providing more than 3.79 billion liters per day. Roughly 14% of public water supply is obtained from alternative sources, primarily desalinated brackish groundwater (SFWMD 2013).

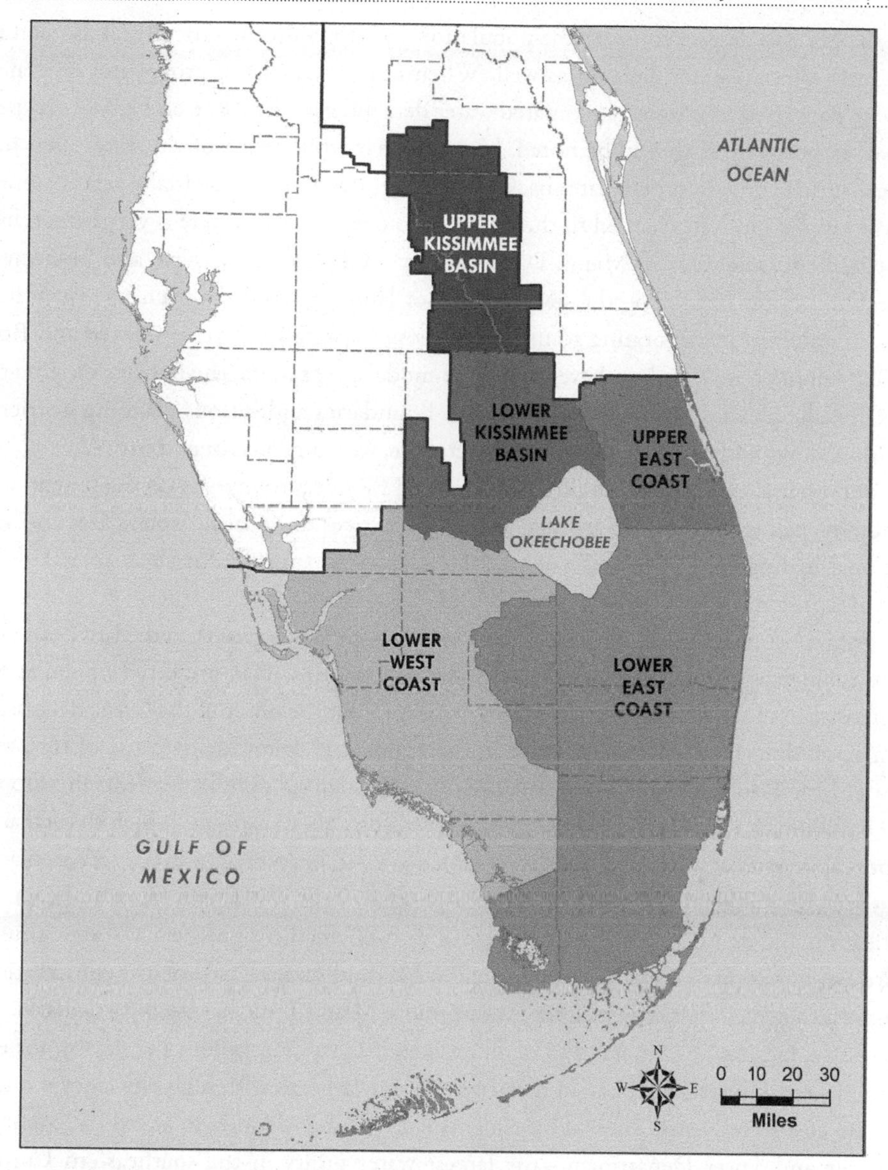

FIGURE 3.5 Water-supply planning areas for the SFWMD (Source: SFWMD 2002).

Whereas most U.S. water management agencies use simple population growth forecasts to estimate future water demand (Billings and Jones 2008), in 2002 the Florida legislature directed the SFWMD to also incorporate land use, as identified in local master plans (Huo et al. 2006). Florida Coastal Everglades (FCE) research has supported this endeavor by assessing the impacts on water demand of changes in land use and land cover. For example, Pokharel (2014) found that in the Redland area of Miami-Dade County between 2001 and 2011, conversion of more than 1,052 ha of agricultural land to residential uses decreased estimated water use by an average of 2,497

L/ha per day, despite considerable population growth. Shifts in agricultural use significantly affected water demand as well. When nearly 1,214 ha of groves and cropland were converted to nurseries, estimated water demand grew by 2,806 and 8,138 L/ha per day, respectively. It should be noted, however, that while much of the water used for agricultural irrigation percolates back into the aquifer, most residential water becomes wastewater and is discharged to the ocean or to deep aquifers; there is very little reuse of treated wastewater in Miami-Dade County. FCE researchers have also pioneered techniques for improving the performance of land use, land cover change quantitative models by incorporating zoning categories (Aldwaik et al. 2015; Onsted and Roy Chowdhury 2014), and we have used these models to examine the impacts on growth of specific policies, such as Urban Growth Boundaries and various planning districts. Addressing another dimension of the water cycle, we have collaborated with researchers from other Long Term Ecological Research (LTER) sites to investigate the potentially homogenizing impacts of yard management practices on local microclimates and regional hydrography (Groffman et al. 2014; Hall et al. 2016; Harris et al. 2012; Polsky et al. 2014; Steele et al. 2014).

Water use is expected to increase with ongoing population growth in the LEC, but not by as much as previously anticipated. The 2005–2006 LEC Plan projected a population increase of more than 31% by 2025 (to 7.3 million), but the onset of the Great Recession in 2008 dramatically slowed growth in the region and statewide. Because of this, the 2013 LEC Plan projected an increase of only 18%, from 5.6 million residents in 2010 to 6.6 million by 2030 (SFWMD 2013). Moreover, contrary to expectations, both total and per capita rates of water use have decreased since 2005, because of both the recession and government policies. Severe drought conditions led the SFWMD to restrict landscape irrigation in the LEC in 2010. In 2011 these limits were further tightened, and agricultural irrigation was cut by 45%. Efforts by the SFWMD and counties to promote conservation have also helped reduce per capita consumption. Miami-Dade has been particularly aggressive, facilitating conservation of an estimated 10 million gallons per day by, for example, repairing leaky pipes and replacing old showerheads with high-efficiency models at no cost to residents. These achievements led Audubon Florida to award the county's Water and Sewer Department—the largest water utility in the southeastern United States—its Excellence in Water Conservation Award in 2013. The SFWMD has scaled back its public outreach activities in the wake of budget cuts in 2011, but it continues to work with hotels and other businesses to reduce demand. Nevertheless, the 2013 LEC Plan predicts increases of 12% in total water demand and 20% in public supply demand by 2030. No changes are anticipated in agricultural demand, because the EAA is a fully developed, stable agricultural area where permitted hectares and cropping practices are not expected to change over the next 20 years (SFWMD 2013).

The Comprehensive Everglades Restoration Plan (CERP) was heralded as a "win-win" strategy for increasing available water supply to meet the future needs of a projected 6 million additional south Florida residents while also restoring water flows to the coastal

Everglades (see Chapters 1 and 2). Planned CERP projects will capture a substantial portion of the surface water currently discharged to tide (an average of 6.4 billion liters per day), clean it, and convey it south to the Everglades. When the plan was negotiated, many environmentalists, National Park Service officials, and academic scientists critiqued it as a water supply boondoggle because it locked in water allocations for existing users and prioritized projects for human use ahead of flow restoration for the Everglades (Grunwald 2006). It also locked in flood protection guarantees, as we will discuss later in the chapter. The 2000 LEC Plan anticipated that timely completion of CERP projects would largely satisfy increased demand by 2020, but implementation has not proceeded at the expected pace (CISRERP 2014).

South Florida is an extraordinarily challenging environment in which to carry out large-scale water storage. Flat topography means there are no river canyons to dam, hot weather subjects surface impoundments to large evaporative losses, and porous bedrock causes extensive underground seepage. CERP relied heavily on two untested technologies to overcome these challenges: massive in-ground reservoirs in mined-out limestone quarries, and 333 aquifer storage and recovery wells. Today the reservoir concept remains hamstrung by the problem of seepage, and studies have determined that, due to hydrogeologic constraints, only 131 aquifer storage and recovery wells at most can feasibly be constructed (USACE and SFWMD 2013). The CERP projects currently authorized or in the pipeline will provide only partial solutions to the problem of water storage (Graham et al. 2015).

Because no "new water" has yet been made available for people or the Everglades, the SFWMD prohibited new allocations from the Everglades and Loxahatchee watersheds in 2007 and from Lake Okeechobee in 2008. Additional future demand must be met by alternative sources, primarily brackish groundwater treated at Miami-Dade County's first reverse-osmosis desalination plant, which opened in Hialeah in 2014. Desalination is extremely costly and energy-intensive, and disposal of the resulting brine—most likely through deep-well injection into the lower Floridan Aquifer—poses environmental challenges. Nevertheless, some critics maintain that, from an ecosystem restoration perspective, water in south Florida is already over-allocated to human users. Unlike in Tampa Bay and other parts of the state, however, over-pumping of groundwater for consumptive use contributes only minimally to water scarcity. The chief culprit, by far, is flood control. Thus, water flows to the Everglades can be significantly increased only by completing hydrologic restoration projects.

The Water Cycle and Climate Change

South Florida is extremely vulnerable to climate change because of its maritime subtropical location within the major Atlantic hurricane pathways, low relief, high population density, and large tourism- and real-estate–based economy. The manifestations of climate change that are most significant for south Florida include higher temperatures, increased uncertainty in rainfall, increased frequency and uncertainty of extreme weather events, and SLR. All are all highly likely to affect the region (Field et al. 2014; Obeysekera et al. 2011). These "pulses and presses" (*sensu* Collins et al. 2011) have cascading effects on ecosystem structure and function and on the availability of natural resources and ecosystem services (see Chapter 7). They also exacerbate the challenges of sustainable water management in south Florida in a variety of ways (Berry et al. 2011; Obeysekera et al. 2006).

Given current rates of greenhouse gas emissions, air temperatures are expected to increase by 2°C to 4°C globally during the 21st century (Field et al. 2014). Higher temperatures, particularly during the dry season, can increase freshwater losses via ET, with significant impacts on agriculture, horticulture, and the Everglades. Obeysekera et al. (2011) analyzed maximum and minimum temperature data from 17 weather stations across the state between 1892 and 2007 and found that minimum temperatures have increased since 1950, but not consistently over all stations because of intervening factors such as inter-decadal climate cycles, land-cover change, and the urban heat-island effect. Future temperature increases in Florida are thus not expected to be uniform and steady.

Predicting precipitation is typically a far more complex task than predicting temperature because of the large number of local, regional, and global influences at play. Rising greenhouse gas emissions worldwide continuously escalate atmospheric heat storage, altering the energy balance and consequently affecting pressure, winds, and ocean currents in unpredictable ways. Climate teleconnection patterns that influence rainfall in south Florida, such as the ENSO, AMO, and Pacific Decadal Oscillation, also appear to be changing in periodic frequency (see Chapter 4). Obeysekera et al. (2011) found very poor agreement on precipitation predictions for south Florida among major general circulation/global climate models. Nevertheless, there is widespread agreement that rainfall patterns are already becoming more variable and extreme events such as hurricanes, flooding, and drought are becoming more frequent and intense. As we will discuss later in the chapter, both factors will compound the challenges of water management and climate adaptation.

SEA LEVEL RISE

SLR is the dimension of climate change that has most vividly captured the public imagination in south Florida. Global sea levels rose and fell up to 140 m during the most recent glaciation—the Wisconsin (Lambeck et al. 2002)—but the current rates of SLR are driven by two consequences of anthropogenic climate change: increased melting of land-bound ice and thermal expansion of ocean water (Rahmstorf 2010). The global average rate of SLR has been accelerating over the past century (Church and White 2011; Hansen 2007; Rahmstorf 2010), and global warming of 4°C will likely lead to SLR of at least 1 m by 2100, with the coming centuries promising several more meters. And none of these conservative estimates account for the growing instabilities in the Greenland and

West Antarctic ice sheets. Even if warming is limited to 2°C, global mean sea levels will likely continue to rise an estimated 1.5 to 4.0 m by 2300 due to lag effects.

In south Florida, tidal gauges in Key West reveal a rise of 2.24 mm per year over the past 100 years (http://tidesandcurrents.noaa.gov/sltrends/). These rates have been accelerating recently, though, from 1.2 ± 0.2 mm per year between 1901 and 1990 to 2.6 ± 0.4 mm per year between 1993 and 2015 (Hay et al. 2015; Watson et al. 2015), and the Intergovernmental Panel on Climate Change (IPCC)'s Fifth Assessment Report predicts continued acceleration (Church et al. 2013). Given that most of the densely inhabited zone is 1 m or less above mean sea level, south Florida is literally at Ground Zero for SLR. To make matters worse, because of the region's porous bedrock, dikes and sea walls will not hold back the rising seas. It is hardly surprising that the 2014 National Climate Assessment identified Miami as one of the cities most vulnerable to SLR worldwide (Carter et al. 2014).

Most stormwater removal in the region relies on gravity-driven canals. SLR is already reducing the capacity of these canals in the lowest-lying areas, resulting in the ever-more-frequent phenomenon of tidal "sunny-day flooding" in city streets. The "King Tide" on Miami Beach in October 2014 was the focus of several public events and considerable attention from local and national officials, scientists, and news media, but it caused no street flooding that year thanks to the city's timely deployment of new high-capacity pumps that discharge stormwater into Biscayne Bay. The previous week, Miami Beach hosted the Sixth Annual Southeast Florida Regional Climate Leadership Summit, where discussions focused heavily on pumps and other infrastructural adaptations to SLR. These types of engineered solutions are costly and energy-intensive, however, and their benefits may be short-lived, particularly if SLR rates continue to increase (Wdowinski et al. 2016). Discharging untreated stormwater directly to Biscayne Bay, moreover, poses a public health risk due to high levels of fecal bacteria (Gidley et al. 2016; Wendel 2016).

The Everglades landscape was significantly altered by SLR in previous millennia. The presence of freshwater peats and marls at the bottom of sediment cores demonstrates that the Everglades once extended beyond its current coastline in the southwest and across Florida Bay (Enos and Perkins 1979; Wachnika and Wingard 2015; Willard and Berhnardt 2011). After rising very quickly from 21,000 to 6,000 years ago, rates of SLR during the last 6,000 years, except for the last 150 years, have been very low. This stability allowed the extensive Holocene freshwater system to become estuarine mangroves in the southwest and marine seagrasses in Florida Bay (Enos and Perkins 1979; Wachnika and Wingard 2015; Wanless and Tagett 1989; Willard and Berhnardt 2011). During the 20th century, the acceleration of SLR to 2 to 3 mm per year coincided with drainage and flood control, which dramatically reduced freshwater inflows to the coastal Everglades. Fresh groundwater typically flows from land to sea in response to gravity. But if groundwater levels drop below sea level as a result of pumping, water diversions, or SLR, seawater intrudes into the aquifer and sinks below the less-dense fresh water. The displaced fresh water can no longer flow to the coastline and instead discharges landward. In the Everglades, sea water has intruded as far as 15 km and 30 km inland in Taylor and Shark sloughs, respectively, effectively halting the flow of fresh groundwater to Florida Bay (Price et al. 2006). This has exacerbated coastal erosion by collapsing freshwater peats, replacing sawgrass marshes with mangroves, and threatening hardwood hammocks (Davis et al. 2005; Gaiser et al. 2006; Ross et al. 2000; Saha et al. 2011).

The combination of SLR and flood control is accelerating saltwater intrusion into the Biscayne Aquifer as well (Dausman and Langevin 2005), diminishing its freshwater resources by 12% to 17% (Blanco et al. 2013). The City of Hallandale Beach, in Broward County, was already forced to close four of its six wells and drill new wells further inland (Berry et al. 2011). In Miami-Dade County, the extent of intrusion into the Biscayne Aquifer has not changed much from 1984 to 2011, mainly due to construction of gates at the mouths of canals to maintain higher freshwater levels upstream (Fig. 3.6). Nevertheless, continued SLR will eventually threaten several municipal wellfields, including the Miami Springs-Hialeah-Preston plant, particularly if it is accompanied by continued groundwater withdrawals (Hughes and White 2016).

The challenge of ensuring water supply in south Florida will be exacerbated by climate change in other ways as well. Higher temperatures will increase ET losses from surface sources, including planned CERP reservoirs, and droughts will be more frequent and longer in duration (Berry et al. 2011). Meanwhile, as we have noted, municipal water demand is expected to increase. Continued progress in implementing hydrologic restoration projects to increase freshwater inputs to the coastal Everglades is thus urgently needed to protect municipal water supplies. As we will discuss in the next section, flow restoration is necessary also to preserve freshwater habitats and stabilize the Everglades coastline.

Everglades Ecohydrology

FRESHWATER AQUATIC COMMUNITIES

In Everglades wetlands, seasonal and yearly rainfall patterns act in tandem with the corrugated ridge and slough topography and the karstic geology to dictate the depth, extent, and duration of marsh flooding. These patterns affect aquatic organisms through four principal mechanisms:

1. Their growth and reproduction is influenced by water levels.
2. They can die from desiccation.
3. They can become spatially concentrated and thus more susceptible to predation as water levels drop.
4. They can redistribute themselves spatially via dispersal as water levels vary.

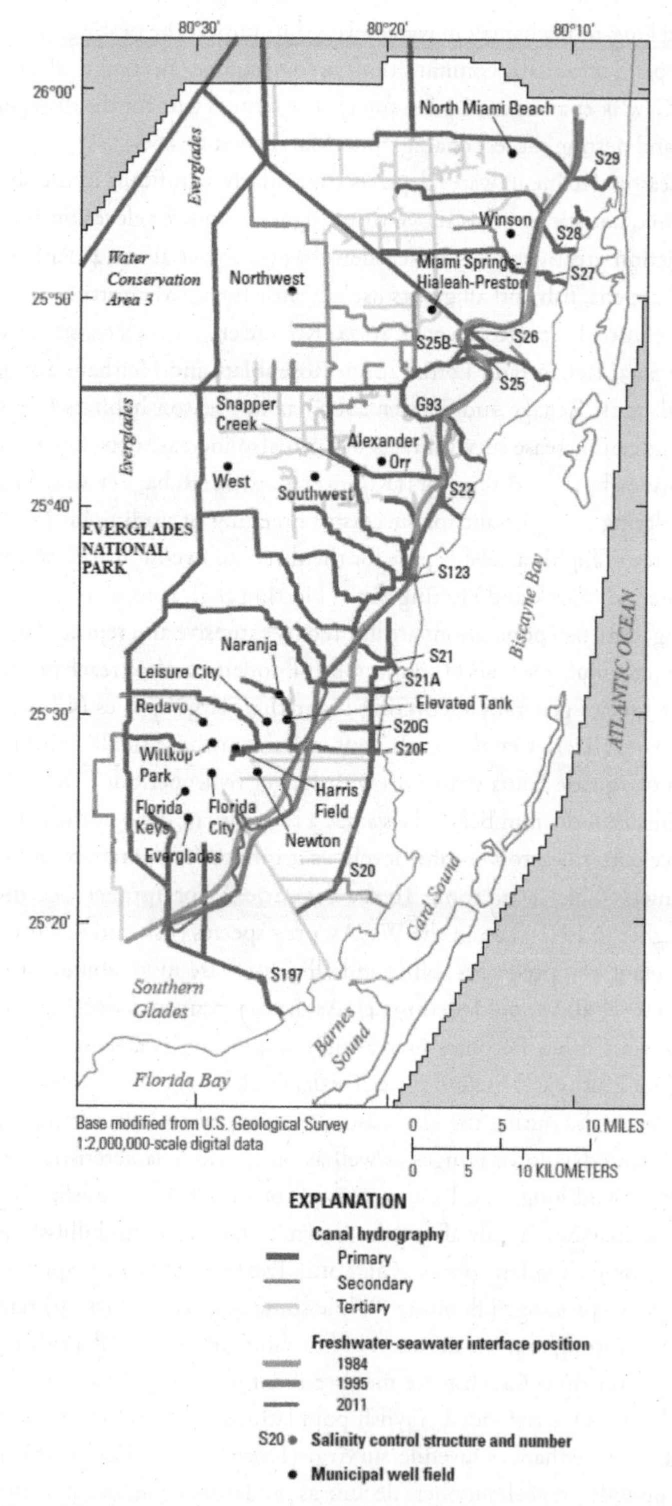

FIGURE 3.6 Extent of saltwater intrusion into the Biscayne Aquifer in 1984, 1995, and 2011 and its relationship to municipal well fields (Hughes and White 2016).

Seasonal and long-term changes in water levels thus impact the nesting success of wading birds that depend on aquatic communities as a food source (Beerens et al. 2011; Frederick et al. 2009; Gawlik et al. 2002; Ogden 1994). The same is true for the distribution, abundance, size, and nesting success of alligators (Mazzotti et al. 2009).

The dry-season decline in water depth is particularly significant in the short term. As habitats shrink, aquatic animals move from higher- to lower-elevation habitats, where spatial restriction greatly increases their densities (Cook et al. 2014; Parkos et al. 2011). When this happens, fish and alligators use alligator holes (Mazzotti et al. 2009; Parkos et al. 2011), solution holes (Kobza et al. 2004; Rehage et al. 2014), coastal creeks (Boucek and Rehage 2013; Rehage and Loftus 2007; Rosenblatt and Heithaus 2011), and canals (Parkos et al. 2011; Rehage and Trexler 2006) as dry-season habitats (see Chapter 4). These habitats can increase survival rates but can also increase exposure to predation by non-native fishes, birds, and alligators (Kobza et al. 2004; Rehage et al. 2014). The timing of seasonal drying is important for successful breeding of wading birds, which need a constant source of highly available prey for the three- to five-month duration of their reproductive period (Cook and Herring 2007; Herring et al. 2010, 2011).

In the long term, fish populations are limited by extensive and repeated dry conditions since they require multiple years of continuously flooded marsh to reach peak abundances, with smaller species generally recovering faster than large species (Chick et al. 2004; Ruetz et al. 2005; Trexler et al. 2005). Thus, drier areas have smaller and more distinct populations of aquatic fauna than sites with longer hydroperiods (Trexler et al. 2005). Drying frequency—the number of days since a site was last dry—is the key determinant of abundance patterns across trophic levels, as it influences the rate of predation of primary consumers (Chick et al. 2004; Trexler et al. 2005). For instance, because drying frequency is higher in ENP than in the WCAs, prey species (fish and macroinvertebrates) are less abundant and predators (birds and alligators) are more abundant (Chick et al. 2004; Mazzotti et al. 2009; Ogden 1994). As drying frequency decreases, predation by larger piscivorous fishes becomes increasingly important as a structuring mechanism (Kushlan 1976; Loftus and Eklund 1994; Trexler et al. 2005).

As water levels rise during the wet season, recovery of aquatic communities depends on dispersal from dry-down refuges as well as on species characteristics such as reproductive strategy and longevity. Eastern mosquitofish, Florida flagfish, and other small fish recolonize marshes rapidly after a dry down, whereas bluefin killifish take weeks or months to become abundant (Goss et al. 2014; Trexler et al. 2005). Species also vary in their responses to prolonged flooding. While some (e.g., mosquitofish) remain consistently abundant throughout the duration of flooding, others (e.g., flagfish and marsh killifish) decline over time. Crayfish are more resilient to drying than their fish predators; thus, after the marsh is reflooded crayfish populations can get a temporary boost when reduced predation enhances juvenile survival (Dorn and Trexler 2007). But over extended wet periods, crayfish numbers decline as predatory fishes become more abundant (Kellogg and Dorn 2012). This disturbance-related pulsing of prey abundance is believed to be responsible for the exceptionally large nesting aggregations of wading birds often observed within a year or two of droughts (Frederick and Ogden 2001).

COASTAL AQUATIC COMMUNITIES

In estuarine regions of the southern Everglades, faunal movements are strongly influenced by variations in salinity that result from intra- and inter-annual variation in rainfall and freshwater flows. As noted previously, freshwater flows are heavily influenced by water management operations, which have greatly reduced total surface water flow south from Lake Okeechobee. Additionally, managers typically hold water in the WCAs during the dry-season months as a conservation measure and release it into ENP only some two to four months into the rainy season. Releases into ENP are also limited by policies protecting the endangered Cape Sable seaside sparrow.

Estuarine species fall into two general functional groups: freshwater and euryhaline (Elliot et al. 2007). Freshwater species include American alligators (Rosenblatt and Heithaus 2011), fishes dominated by centrarchids (Boucek and Rehage 2013), and macroinvertebrates such as riverine grass shrimp (McCarthy et al. 2012). Freshwater discharge affects habitat use more directly for this species group because they suffer physiologic osmotic stress at higher salinities. Florida largemouth bass that inhabit upper estuarine reaches, for instance, can experience high mortality in years with severe dry downs (Lee 2016). Thus, freshwater species track the oligohaline habitat as it seasonally expands and contracts. Most alligators avoid elevated salinities year-round, but a small number make forays into high-salinity regions of the estuaries, and even into the coastal ocean, to feed before returning to lower-salinity waters. The movement of these "commuting" individuals fluctuates, becoming more pronounced as low-salinity water pushes further seaward in the wet season (Rosenblatt and Heithaus 2011). It is highly likely that the spatial extent of habitat for the alligator and other freshwater taxa in the coastal Everglades has contracted landward with decreased freshwater flows, and that it will continue to change in the future in response to SLR and hydrologic restoration (see Chapter 8).

Euryhaline species include the American crocodile (Mazzotti et al. 2009), teleost fishes such as snook and mojarras (Boucek and Rehage 2013), and elasmobranchs, most notably juvenile bull sharks (Heithaus et al. 2009). This group is less directly constrained by salinity than are freshwater species, but rainfall patterns and freshwater flows may still play an important role indirectly. Thus, restoration is likely to affect the behavior and habitat use patterns of these species as well. Many Everglades euryhaline species evolved in the tropics and colonized south Florida through northerly range expansions from the Caribbean. In the tropics, while temperature and photoperiod vary little during the year, the hydrologic regime varies strongly, with distinct high and low rainfall periods and corresponding freshwater flows (Junk 1989; Pringle 2001). Consequently, tropical species have evolved life histories, spawning patterns, and feeding strategies that are centered completely on this hydrologic predictability (Gillson 2011; Robins et al. 2005; Welcomme 2001). For instance, snook, a common Everglades estuarine resident, synchronizes its spawning migrations with the onset of the wet season (Andrade et al. 2013). Both snook and, to a lesser extent, juvenile bull sharks migrate to the uppermost reaches of the estuary in the dry season to take advantage of high concentrations of prey forced out of drying Everglades marshes (Boucek and Rehage 2013; Matich and Heithaus 2014). Not only is this foraging opportunity facilitated by regional rainfall patterns and inundation of marshes, but the hydrologic cycle likely provides the cue for these species. Changes in salinity affect the habitat use patterns of bull sharks only minimally in the Shark River Estuary (e.g., Heithaus et al. 2009) but critically in the Caloosahatchee River, where anthropogenic modifications to freshwater flows are greater and tend to occur in large pulses (Simpfendorfer et al. 2005). Freshwater flows also affect the vulnerability of snook populations to extreme cold events, which can reduce abundance by up to 95% (Boucek et al. 2016; Stevens et al. 2016). In the Shark River, during periods of below-average flows, snook tend to concentrate in the upstream oligohaline zone, which is colder and therefore a higher-risk habitat than downstream areas (Boucek et al. 2017).

One of the major ecosystem services provided by Everglades estuaries is recreational fisheries. Angling opportunities are influenced by fish migration patterns, which are tied to the hydrologic cycle (Boucek and Rehage 2013, 2015). Vast expanses of marsh are inundated in the wet season, creating a high-quality, productive environment for many fish that inhabit these marshes. Catch rates of Florida largemouth bass, in particular, are driven by variation in freshwater flows. This temperate large-bodied fish, common throughout Florida, is a subspecies of the Northern largemouth bass that inhabits the Mississippi drainage (Philipp et al. 1983). In the Everglades, nearly all of its growth occurs in these inundated marshes, and monthly survival is strongly affected by the duration of marsh flooding (Lee 2016). When these marshes dry, the bass are forced into the upper estuary and other deep-water habitats. Because these populations are subsidized by marsh habitats, bass abundance in the estuary during the dry season far exceeds what could be sustained on estuarine productivity alone. This creates excellent opportunities for anglers (Boucek and Rehage 2015).

Although bass concentrations are primarily determined by intra-annual variation in rainfall and upstream water management, during the dry season inter-annual variation can play a larger role. Throughout several consecutive wet years in the mid-1990s, for example, bass catches in Everglades estuaries were at record highs (Boucek and Rehage 2015). Several factors were most likely at play: salinity remained low in the estuary during the dry season; the duration of the dry season and thus of stressful crowded conditions was reduced; and marsh water depths in the wet season (a proxy for marsh habitat availability) were high. These favorable conditions persisted for roughly four years (1994–1997), during which time anglers were catching three bass per hour on average, or roughly triple the state's average catch rate. Conversely, extended droughts can have the opposite effect, producing high bass mortalities due to prolonged time spent in the crowded estuary and salinities that can increase to lethal levels. Analysis of a 20-year time series reveals that after droughts, bass catches in the Everglades drop the following year to less than 0.2 fish per hour.

PLANT COMMUNITIES

As with aquatic fauna, the combination of seasonal flooding and irregular topography in the Everglades has produced a mosaic of plant communities with widely differing responses to flooding, drought, fire, and salinity (Myers and Ewel 1990). The combination of reduced freshwater flows and SLR has increased seawater intrusion during spring tides and in the dry season, leading to increased groundwater salinization (Willard and Bernhardt 2011). A likely decrease in annual rainfall due to climate change will have impacts as well, favoring expansion of more xeric plant communities, but because of considerable uncertainty in forecasting future precipitation regimes, these impacts are more difficult to predict than are the impacts of SLR and water management.

Coastal forests include salt-tolerant mangroves, moderately tolerant buttonwood hammocks, and salinity-intolerant hardwood hammocks found on slightly higher ground (Saha et al. 2011). Modeling by Teh et al. (2008) predicted that with increased groundwater salinization, mangroves would expand at the expense of buttonwoods and coastal hammocks. Jiang et al. (2012) predicted similar effects from soil water salinization. These predictions are supported by numerous anecdotal observations of mangrove encroachment inland and by FCE research. A 50-year data analysis of coastal southeastern vegetation found that the boundary between mixed-graminoid mangrove and sawgrass communities has shifted 3.3 km inland since the mid-1940s and the low-productivity hypersaline white zone has shifted 1.5 km inland on average, with greater incursion in areas cut off from upstream freshwater sources by roads or levees (Ross et al. 2000). Similar patterns of SLR transgression have also been observed in Shark River Slough, where mangroves first appeared around Mahogany Hammock two decades ago and have since grown in number. Based on data from aerial resistivity surveys and from groundwater wells along Shark River Slough, Saha et al. (2011) linked this encroachment to groundwater salinization.

Saha et al. (2011) hypothesized that hardwood communities found in areas of higher elevation depend on a rain-derived freshwater "lens" that, due to its lower density, perches above the saline water table. Saha et al. (2015) confirmed this freshwater lens hypothesis in a field study using highly accurate elevation data to analyze the stable isotopic composition of various water sources to coastal plant communities at different elevations. While they found negligible elevation differences between buttonwoods and hardwood hammocks, the less salinity-tolerant hardwood species were located further from seawater pathways. A similar study is underway in pine rocklands, hardwood hammocks, buttonwood hammocks, and mangrove communities in the Florida Keys. Mangrove encroachment is expected to continue with ongoing salinization of the coastal rhizosphere.

However, Koch et al. (2015) found no evidence that the rate of mangrove expansion can keep up with the rate of SLR predicted over the next half-century under a global warming scenario of +1.5°C. At present, soil elevation in south Florida mangrove forests increases by 0.9 to 2.5 mm per year. Peat accumulation would thus have to increase twofold to fourfold to accommodate this rate of SLR. These studies all confirm the critical need to restore freshwater flows to the coastal Everglades so as to maintain hydrologic conditions for the diverse plant and animal communities for as long as possible. Increased flows are critical for limiting saltwater intrusion, facilitating mangrove peat accumulation, and stabilizing coastlines.

Human Management of the Water Cycle

The backbone of south Florida's water management system is known colloquially as the C&SF Flood Control Project, but Congress authorized it in 1948 for "flood control and other purposes"—namely agricultural irrigation, municipal and industrial water supply, preservation of fish and wildlife, water supply to ENP, prevention of saltwater intrusion, navigation, and recreation. The project was politically contentious from the outset. Urban coastal residents accused water managers of benefiting EAA farmers at their expense, and the U.S. Fish & Wildlife Service and National Park Service worried that wildlife preservation would take a backseat to flood control and water supply (Blake 1980; Godfrey and Catton 2011; McCally 1999). Fifty years later, CERP promised to meet the needs of the (vastly expanded) urban and agricultural sectors alike, while also rehabilitating natural systems, by "growing the water pie" through increased retention of floodwaters. Yet these concerns persist. The C&SF system has prevented catastrophic flooding for almost 70 years, but with unintended and devastating consequences. Our research suggests that adequately fulfilling all C&SF Project mandates simultaneously is impossible in practice, and as a result water managers are perpetually forced to make politically painful tradeoffs among competing users, including the Everglades (Schwartz 2014).

The hydraulic infrastructure of south Florida is jointly managed by the USACE and the SFWMD. The USACE (with input from the SFWMD) controls water levels in Lake Okeechobee and the WCAs, including releases into ENP through the S-12 gates under Tamiami Trail. The SFWMD moves water through the rest of the system, and allocates water to permitted users. The system's lynchpin is Lake Okeechobee. It is no longer the "liquid heart" of the Greater Everglades watershed. Rather, it is Florida's largest multipurpose reservoir, operated in accordance with a regulation schedule that specifies seasonally varying minimum and maximum water levels. The intrinsic challenge of managing a multipurpose reservoir is the tradeoff between holding water for supply and maintaining capacity to store floodwaters (EarthTech et al. 2005). Water users typically seek to have lake stages maintained as high as possible to ensure sufficient supply during droughts, but this reduces retention capacity. Lake Okeechobee is the primary water supply source for farms and three small municipalities in the EAA, and the backup source for the urban tri-county area. During wet periods, in addition to rainfall the lake receives flows and floodwaters from several basins to the north and west of the lake, comprising 14,500 km^2, as well as from the EAA through the controversial practice of "backpumping" (discussed later in this section and in Chapter 5).

The challenges of negotiating this tradeoff are compounded by the lake's hydrologic modifications through the C&SF Project. Due to channelization of the Kissimmee River, floodwaters that formerly lingered for six to eight months in this upstream floodplain, where it was also cleaned by riparian wetlands, now flow into Lake Okeechobee in one month. Construction of the Herbert Hoover Dike reduced the lake's spatial extent by nearly half and eliminated its ability to overflow the southern shore. Because the dike lacks a conventional spillway, the lake's outflow capacity is now only one-sixth of its inflow capacity (USACE 2015). During extremely wet periods lake levels are lowered through releases east to the St. Lucie River and west to the Caloosahatchee River, which severely damages these estuarine and coastal ecosystems. Huge pulses of turbid, tannic freshwater polluted with nutrients and bacteria make their way to east- and west-coast estuaries, where they trigger harmful algal blooms and decimate oysters and seagrass beds by lowering salinities. Conversely, the Caloosahatchee Estuary is affected even more severely by elevated salinities during prolonged droughts, when Lake Okeechobee holds insufficient water to meet all needs and flows to the Caloosahatchee River are reduced or halted. Heavy rainfall raises water levels not only in the lake but also in the EAA and WCAs, which limits the option to discharge the lake's water southward.

Operational flexibility is further constrained by the diverse purposes for which Lake Okeechobee and downstream landscapes are managed. Excessive discharges strain the constructed treatment wetlands that have been built south of the EAA for nutrient removal (see Chapter 5). Prolonged high levels harm the lake's littoral marshes and submerged vegetation, adversely impacting populations of wading birds, including the endangered Everglade snail kite, as well as the largemouth bass that support a nationally renowned sport-fishing industry. High water levels in the WCAs threaten federally protected species as well as tree islands, which are important cultural resources for the Miccosukee Tribe and ecologically critical to upland Everglades species. As we discuss further, sustained inundation of EAA fields is ruled out by flood protection entitlements for EAA landowners.

Another daunting challenge is posed by the hazardous condition of the 230-km Hoover Dike. In places, the dike is threatened because of both construction methods and operational decisions. When the USACE built portions of the dike in the 1930s, then extended it the rest of the way around Lake Okeechobee in the 1950s and 1960s, they used mainly uncompacted fill from canal dredging and built it on top of muck soils and permeable bedrock (Bromwell 2006). The dike was designed as a flood control levee, but when the USACE raised the maximum Lake Okeechobee stage 0.46 m to boost EAA water supply in 1978, and maintained it there through 1992, they increased the hydraulic loading beyond what the levee was built to withstand (Bromwell et al. 2006; Godfrey and Catton 2011). This accelerated internal erosion of the levee, eventually creating an unacceptably high risk of catastrophic breaching (USACE 2015). The USACE began reporting areas of erosion in the mid-1980s. They completed an engineering evaluation in 2000 and began repairs in 2005 on the most compromised section, near the EAA town of Belle Glade. Hurricane Katrina prompted increased scrutiny of flood control infrastructure nationwide, and a second evaluation in 2007 ranked the dike in the Corps' lowest dam safety category: "critically near failure or extreme high risk" (USACE 2015). The dike rehabilitation project was revised and expanded, and structural improvements costing $600 million have reduced the Belle Glade section's failure risk to tolerable levels. A second phase is projected to be completed in 2025 at a total cost of $1.5 billion, but in the meantime, a new USACE study in 2015 gave the dike an "urgent and compelling" risk ranking (USACE 2015).

This study identified a second failure mode as well. If a severe storm were to hit Lake Okeechobee when the water level is already elevated, high-speed winds across the lake's large surface area could generate sufficient storm surge to overtop the levee (USACE 2015). The USACE deemed this a non-critical threat due to the low probability of high loading and a severe storm occurring simultaneously. As we noted already, though, climate change is increasing the frequency and unpredictability of extreme weather events. While managers have incorporated increasingly sophisticated meteorological data and climate modeling into operational decision making since 2000, they remain unable to predict future rainfall patterns with any certainty (Cadavid et al. 2006; Obeysekera et al. 2006). Yet, as the saying goes, hindsight is 20/20. For example, the Lake Okeechobee regulation schedule was revised in 2000 to address ecological concerns but, according to a SFWMD official, after a prolonged wet period in 2003–2005 it was denounced by many stakeholders for holding stages too high. If those years had brought drought instead, "everybody would have thought it was the best regulation schedule since sliced bread" (anonymous). Stakeholders tend to blame the lake schedule for the impacts of any recent flood or drought, but water managers must strive to balance the harms and benefits incurred by their decisions and cannot easily alter schedules in response to individual events.

To reduce the risks of dike failure from both erosion and overtopping, in 2008 the USACE lowered the lake's maximum stage by 0.38 m (to 5.26 m National Geodetic Vertical Datum [NGVD] above sea level). This increased the need for emergency releases. Climate change appears to be magnifying the problem as well: 2013 had the wettest start to its wet season in 45 years. Lake releases continued full force from May through October, triggering water safety advisories and devastating the tourism and fishing industries that are the backbone of local economies in the Stuart and Fort Myers areas. Thousands of coastal residents took to the streets in protest, dubbing this the "Lost Summer." Before the estuaries could recover ecologically or economically, a powerful El Niño pattern in late 2015 brought record dry-season rainfall beginning in November and peaking in

January 2016 with 23 cm (476% above average). Releases began on February 1 and by mid-June their volume had already surpassed the 2013 total of 515.2 billion liters. In May an 85-km^2 blue-green algae bloom was discovered on Lake Okeechobee near the St. Lucie locks and was found to contain high levels of toxic microcystin bacteria. Heavy rainfall compelled the USACE to continue releases to the river, where the toxic lake water mixed with nutrients from local basin runoff. With an added boost from high temperatures and lowered salinities, the bloom grew explosively, spreading beyond the inlet to the Atlantic Ocean for the first time in memory and prompting authorities to close all Treasure Coast beaches. Residents complained of an unusually foul odor and suffered respiratory and gastrointestinal distress. Governor Rick Scott declared a state of emergency for the affected counties. The story broke in national media outlets just before the July 4 holiday weekend, with the *Washington Post* comparing the blooms to "chunky guacamole" and "a festering, infected sore" (Mettler 2016). The blooms finally began to dissipate toward the end of the month as drier weather allowed the USACE to reduce releases, but the local tourism industry had already sustained heavy losses (Cahyanto 2016).

The disaster sparked unprecedented levels of political mobilization in the coastal communities. Real estate agents, boat captains, residents, and municipal officials joined forces with environmentalists, taking their fight to the streets, social media platforms, the statehouse, and Congress. Their complaint was much the same as in 1930s and 1940s: flood control operations benefit EAA landowners at the expense of estuarine ecosystems. This argument received a symbolic boost when 30 cm of rain flooded the EAA in January 2016, causing heavy crop losses and shutting down the area's four sugar mills for over two weeks. As required by CERP's flood protection mandate, the SFWMD initiated backpumping into Lake Okeechobee to drain fields and prevent flooding in the lakeside towns. This happened one day before the USACE began releases to the St. Lucie River.

The crux of the problem is the lake's limited outflow capacity. Some estuary advocates expressed hope that the Lake Okeechobee regulation schedule would be revised to permit more floodwater retention after dike repairs are completed (a move that would likely be opposed for reasons of both dam safety and lake ecology). But their primary demand was for increased storage and conveyance capacity south of the lake. CERP calls for construction of aboveground reservoirs to store 444,052,800 m^3 of water in the EAA and 308,370,000 m^3 north of the lake, but to date no planning has been undertaken for the latter, and the former remains mired in controversy. The USACE did begin building a reservoir on EAA farmland that was acquired by the state in 2006, but construction was halted in 2009 after Governor Charlie Crist announced a deal to purchase all land and assets of the U.S. Sugar Corporation, one of three very large agribusiness operations in the EAA. Because many Everglades professionals identified the lack of willing land sellers in the EAA as a key obstacle to hydrologic restoration (Schwartz 2014), this deal was widely celebrated as a potential game-changer. The SFWMD promptly launched a planning process with extensive public participation to reconceptualize strategies for storage, treatment, and conveyance of water both north and south of Lake Okeechobee. Due to the economic recession and political resistance, however, the buyout was reduced from 75,676 to 10,845 ha. When the state initiated a new set of nutrient removal projects to comply with ongoing federal water quality litigation in 2012, the U.S. Sugar Corporation lands and the EAA reservoir site were both repurposed to create shallow reservoirs for attenuating pulse flows to these constructed treatment marshes and to expand the marshes (see Chapter 5). Thus, none of this land will be used for CERP water storage.

After the "Lost Summer" of 2013, the state legislature tapped the University of Florida Water Institute to holistically evaluate water storage needs and potential strategies. Building on the SFWMD's 2009 planning exercise, the study found that completion of existing CERP/CEPP projects would reduce Lake Okeechobee releases by less than 55% and would meet less than 75% of Everglades dry-season demand, whereas an additional 1,233,480,000 m^3 of storage could achieve 90% of both targets (Graham et al. 2015). The authors of this study called on state and federal authorities to begin strategic planning to determine the most effective configurations both north and south of the lake to achieve this target. Later that year, restoration advocates lobbied the state aggressively but unsuccessfully to exercise an option, included in the 2010 U.S. Sugar Corporation deal, to purchase another 18,939-ha parcel for construction of a new deep storage reservoir in the EAA. During the 2016 estuary crisis, advocates renewed the call for state to "buy the land and send the water south," but Governor Scott and SFWMD officials—along with EAA landowners and officials—contended that further EAA land acquisition was unnecessary and that additional storage should be sited north of Lake Okeechobee. And political conflict continues to thwart the quest for expanded water storage and conveyance capacity. Unless the land acquisition impasse is resolved, a technological "silver bullet" is devised to supplement aquifer storage and recovery, or flood protection and water supply entitlements are substantially reduced, ecosystem restoration will remain hobbled by the demands of flood control and an ironic lack of sufficient water for the Everglades.

Conclusions

The Everglades is an iconic landscape in a variety of ways, as we saw in Chapter 2, and water is the defining element of its iconicity. The Greater Everglades watershed, which was once one of the planet's largest wetland landscapes, was shaped by variable rainfall and uninterrupted surface water flows that gave rise to its characteristic geomorphology and plant and animal communities. Seasonal inundation is the sine qua non of ecological productivity in the Everglades but also the primary obstacle to the modes of human productivity pursued by Euro-American settlers. For over a century, efforts to control hydrologic variability have determined our relationship with the Everglades.

Hydrologic modification and multipurpose water management on an epic scale have yielded enormous benefits in flood protection and agricultural production, but with unintended and profound impacts on the structure, function, and ecosystem services of the Everglades. Flood control operations result in too much water in the wrong places (released to tide via coastal estuaries) and not enough water to recharge the Biscayne Aquifer and hydrate the coastal Everglades. These impacts are increasingly exacerbated by anthropogenic climate change, and the combination of decreased freshwater flows and rising sea levels is transforming Everglades landscapes by raising salinities, accelerating peat loss, and altering species dynamics. The viability of south Florida's urban agglomerations and agricultural industry is threatened by the water control system that brought them into being, and SLR and climate change are magnifying the challenges of operating this system. How homeowners, policy makers, flood insurance providers, and real estate developers respond to climate change will play an ever-more-dominant role in shaping patterns of water and land use.

These daunting challenges cannot be met without hydrologic restoration to increase floodwater retention and southward flows. The state and federal governments have worked together for decades to "get the water right," overcoming tremendous biophysical and institutional obstacles to achieve some remarkable successes. Everglades scientists largely agree on key restoration goals and viable strategies going forward, but continued progress depends on resolution of longstanding sociopolitical conflicts and on the capacity of south Florida residents and water managers to adapt to an environment of growing climatic and sociopolitical uncertainty.

References

Abtew, W. 1996. Evapotranspiration measurements and modeling for three wetland systems in south Florida. *JAWRA Journal of the American Water Resources Association* 32: 465–473.

Abtew, W. 2001. Evaporation estimation for Lake Okeechobee in south Florida. *Journal of Irrigation and Drainage Engineering* 127(3): 140–147.

Abtew, W. 2004. Evapotranspiration in the Everglades: comparison of Bowen Ratio and model estimates. Technical Paper EMA # 417. West Palm Beach: South Florida Water Management District.

Abtew, W., J. Obeysekera, M. Irizarry-Ortiz, D. Lyons, and A. Reardon. 2003. Evapotranspiration estimation for south Florida. World Water & Environmental Resources Congress 2003: 1–9.

Aldwaik, S.Z., J.A. Onsted, and R.G. Pontius, Jr. 2015. Behavior-based aggregation of land categories for temporal change analysis. *International Journal of Applied Earth Observation and Geoinformation* 35: 229–238.

Arguez, A., I. Durre, S. Applequist, M. Squires, R. Vose, X. Yin, and R. Bilotta. 2010. NOAA's U.S. Climate Normals (1981–2010). NOAA National Climatic Data Center. DOI:10.7289/V5PN93JP.

Barr, J., M. DeLonge, and J. Fuentes. 2014. Seasonal evapotranspiration patterns in mangrove forests. *Journal of Geophysical Research: Atmospheres* 119(7): 3886–3899.

Beerens, J.M., D.E. Gawlik, G. Herring, and M.I. Cook. 2011. Dynamic habitat selection by two wading bird species with divergent foraging strategies in a fluctuating wetland. *Auk* 128: 651–662.

Berry, L., F. Bloetscher, N. Hernández Hammer, M. Koch-Rose, D. Mitsova-Boneva, J. Restrepo, T. Root, and R. Teegavarapu. 2011. Florida Water Management and Adaptation in the Face of Climate Change. Florida Climate Change Task Force. Unpublished white paper.

Billings, B., and C. Jones, C. 2008. *Forecasting urban water demand*, 2nd ed. Neverer, Colorado: American Waterworks Association.

Blake, N.M. 1980. *Land Into Water—Water Into Land: A History of Water Management in Florida*. Gainesville: University Press of Florida.

Blanco, R.I., G.M. Naja, R. Rivero, and R.M. Price. 2013. Spatial and temporal changes in groundwater salinity in south Florida. *Applied Geochemistry* 38: 48–58.

Boucek, R.E., E.E. Gaiser, H. Liu and J.S. Rehage. 2016. A review of subtropical community resistance and resilience to extreme cold spells. *Ecosphere* 7(10): e01455.

Boucek, R.E., M.R. Heithaus, R. Santos, P. Stevens, and J.S. Rehage. 2017. Can animal habitat use patterns influence population resistance to climate disturbance? A sub-tropical sportfish case study. *Global Change Biology* 23(10): 4045–4057.

Boucek, R.E., and J.S. Rehage. 2013. No free lunch: displaced marsh consumers regulate a pretty subsidy to an estuarine consumer. *Oikos*. doi:10.1111/j.1600-0706.2013.20994.x

Boucek, R.E., and J.S. Rehage. 2015. A tale of two fishes: using recreational angler records to examine the link between fish catches and floodplain connections in a subtropical coastal river. *Estuaries and Coasts* 38(S1): 124–135.

Bromwell, L.G., R.G. Dean, and S.G. Vick. 2006. Report of Expert Review Panel: Technical Evaluation of Herbert Hoover Dike, Lake Okeechobee, Florida. BCI Engineers & Scientists, Inc. Unpublished report prepared for the South Florida Water Management District.

Cadavid, L.G., C.J. Neidrauer, J.T.B. Obeysekera, E.R. Santee, P. Trimble, and W. Wilcox. 2006. Lake Okeechobee Operations by Means of the Water Supply and Environment (WSE) Regulation Schedule. West Palm Beach: South Florida Water Management District. Unpublished report.

Cahyanto, I. 2017. *The Impacts of 2016 Harmful Algal Bloom (HAB) Outbreak on Florida Tourism*. Project report. Spearfish, South Dakota: University of Florida Tourism Crisis Management Initiative and Black Hills State University.

Carter, L.M., J.W. Jones, L. Berry, V. Burkett, J.F. Murley, J. Obeysekera, P.J. Schramm, and D. Wear. 2014. Southeast and the Caribbean. In *Climate Change Impacts in the United States: The Third National Climate Assessment*, edited by J.M. Melillo, T.C. Richmond, and G.W. Yohe, 396–417. Washington, DC: U.S. Global Change Research Program.

Chick, J.H., C.R. Ruetz III, and J.C. Trexler. 2004. Spatial scale and abundance patterns of large fish communities in freshwater marshes of the Florida Everglades. *Wetlands* 24: 652–664.

Childers, D.L., J.N. Boyer, S.E. Davis, C.J. Madden, D.T. Rudnick and F.H. Sklar. 2006. Relating precipitation and water management to nutrient concentration patterns in the oligotrophic "upside down" estuaries of the Florida Everglades. *Limnology and Oceanography* 51(1): 602–616.

Church, J.A., P.U. Clark, A. Cazenave, J.M. Gregory, S. Jevrejeva, A. Levermann, M.A. Merrifield, G.A. Milne, R.S. Nerem, P.D. Nunn, A.J. Payne, W.T. Pfeffer, D. Stammer, and A.S. Unnikrishnan. 2013. Sea level change. In *Climate Change 2013: The Physical Science Basis*. Working Group I Contribution to the Fifth Assessment Report of the Intergovernmental Panel on Climate Change, edited by T.F. Stocker, D. Qin, G.-K. Plattner, M.M.B. Tignor, S.K. Allen,

J. Boschung, A. Nauels, Y. Xia, V. Bex, and P.M. Midgley, 1137–1216. New York: Cambridge University Press.

Church, J.A., and N.J. White. 2011. Sea level rise from the late 19th to the early 21st century. *Survey Geophysics* 32: 585–602.

Collins, S.L., S.R. Carpenter, S.M. Swinton, D.E. Orenstein, D.L. Childers, T.L. Gragson, N.B. Grimm, J.M. Grove, S.L. Harlan, J.P. Kaye, A.K. Knapp, G.P. Kofinas, J.J. Magnuson, W.H. McDowell, J.M. Melack, L.A. Ogden, G.P. Robertson, M.D. Smith, and A.C. Whitmer. 2011. An integrated conceptual framework for long-term social-ecological research. *Frontiers in Ecology and the Environment* 9(6): 351–357.

Committee on Independent Scientific Review of Everglades Restoration Progress (CISRERP). 2014. *Progress Toward Restoring the Everglades: The Fifth Biennial Review*. Washington, DC: National Academies Press.

Cook, M.I., E.M. Call, R.M. Kobza, S. Hill, and C. Saunders. 2014. Seasonal movements of crayfish in a fluctuating wetland: implications for restoring wading bird populations. *Freshwater Biology* 59: 1608–1621.

Cook, M.I., and H. Herring. 2007. Food availability and White Ibis reproductive success: an experimental study. In *South Florida Wading Bird Report*, Volume 13, edited by M.I. Cook and H.K. Herring, 42–45. West Palm Beach: South Florida Water Management District.

Cunningham, K.J. 2004. Application of ground-penetrating radar, digital optical borehole images, and cores for characterization of porosity hydraulic conductivity and paleokarst in the Biscayne Aquifer, southeastern Florida, USA. *Journal of Applied Geophysics* 55(1): 61–76.

Dausman, A., and C.D. Langevin. 2005. Movement of the saltwater interface in the surficial aquifer system in response to hydrologic stresses and water-management practices, Broward County, Florida. U.S. Geological Survey Scientific Investigations Report 2004–5256. https://pubs.er.usgs.gov/publication/sir20045256.

Davidov, S. 2013. HighWaterLine project charts the impact of sea level rise in Miami. *Miami New Times*. http://www.miaminewtimes.com/arts/highwaterline-project-charts-the-impact-of-sea-level-rise-in-miami-6494902.

Davis, S.M., D.L. Childers, J.J. Lorenz, H.R. Wanless, and T.E. Hopkins. 2005. A conceptual model of ecological interactions in the mangrove estuaries of the Florida Everglades. *Wetlands* 25(4): 832–842.

DiFrenna, V.J., R.M. Price, and M.R. Savabi. 2008. Identification of a hydrodynamic threshold in karst rocks from the Biscayne Aquifer, south Florida, USA. *Hydrogeology Journal* 16(1): 31–42.

Dorn, N.J., and J.C. Trexler. 2007. Crayfish assemblage shifts in a large drought-prone wetland: the roles of hydrology and competition. *Freshwater Biology* 20(1): 113–115.

Douglas, E.M., J.M. Jacobs, D.M. Sumner, and R.L. Ray. 2009. A comparison of models for estimating potential evapotranspiration for Florida land cover types. *Journal of Hydrology* 373: 366–376.

EarthTech, Hydrologics and SynInt, Inc. 2005. Task 3 Final Report: Water Management Operations Rules Inventory and Development of Preliminary Optimization-Based Model of Current Operating Rules. Jupiter, Florida: EarthTech. Unpublished report prepared for the U.S. Army Corps of Engineers Jacksonville District.

Elliot, M., A.K. Whitfield, I.C. Potter, S.J.M. Blaber, D.P. Cyrus, F.G. Nordlie, and T.D. Harrison. 2007. The guild approach to categorizing estuarine fish assemblages: a global review. *Fish and Fisheries* 8: 241–268.

Enos, P., and R.D. Perkins. 1979. Evolution of Florida Bay from island stratigraphy. *Geological Society of America Bulletin* 90(1): 59–83.

Field, C.B., V.R. Barros, K.J. Mach, M.D. Mastrandrea, M. van Aalst, W.N. Adger, D.J. Arent, J. Barnett, R. Betts, T.E. Bilir, J. Birkmann, J. Carmin, D.D. Chadee, A.J. Challinor, M. Chatterjee, W. Cramer, D.J. Davidson, Y.O. Estrada, J.-P. Gattuso, Y. Hijioka, O. Hoegh-Guldberg, H.Q. Huang, G.E. Insarov, R.N. Jones, R.S. Kovats, P. Romero-Lankao, J.N. Larsen, I.J. Losada, J.A. Marengo, R.F. McLean, L.O. Mearns, R. Mechler, J.F. Morton, I. Niang, T. Oki, J.M. Olwoch, M. Opondo, E.S. Poloczanska, H.-O. Pörtner, M.H. Redsteer, A. Reisinger, A. Revi, D.N. Schmidt, M.R. Shaw, W. Solecki, D.A. Stone, J.M.R. Stone, K.M. Strzepek, A.G. Suarez, P. Tschakert, R. Valentini, S. Vicuña, A. Villamizar, K.E. Vincent, R. Warren, L.L. White, T.J. Wilbanks, P.P. Wong, and G.W. Yohe. 2014. Technical summary. In *Climate Change 2014: Impacts, Adaptation, and Vulnerability. Part A: Global and Sectoral Aspects.* Contribution of Working Group II to the Fifth Assessment Report of the Intergovernmental Panel on Climate Change, edited by C.B. Field, V.R. Barros, D.J. Dokken, K.J. Mach, M.D. Mastrandrea, T.E. Bilir, M. Chatterjee, K.L. Ebi, Y.O. Estrada, R.C. Genova, B. Girma, E.S. Kissel, A.N. Levy, S. MacCracken, P.R. Mastrandrea, and L.L. White, 35–94. Cambridge and New York: Cambridge University Press.

Frederick, P.C., D.E. Gawlik, J.C. Ogden, M.I. Cook, and M. Lusk. 2009. The white ibis and wood stork as indicators for restoration of the Everglades ecosystem. *Ecological Indicators* 9: S83–S95.

Frederick, P.C., and J.C. Ogden. 2001. Pulsed breeding of long-legged wading birds and the importance of infrequent severe drought conditions in the Florida Everglades. *Wetlands* 21: 484–491.

Gaiser, E.E., D.L. Childers, R.D. Jones, J.H. Richards, L.J. Scinto, and J.C. Trexler. 2006. Periphyton responses to eutrophication in the Florida Everglades: cross-system patterns of structural and compositional change. Pt. 2, *Limnology and Oceanography* 51(1): 617–630.

Gawlik, D.E. 2002. The effects of prey availability on the numerical response of wading birds. *Ecological Monographs* 72: 329–346.

German, E.R. 2000. Regional evaluation of evapotranspiration in the Everglades: U.S. Geological Survey Water-Resources Investigations Report 00–4217. https://pubs.er.usgs.gov/publication/wri004217.

Gidley, M., H. Briceño, A. Serna, E. Kelly, and C. Sinigalliano. 2016. Characterizing Microbial Water Quality of Extreme Tide Floodwaters Discharged from an Urbanized Subtropical Beach: Case Study of Miami Beach with Implications for Sea Level Rise and Public Health. Paper presented at the AGU Ocean Science Meeting, February 21–26, New Orleans, Louisiana.

Gillson, J. 2011. Freshwater flow and fisheries production in estuarine and coastal systems: where a drop of rain is not lost. *Reviews in Fisheries Science* 19(3): 168–186.

Godfrey, M.C., and T. Catton. 2011. *River of Interests: Water Management in South Florida and the Everglades, 1948–2010*. Washington, DC: Historical Research Associates.

Goss, C., W.F. Loftus, and J.C. Trexler. 2014. Seasonal fish dispersal in ephemeral wetlands of the Florida Everglades. *Wetlands* 34(Suppl 1): 147–157.

Graham, W.D., M.J. Angelo, T.K. Frazer, P.C. Frederick, K.E. Havens, and K.R. Reddy. 2015. Options to Reduce High Volume Freshwater Flows to the St. Lucie and Caloosahatchee Estuaries and Move More Water from Lake Okeechobee to the Southern Everglades. Gainesville: University of Florida Water Institute. Unpublished document.

Groffman, P.M., J. Cavender-Bares, N.D. Bettez, J.M. Grove, S.J. Hall, J.B. Heffernan, S.E. Hobbie, K.L. Larson, J.L. Morse, C. Neill, K. Nelson, J. O'Neill-Dunne, L. Ogden, D.E. Pataki, C. Polsky, R. Roy Chowdhury, and M. Steele. 2014. Ecological homogenization of urban America. *Frontiers in Ecology & Environment* 12(1): 74–81.

Grunwald, M. 2006. *The Swamp: The Everglades, Florida and the Politics of Paradise*. New York: Simon & Schuster.

Hall, S.J., J. Learned, B. Ruddell, K.L. Larson, J. Cavender-Bares, N.D. Bettez, P.M. Groffman, J.M. Grove, J.B. Heffernan, S.E. Hobbie, K. Larson, J.L. Morse, C. Neill, K.C. Nelson, J. O'Neil-Dunne, L.A. Ogden, D.E. Pataki, W.D. Pearse, C. Polsky, R. Roy Chowdhury, M.K. Steele, and T. Trammell. 2016. Convergence of microclimate in residential landscapes across diverse cities in the United States. *Landscape Ecology* 31(1): 101–117.

Hansen, J. 2007. Scientific reticence and sea level rise. *Environmental Research Letters* 2: 024002.

Harris, E.M., C. Polsky, K.L. Larson, R. Garvoille, D. G. Martin, J. Brumand, and L. Ogden. 2012. Heterogeneity in residential yard care: evidence from Boston, Miami, and Phoenix. *Human Ecology* 40(5): 735–749.

Hay, C.C., E. Morrow, R.E. Kopp, and J.X. Mitrovica. 2015. Probabilistic reanalysis of twentieth-century sea-level rise. *Nature* 517(7535): 481–484.

Heithaus, M.R., B. Delius, A.J. Wirsing, and M.M. Dunphy-Daly. 2009. Physical factors influencing the distribution of a top predator in a subtropical oligotrophic estuary. *Limnology and Oceanography* 54: 472–482.

Herring, G., M.I. Cook, D.E. Gawlik, and E.M. Call 2011. Food availability is expressed through physiological stress indicators in nestling White Ibis: a food supplementation experiment. *Functional Ecology* 25: 682–690.

Herring, G., D.E. Gawlik, M.I. Cook, and J.M. Beerens. 2010. Sensitivity of nesting Great Egrets (*Ardea alba*) and White Ibises (*Eudocimus albus*) to reduced prey availability. *Auk* 127: 660–670.

Hughes, J.D., and J.T. White. 2016. Hydrologic Conditions in Urban Miami-Dade County, Florida, and the Effect of Groundwater Pumpage and Increased Sea Level on Canal Leakage and Regional Groundwater Flow. USGS Scientific Investigations Report 2014–5162. https://pubs.er.usgs.gov/publication/sir20145162.

Huo, J., A. Mittl, G. Reilly, and R. Sosnowski. 2006. Application of GIS to plan long range water supply facilities by linking land use and water billing data of city of Cape Coral in southwest Florida. Water Environment Foundation WEFTEC: 1319–1327.

Jiang, J., D.L. DeAngelis, T.J. Smith III, S.Y. Teh, and H. Koh. 2012. Spatial pattern formation of coastal vegetation in response to external gradients and positive feedbacks affecting soil porewater salinity: a model study. *Landscape Ecology* 27(1): 109–119.

Kellogg, C.M., and N.J. Dorn. 2012. Consumptive effects of fish reduce wetland crayfish recruitment and drive species turnover. *Oecologia* 168: 1111–1121.

Kisekka, I., K.W. Migliaccio, M.D. Dukes, B. Schaffer, and J.H. Crane. 2010. Evapotranspiration-based irrigation scheduling and physiological response in a Carambola (*Averrhoa carambola* L.) orchard. *Applied Engineering in Agriculture* 26(3): 373–380.

Kisekka, I., K.W. Migliaccio, M.D. Dukes, B. Schaffer, and J.H. Crane. 2012. Evapotranspiration-Based Irrigation Scheduling for Agriculture. University of Florida IFAS Extension Publication #AE457.

Kobza, R.M., J.C. Trexler, W.F. Loftus, and S.A. Perry. 2004. Community structure of fishes inhabiting aquatic refuges in a threatened karst wetland and its implications for ecosystem management. *Biological Conservation* 116(2): 153–165.

Koch, M.S., C. Coronado, M.W. Miller, D.T. Rudnick, E. Stabenau, R.B. Halley, and F.H. Sklar 2015. Climate change projected effects on coastal foundation communities of the Greater Everglades using a 2060 Scenario: need for a new management paradigm. *Environmental Management* 55(4): 857–875.

Kushlan, J.A. 1976. Wading bird predation in a seasonally fluctuating pond. *Auk* 93: 464–476.

Lagomasino, D. 2014. Ecohydrology, Evapotranspiration and Hydrogeochemistry of Carbonate Mangrove Wetlands. FIU Electronic Theses and Dissertations, Paper 1258. http://digitalcommons.fiu.edu/etd/1258.

Lambeck, K., T.M. Esat, and E. Potter. 2002. Links between climate and sea levels for the past three million years. *Nature* 419: 199–206.

Lee, J.A. 2016. Survival of Florida Largemouth Bass in a coastal refuge habitat across years of drying severity. Master thesis, Florida International University, Miami, FL.

Lee, T.M., L.A. Sacks, and A. Swancar. 2014. Exploring the long-term balance between net precipitation and net groundwater exchange in Florida seepage lakes. *Journal of Hydrology* 519: 3054–3068.

Light, S.S., and W. Dineen. 1994. Water control in the Everglades: a historical perspective. In *Everglades: The Ecosystem and Its Restoration*, edited by S.M. Davis and J.C. Ogden, 47–84. Delray Beach, FL: St. Lucie Press.

Loftus, W.F., and A.M. Eklund. 1994. Long-term dynamics of an Everglades small-fish assemblage. In *Everglades: The Ecosystem and Its Restoration*, edited by S.M. Davis and J.C. Ogden, 461–483. Delray Beach, Florida: St. Lucie Press.

Manda, A.K., and M.R. Gross. 2006. Estimating aquifer-scale porosity and the REV for karst limestones using GIS-based spatial analysis. *Geological Society of America Special Papers* 404: 177–189.

Matich, P., and M.R. Heithaus. 2014. Multi-tissue stable isotope analysis and acoustic telemetry reveal seasonal variability in the trophic interactions of juvenile bull sharks in a coastal estuary. *Journal of Animal Ecology* 83(1): 199–213.

Mazzotti, F.J., G.R. Best, L.A. Brandt, M.S. Cherkiss, B.M. Jeffery, and K.G. Rice. 2009. Alligators and crocodiles as indicators for restoration of Everglades ecosystems. *Ecological Indicators* 9(6): S137–S149.

McCally, D. 1999. *The Everglades: An Environmental History*. Gainesville: University Press of Florida.

McCarthy, L.C., W.F. Loftus, and J.S. Rehage. 2012. Shrimp along an Everglades estuarine gradient: do multiple species have similar trophic function? *Bulletin of Marine Science* 88(4): 843–861.

McPherson, B.F., and R. Halley. 1997. The south Florida environment: a region under stress. U.S. Geological Survey Circular 1134.

McVoy, C.W., W. Park Said, J. Obeysekera, J. VanArman, and T.W. Dreschel. 2011. *Landscapes and Hydrology of the Predrainage Everglades*. Gainesville: University Press of Florida.

Mettler, K. 2016. This disgusting, "guacamole-thick" goop is invading Florida's coastline. *Washington Post*, 1 July 2016.

Miller, J.A. 1997. Hydrogeology of Florida. In *The Geology of Florida*, edited by A.F. Randazzo and D.S. Jones, 69–88. Gainesville: University Press of Florida.

Moses, C.S., W.T. Anderson, C. Saunders, and F. Sklar. 2013. Regional climate gradients in precipitation and temperature in response to climate teleconnections in the Greater Everglades ecosystem of South Florida. *Journal of Paleolimnology* 49(1): 5–14.

Myers, R.L., and J.J. Ewel. 1990. Problems, prospects, and strategies for conservation. In *Ecosystems of Florida*, edited by R.L. Myers and J.J. Ewel, 765. Gainesville: University of Florida Press.

Nuttle, W.K., J.W. Fourqurean, B.J. Cosby, J.C. Zieman, and M.B. Robblee. 2000. Influence of net freshwater supply on salinity in Florida Bay. *Water Resources Research* 36(7): 1805–1822.

Obeysekera, J., M. Irrizarri, J. Park, J. Barnes, and T. Dessalegne. 2011. Climate change and its implications for water resources management in south Florida. *Stochastic Environmental Research and Risk Assessment* 25(4): 495–516.

Obeysekera, J., P. Trimble, C. Neidrauer, C. Pathak, J. VanArman, T. Strowd, and C. Hall. 2006. Consideration of Long-Term Climatic Variability in Regional Modeling for SFWMD Planning & Operations. Appendix 2-2 of the 2007 South Florida Environmental Report. West Palm Beach: South Florida Water Management District. Unpublished draft document.

Ogden, J.C. 1994. A comparison of wading bird nesting dynamics, 1931–1946 and 1974–1989 as an indication of changes in ecosystem conditions in the southern Everglades. In *Everglades: The Ecosystem and Its Restoration*, edited by S. Davis and J.C. Ogden, 533–570. Delray Beach, Florida: St. Lucie Press.

Onsted, J.A., and R. Roy Chowdhury. 2014. Does zoning matter? A comparative analysis of landscape change in Redland, Florida using cellular automata. *Landscape and Urban Planning* 121: 1–18.

Parkos, J.J. III, C.R. Ruetz III, and J.C. Trexler. 2011. Disturbance regime and limits on benefits of refuge use for fishes in a fluctuating hydroscape. *Oikos* 120(10): 1519–1530.

Philipp, D.P., W.F. Childers, and G.S. Whitt. 1983. A biochemical genetic evaluation of the northern and Florida subspecies of largemouth bass. *Transactions of the American Fisheries Society* 112(1): 1–20.

Pokharel, S. 2014. Analysis of land use change as a method of predicting water demands in an urbanizing environment: Redland, *Miami-Dade County, Florida*. Master's thesis, Florida International University, FIU Electronic Theses and Dissertations.

Polsky, C., J. Morgan Grove, C. Knudson, P.M. Groffman, N. Bettez, J. Cavender-Bares, S. Hall, J. Heffernan, S. Hobbie, K. Larson, J. Morse, C. Neill, K. Nelson, L. Ogden, J. O'Neill-Dunne, D. Pataki, R. Roy Chowdhury, and M. Steele. 2014. Assessing the homogenization of urban land management with an application to US residential lawncare. *Proceedings of the National Academy of Sciences of the USA* 111(12): 4432–4437.

Price, R.M., P.K. Swart, and J.W. Fourqurean. 2006. Coastal groundwater discharge-an additional source of phosphorus for the oligotrophic wetlands of the Everglades. *Hydrobiologia* 569(1): 23–36.

Price, R.M., P.K. Swart, and H.E. Willoughby. 2008. Seasonal and spatial variation in the stable isotopic composition (δ^{18}O and δD) of precipitation in south Florida. *Journal of Hydrology* 358(3-4): 193–205.

Pringle, C.M. 2001. Hydrologic connectivity and the management of biological reserves: a global perspective. *Ecological Applications* 11(4): 981–998.

Rahmstorf, S. 2010. A new view on sea level rise. *Nature Reports Climate Change* 1004: 44–45.

Rehage, J.S., S.E. Liston, K.J. Dunker, and W.F. Loftus. 2014. Fish community responses to the combined effects of decreased hydroperiod and nonnative fish invasions in a karst wetland: are Everglades solution holes sinks for native fishes? *Wetlands* 34(Suppl 1): 159–173.

Rehage, J.S., and W.F. Loftus. 2007. Seasonal fish community variation in headwater mangrove creeks in the southwestern everglades: an examination of their role as dry-down refuges. *Bulletin of Marine Science* 80: 625–645.

Rehage, J.S., and J.C. Trexler. 2006. Assessing the net effect of anthropogenic disturbance on aquatic communities in wetlands: community structure relative to distance from canals. *Hydrobiologia* 569(1): 359–373.

Robins, J.B., I.A. Halliday, J. Staunton-Smith, D.G. Mayer, and M.J. Sellin. 2005. Freshwater-flow requirements of estuarine fisheries in tropical Australia: a review of the state of knowledge and application of suggested approach. *Marine and Freshwater Research* 56(3): 343–360.

Rosenblatt, A.E., and M.R. Heithaus. 2011. Does variation in movement tactics and trophic interactions among American alligators create habitat linkages? *Journal of Animal Ecology* 80(4): 786–798.

Ross, M.S., J.F. Meeder, J.P. Sah, P.L. Ruiz, and G.J. Telesnicki. 2000. The southeast saline Everglades revisited: 50 years of coastal vegetation change. *Journal of Vegetation Science* 11(1): 101–112.

Roy Chowdhury, R., K. Larson, J.M. Grove, C. Polsky, E. Cook J. Onsted, and L. Ogden. 2011. A multi-scalar approach to theorizing socio-ecological dynamics of urban residential landscapes. *Cities and the Environment (CATE)* 4(1): Article 6.

Ruetz, C.R., J.C. Trexler, F. Jordan, W.F. Loftus, and S.A. Perry. 2005. Population dynamics of wetland fishes: spatio-temporal patterns synchronized by hydrological disturbance? *Journal of Animal Ecology* 74: 322–332.

Saha, A.K., C.S. Moses, R.M. Price, V. Engel, T.J. Smith III, and G. Anderson. 2012. A hydrological budget (2002–2008) for a large subtropical wetland ecosystem indicates marine groundwater discharge accompanies diminished freshwater flow. *Estuaries and Coasts* 35(2): 459–474.

Saha, A.K., S. Saha, J. Sadle, J. Jiang, M.S. Ross, R.M. Price, L.S.L. Sternberg, and K.S. Wendelberger. 2011. Sea level rise and South Florida coastal forests. *Climate Change* 107(1-2): 81–108.

Saha, S., J. Sadle, C. van der Heiden, and L. Sternberg. 2015. Salinity, groundwater, and water uptake depth of plants in coastal uplands of Everglades National Park (Florida, USA). *Ecohydrology* 8(1): 128–136.

Saha, A.K., L.S.L. Sternberg, M.S. Ross, and F. Miralles-Wilhelm. 2010. Water source utilization and foliar nutrient status differs between upland and flooded plant communities in wetland tree islands. *Wetlands Ecology and Management* 18(3): 343.

Schedlbauer, J.L., S.F. Oberbauer, G. Starr, and K.L. Jimenez. 2011. Controls on sensible heat and latent energy fluxes from a short-hydroperiod Florida Everglades marsh. *Journal of Hydrology* 411(3-4): 331–341.

Schwartz, K.Z.S. 2014. The anti-politics of biopolitical disaster on Florida's coasts. Paper presented at the Western Political Science Association conference, Seattle, Washington.

Shih, S.F., G.S. Rahi, and D.S. Harrison. 1982. Evapotranspiration studies on rice in relation to water use efficiency. *Transactions of the ASAE* 25(3): 0702–0707.

Simpfendorfer, C.A., G.G. Freitas, T.R. Wiley, and M.R. Heupel. 2005. Distribution and habitat partitioning of immature bull sharks (*Carcharhinus leucas*) in a southwest Florida estuary. *Estuaries* 28: 78–85.

South Florida Water Management District (SFWMD). 2013. Lower East Coast Water Supply Plan Update. West Palm Beach: South Florida Water Management District. Unpublished document.

South Florida Water Management District (SFWMD). 2014. Water Supply Planning. http://www.sfwmd.gov/portal/page/portal/xweb%20-%20release%203%20water%20supply/water%20supply%20planning.

Steele, M.K., J. B. Heffernan, N. Bettez, J. Cavender-Bares, P.M. Groffman, M. Grove, S. Hall, S. E. Hobbie, K. Larson, J. L. Morse, C. Neill, K.C. Nelson, J. O'Neil-Dunne, L. Ogden, D.E. Pataki, C. Polsky, and R. Roy Chowdhury. 2014. Convergent surface water distributions in U.S. cities. *Ecosystems* 17(4): 685–697.

Stevens, P.W., D.A. Blewett, R.E. Boucek, J.S. Rehage, B.L. Winner, J.M. Young, J.A. Whittington, and R. Paperno. 2016. Resilience of a tropical sport fish population to a severe cold event varies across five estuaries in southern Florida. *Ecosphere* 7(8): e01400.

Sullivan, P.L., R.M. Price, V. Engel, and M.S. Ross. The influence of vegetation on the hydrodynamics and geomorphology of a tree island in Everglades National Park (Florida, USA). *Ecohydrology* 7(2): 727–744.

Sullivan, P.L., R.M. Price, F. Miralles-Wilhelm, M.S. Ross, L.J. Scinto, T.W. Dreschel, F.H. Sklar, and E. Cline. 2012. The role of recharge and evapotranspiration as hydraulic drivers of ion concentrations in shallow groundwater on Everglades tree islands, Florida (USA). *Hydrological Processes* 28(2): 293–304.

Teh S.Y., D.L. DeAngelis, L.S.L. Sternberg, F.R. Miralles-Wilhelm, T.J. Smith, and H. Koh. 2008. A simulation model for projecting changes in salinity concentrations and species dominance in the coastal margin habitats of the Everglades. *Ecological Modelling* 213(2): 245–256.

Trexler, J.C., W.F. Loftus, and S. Perry. 2005. Disturbance frequency and community structure in a twenty-five-year intervention study. *Oecologia* 145: 140–152.

Troxler, T.G., D.L. Childers, and C.J. Madden. 2014. Drivers of decadal-scale change in southern Everglades wetland macrophyte communities of the coastal ecotone. *Wetlands* 34(1): 81–90.

Trout, K., and M. Ross. 2006. Estimating evapotranspiration in urban environments. In *Urban Groundwater Management and Sustainability*, edited by J.H. Tellam, M.O. Rivett, R.G. Israfilov, and L.G. Herringshaw, 157–168. Dordrecht, the Netherlands: Springer.

U.S. Army Corps of Engineers (USACE). 2015. Herbert Hoover Dike Dam Safety Modification Study Draft Environmental Impact Statement. Unpublished document. Jacksonville, FL: USACE.

U.S. Army Corps of Engineers (USACE) and SFWMD. 2013. Final Technical Data Report: Comprehensive Everglades Restoration Plan Aquifer Storage and Recovery Pilot Project. Unpublished document. Jacksonville and West Palm Beach, FL: USACE and SFWMD.

Villalobos-Vega, R. 2010. Water Table and Nutrient Dynamics in Neotropical Savannas and Wetland Ecosystems. Ph.D. dissertation, University of Miami. http://scholarlyrepository.miami.edu/oa_dissertations/389

Wachnicka, A.H., and G.L. Wingard. 2015. Biological indicators of changes in water quality and habitats of the coastal and estuarine areas of the Greater Everglades Ecosystem. In *Microbiology of the Everglades Ecosystem*, edited by J.A. Entry, A.D. Gottlieb, K. Jayachandran, and A. Ogram, 127–144. Boca Raton, FL: CRC Press.

Wang, X., L.O. Sternberg, M.S. Ross, and V.C. Engel. 2011. Linking water use and nutrient accumulation in tree island upland hammock plant communities in the Everglades National Park, USA. *Biogeochemistry* 104(1–3): 133–146.

Wanless, H. R., and M.G. Tagett. 1989. Origin, growth and evolution of carbonate mudbanks in Florida Bay. *Bulletin of Marine Science* 44(1): 454–489.

Watson, C.S., N.J. White, J.A. Church, M.A. King, R.J. Burgette, and B. Legresy. 2015. Unabated global mean sea-level rise over the satellite altimeter era. *Nature Climate Change* 5: 565–568.

Wdowinski, S., R. Bray, B.P. Kirtman, and Z. Wu. 2016. Increasing flooding hazard in coastal communities due to rising sea level: case study of Miami Beach, Florida. *Ocean & Coastal Management* 126: 1–8.

Welcomme, R.L. 2001. *Inland Fisheries Ecology and Management*. Oxford, UK: Blackwell Science Ltd.

Wendel, J. 2016. Dirty water: unintended consequence of climate resiliency. *Eos*. https://eos.org/articles/dirty-water-unintended-consequence-of-climate-resiliency.

Willard, D.A., and C.E. Bernhardt. 2011. Impacts of past climate and sea level change on Everglades wetlands: placing a century of anthropogenic change into a late-Holocene context. *Climatic Change* 107(1–2): 59–80.

Winsberg, M.D., and M. Simmons. 2009. An analysis of the beginning, end, length, and strength of Florida's hot season. http://climatecenter.fsu.edu/topics/specials/floridas-hot-season.

Wu, C.-L., and S. Shukla. 2013. Eddy covariance-based evapotranspiration for a subtropical wetland. *Hydrological Processes* 28: 5879–5896.

Zapata-Rios, X., and R.M. Price. 2012. Estimates of groundwater discharge to a coastal wetland using multiple techniques: Taylor Slough, Everglades National Park. *Hydrogeology Journal* 20: 1651–1668.

4

Ecosystem Fragmentation and Connectivity

LEGACIES AND FUTURE IMPLICATIONS OF A RESTORED EVERGLADES

John Kominoski, Jennifer Rehage, and Bill AndersonwithRoss Boucek, Henry Briceño, Mike Bush, Tom Dreschel, Michael Heithaus, Rudolf Jaffé, Laurel Larsen, Philip Matich, Christopher McVoy, Adam Rosenblatt, and Tiffany Troxler

In a Nutshell

- A century of fragmentation and novel hydrologic and ecological connectivity characterize the postdrainage Everglades landscape.
- Human engineering has dramatically reduced the spatiotemporal patterns of connectivity that characterized the predrainage ecosystem, but it has also increased connectivity among disparate ecosystems in novel ways.
- Compartmentalized hydroecological dynamics have shifted the dependent drivers of connectivity from seasonal and local or regional scales to decadal and global scales via teleconnections, temporal correlations in climatic patterns at widely separated geographic locations.
- Long-term, spatially explicit ecological research has addressed questions about ecological connectivity in complex wetland landscapes, while also raising new ones.
- Restoration of the Everglades will test the human–environmental framework that links freshwater–saltwater connectivity to ecosystem resilience throughout this changing coastal landscape.

Introduction

Fragmentation characterizes many ecosystems that have been redesigned by humans, and often these same systems are connected to distant systems in novel ways due to human engineering. Ecosystem connectivity integrates the movement of materials, energy, and organisms through space and time, and modern ecosystems are highly connected through global phenomena such as atmospheric deposition of nutrients, elevated atmospheric greenhouse gases, and climate-related teleconnections. Ecosystems also become disconnected through development of human infrastructure, such as roads, levees, fences, and changes in land use.

The degree to which aquatic ecosystem connectivity is altered by human design and engineering may be characterized by hydrologic regimes (Poff et al. 1997) and hydrologic connectivity (Pringle 2003). In many systems, hydrologic connectivity drives ecological connectivity across multiple temporal and spatial scales. These include seasonal to multi-decadal patterns as well as longitudinal (upstream–downstream), lateral (channel–floodplain), and vertical (atmosphere–surface water or surface water–groundwater) dimensions. By *hydrologic connectivity* we refer to the water-mediated transfer of matter, energy, and organisms within and across elements of the hydrologic cycle (*sensu* Pringle 2001). These concepts of connectivity are analogous to landscape connectivity, which first developed to examine landscape structure in the context of patchy habitats and organismal movements (Merriam 1984). Ecological connectivity, both within and across ecosystem boundaries, is a key component of the ecological integrity; connectivity of ecosystems is a fundamental concern for the long-term protection, management, and sustainability of ecosystems and landscapes (Bracken et al. 2013; Fullerton et al. 2010; Pringle 2003). Land use change disrupts hydrologic and ecological connectivity across landscapes, but it also produces new forms of connectivity and increases the importance of global drivers of connectivity over local/regional drivers. In this chapter we illustrate:

1. How predrainage ecological connectivity in the Everglades has been greatly reduced
2. How organic matter tracers and animal movements provide evidence of reduced connectivity
3. The (dis)connectivity of the canal system
4. The increased importance of teleconnections to Everglades connectivity
5. Future Everglades connectivity through freshwater restoration and sea level rise (SLR).

Changes to Everglades Connectivity

The Greater Everglades is a globally iconic and regionally important landscape containing both fragmented and connected ecosystems. The world's largest restoration effort is under way to enhance ecological connectivity in the Everglades in the context of global changes in climate and teleconnections that offer opportunities for and challenges to maintaining long-term ecologic connectivity. The expansive, predrainage Everglades was characterized by a high degree of hydrologic connectivity relative to the largely fragmented modern landscape (McVoy et al. 2011; see Chapter 1). In the predrainage Everglades, hydrologic connectivity occurred by southward sheet flow that connected the entire watershed and by the pronounced seasonality in rainfall, which regulated the spatiotemporal dynamics of connectivity (McVoy et al. 2011). Hydrologic connectivity also originated from strong linkages to water sources north of the Everglades and to the large freshwater outflows to the south along both coasts in the southern tip of the peninsula.

Although some ecological connectivity remains in the engineered Everglades, alterations to the system have resulted in fragmentation and changes to predrainage flow patterns. For instance, canals and levees now bisect wetlands, dramatically reducing hydrologic connectivity across the landscape. But canals also act to increase spatial and temporal connectivity among previously disparate regions such as (1) Lake Okeechobee and the west and east coasts (via lake discharges to the Caloosahatchee and St. Lucie estuaries); (2) the central Everglades and the urban southeastern corridor; (3) the Everglades Agricultural Area (EAA) and Florida Bay; and (4) the Redland Agricultural Area in southwest Miami-Dade County and Biscayne Bay. Hydrologic engineering and water management have decoupled inundation patterns from the seasonality of rainfall, leaving some areas of the landscape drier and others wetter than during predrainage periods virtually all the time. These changes in water regime have large implications for the associated biological communities and ecosystems (see Chapter 3). Water management effectively has changed freshwater and marine hydroperiods, changing both the source and residence time of water throughout the Everglades landscape.

The postdrainage Everglades is most characterized by changes in vegetation communities within ridge and slough habitats that are linked to reduced water flow, altered water levels, and accelerated peat loss. This modification in topography has resulted in a reduced differentiation of ridge and slough flow regimes and increased resistance to sediment transport within former sloughs. Within sloughs, increases in vegetation stem densities have also significantly reduced water flow (Harvey and Larsen 2002). The differentiation of flow regimes has been further reduced by peat oxidation, which has reduced elevation differences between ridges and sloughs. This loss of peat has effectively flattened the "corrugated" ridge and slough microtopography across much of Everglades National Park (see Fig. 1.1 in Chapter 1). When anthropogenic drainage exposed formerly inundated anaerobic peat soils to aerobic conditions, peat loss began to occur, both by slow, continual microbial oxidation and by occasional peat fires. The greatest losses of peat thickness, and of elevation, occurred where the peat was thickest—in Shark River Slough. This loss resulted in increased "dishing" of the topography perpendicular to flow; that is, greater elevation differences within Shark River Slough relative to the flanking marshes (the present-day "marl prairies") to the east and west. Prior to the drainage-induced peat loss, the transverse cross-section of this ground surface was much flatter (see Fig. 1.1 in Chapter 1; McVoy et al. 2011). These changes in topography have ramifications for the opportunities to restore hydrologic connectivity.

Despite dramatic changes to the predrainage connectivity of the Everglades, some aspects of spatial and temporal connectivity have remained largely intact (climatic teleconnections) or have even increased (enhanced marine transgression due to accelerating SLR). Teleconnections are temporal correlations in climatic patterns at widely separated geographic locations. Teleconnections from the El Niño-Southern Oscillation (ENSO) and the Atlantic Multi-decadal Oscillation (AMO) affect patterns of precipitation and temperature regimes across the south Florida peninsula (Childers et al. 2006; Enfield et al. 2001; Moses et al. 2013). However, changes in climate and land use for urbanization and agriculture have increased the importance of these teleconnections as drivers of connectivity. Although Everglades restoration should increase the spatiotemporal extent of freshwater connectivity to the southern Everglades, storm surges and marine transgression due to SLR will continue to increase the connectivity of coastal wetlands with the ocean.

Our research focus in the Everglades has focused on the most hydrologically altered portion of the remaining landscape—Everglades National Park (ENP). This region is considerably drier than under predrainage conditions, with shallower water depths and shortened hydroperiods (Fig. 4.1). Overall flow volumes and hydroperiods have been reduced by water management and engineered infrastructure (e.g., Tamiami Trail) that have reduced southwestern sheet flow. Through spatially resolved long-term research, we have elucidated the critical links integrating fresh and marine water supplies, organic matter and consumer dynamics, and ecosystem resilience. These important findings are informing a more effective Everglades restoration process (see Chapter 8).

Types of Ecological Connectivity

Changes in ecological connectivity can be parsed into changes in *structural, functional,* and *process* connectivity, which are interrelated across space and time (Fig. 4.2). Structural connectivity is a type of spatial connectivity that describes the physical adjacency or contiguity of landscape features, such as ridges and sloughs (Calabrese and Fagan 2004; Tischendorf and Fahrig 2000). It establishes the physical template that regulates functional connectivity, which is a spatiotemporal attribute that describes the rate at which water, dissolved or suspended matter, and organisms are able to move through the landscape (i.e., fluxes, flows; Ali and Roy 2010; Wainwright et al. 2011). Process connectivity is also a spatiotemporal attribute that describes the dynamic linkages among variables as characterized by feedbacks, forcings, or synchronicities (Miller et al. 2012; Ruddell and Kumar 2009). For example, surface water flow would be dynamically connected to

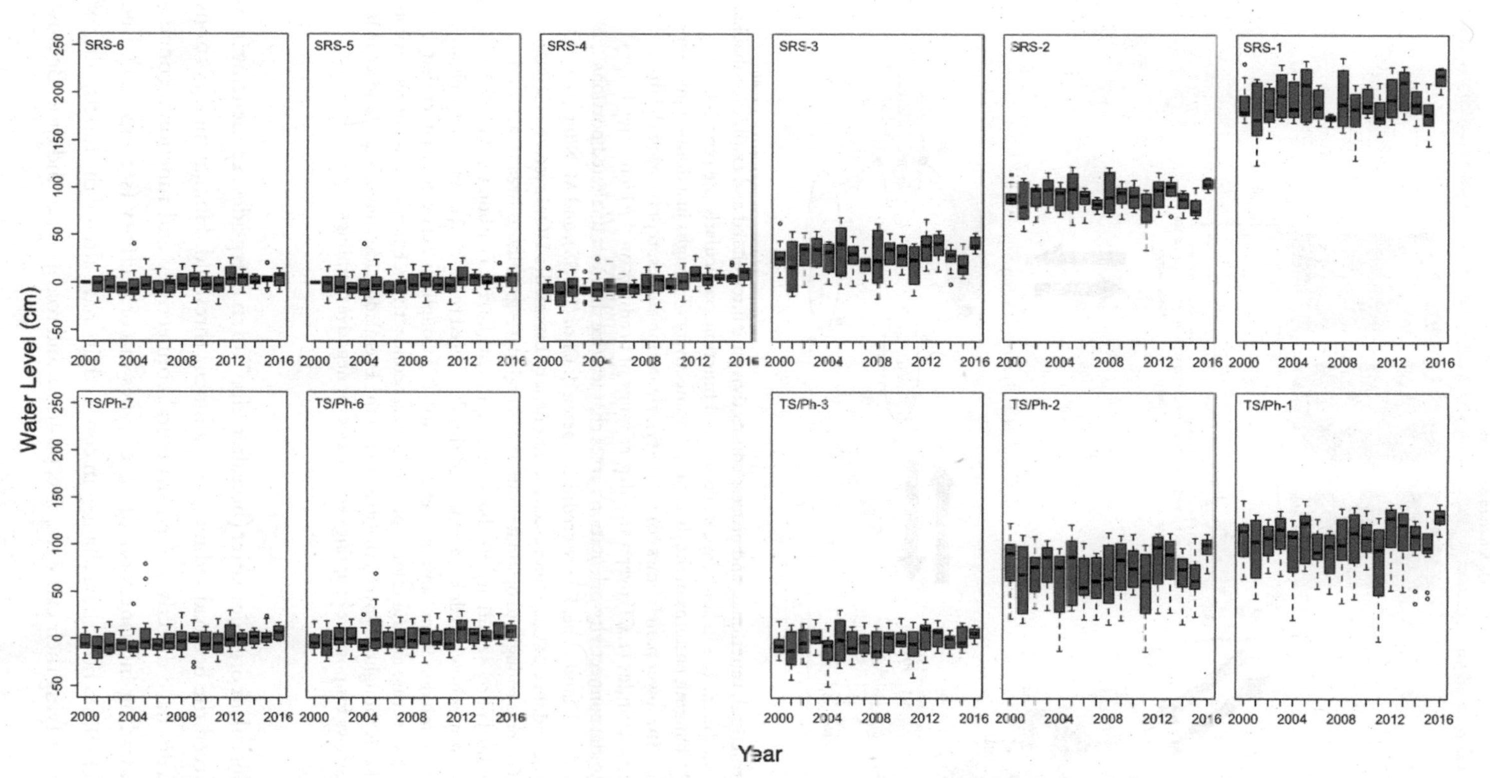

FIGURE 4.1 Water levels and hydroperiod (level > 0) from 2002 to 2012 at freshwater and estuarine sites along the Shark River Slough (SRS) and Taylor Slough/Panhandle (TS/Ph) transects in Everglades National Park, Florida, USA. Freshwater sites include SRS 1 and TS/Ph 1 & 2, and the estuarine sites are SRS 4 & 5 and TS/Ph 6 & 7. Declines in water levels and reductions in hydroperiod are evident at SRS 1, which is immediately downstream of Tamiami Trail and at TS/Ph 1 and 2.

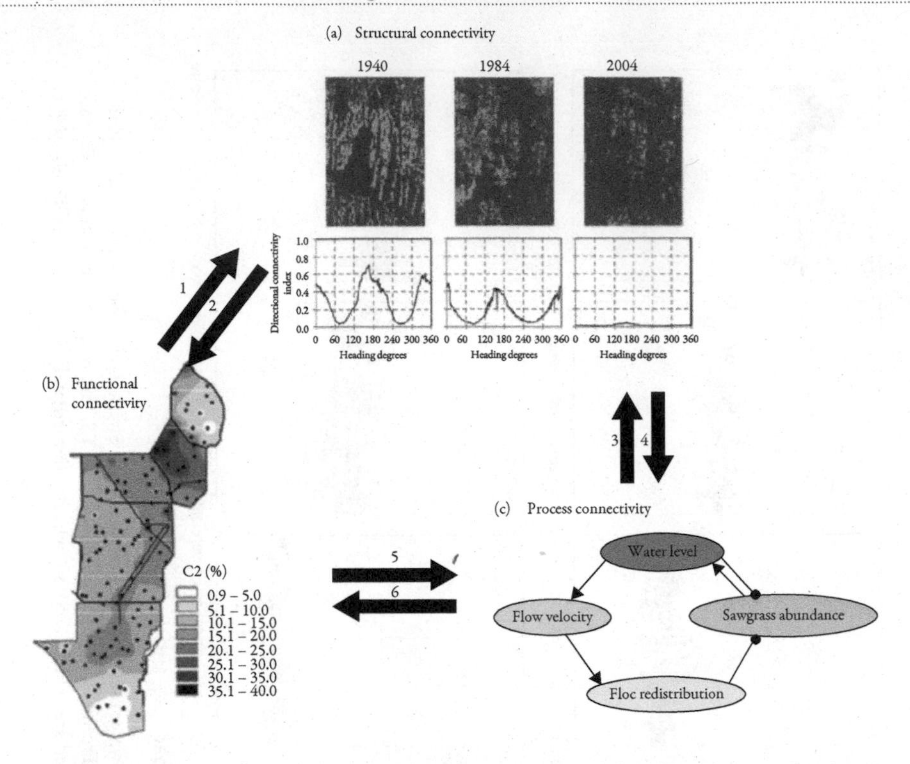

FIGURE 4.2 Structural, functional, and process connectivity are interrelated and exhibit feedbacks that can lead to nonlinearities in landscape structure and function, particularly the functioning of ecotones. (A) Changing patterns in the directional connectivity of sloughs in a drained portion of the Everglades. The lower panel shows connectivity-orientation curves, representing the directional connectivity index (DCI) over a 360-degree range of headings on the landscape. Large-amplitude sinusoidal connectivity-orientation curves characterize the most unaltered portions of the ridge and slough landscape. Figure credit: L. Larsen, J. Choi (USGS), and M. Nungesser (SFWMD). (B) Abundance of C2, a fluorescent component of Everglades DOM resolved through EEM-PARAFAC that is thought to originate from the Everglades Agricultural Area (EAA) and that exhibits minimal photosensitivity. Its abundance represents fluxes of carbon from the EAA. The figure demonstrates how canals have enhanced this flux-based functional connectivity. Figure reproduced with permission from Yamashita et al. (2010). (C) Couplings between variables that govern ecosystem structure and function represent process connectivity. Here, a sediment redistribution feedback that is thought to govern the structure of the Everglades ridge and slough landscape is depicted using arrows for positive couplings and circles for negative couplings.

transport of flocculent organic matter (hereafter "floc") in an Everglades experiencing flow pulses that exceed the bedload sediment entrainment threshold. Human modifications to the Everglades have directly altered structural connectivity and functional connectivity, with cascading influences on all other types of connectivity (see Fig. 4.2). The Everglades exemplifies how these changes in connectivity have altered the functioning of ecotones by modifying their existence, size, position, and role in space and time. Because of feedbacks among the different types of connectivity, changes in connectivity and their ecological consequences can be rapid and can exhibit strong nonlinearity.

STRUCTURAL CONNECTIVITY

The ridge and slough mosaic was historically one of the dominant landscape patterns in the freshwater Everglades. This mosaic was characterized by elevated peat ridges vegetated by sawgrass and elongated parallel to flow, interspersed among deeper, more open, and interconnected sloughs. This "corrugated" landscape and its assortment of habitat types sustained high fish and wading bird diversity (Ogden 2005; see Chapter 2). The interconnectedness of the sloughs also likely made them important fish dispersal pathways (Sokol et al. 2014; Yurek et al. 2013). Since creation of the Water Conservation Areas (WCAs; see Chapter 1), this ridge and slough landscape has changed substantially, with the encroachment of sawgrass ridges into sloughs through much of its extent but with a loss of ridges at the downstream, deeper ends of WCA impoundments.

Changes in slough connectivity across the system and over decades have been quantified within multi-km^2 footprints from digitized imagery with a lacunarity index (Wu et al. 2006) and a directional connectivity index (DCI; Larsen et al. 2012). The lacunarity index is used to quantify the "gappiness" of landscape texture based on the frequency distribution of landscape elements in a sliding box. The DCI, which ranges from zero (no connectivity) to one (complete connectivity), is sensitive to the direction of historical flows (Larsen et al. 2012). Sinusoidal DCI curves suggest strongly directional and connected sloughs, whereas flat, low, and/or noisy DCI curves are associated with degraded landscapes (see Fig. 4.2a).

Structural connectivity can change dramatically in just two decades in the Everglades. We simulated changes in slough connectivity using an ecogeomorphic model of the ridge and slough landscape (Larsen and Harvey 2010) and found that after a small decrease in water level or flow velocity, slough connectivity may be lost in just a matter of years (Larsen et al. 2012). Threshold-type changes in slough connectivity followed by more gradual decreases in slough areal coverage are measured after hydrologic perturbations, suggesting that structural connectivity metrics such as the DCI may have utility as an early warning signal of imminent landscape change or, conversely, as a performance metric of the success of restoration efforts.

FUNCTIONAL AND PROCESS CONNECTIVITY

Functional and process connectivity assessments in the Everglades have typically sought to achieve a process-level understanding of hydroecological processes in the system and to establish baseline measurements of system functioning in order to better assess and predict responses to future management decisions. Because they focus on fluxes, these assessments are relevant to the functioning of ecotones, including the surface water–groundwater interface, coastal ecotones, and ecotones separating discrete vegetation patches. Among the first functional connectivity assessments were studies of water fluxes between surface water and groundwater in order to quantify magnitudes of hydrologic exchange between canals and the surficial aquifer (Harvey et al. 2004; Price et al. 2003), the intrusion of salt water into coastal wetlands (Price et al. 2006), or time scales for remediation of legacy contaminations of mercury (Harvey et al. 2002) and phosphorus (Harvey et al. 2006; Larsen et al. 2014).

Process connectivity in the Everglades has been approached by studies of the interactions among water, vegetation, and nutrient biogeochemistry. For example, differences in transpiration-driven pumping among vegetation types or different rates of exchange across the groundwater–surface water interface can lead to heterogeneities in nutrient retention (Givnish et al. 2008; Ross et al. 2006). These studies suggested that differences in transpiration rates between tree islands and the surrounding marsh are substantial and may lead to high preferential retention of phosphorus on tree islands (Gaiser et al. 2012; Sullivan et al. 2011, 2014; Troxler-Gann et al. 2005; Wetzel et al. 2005). On the other hand, different rates of vertical hydrologic exchange between ridges and sloughs—resulting from their different elevations—may be responsible for the slight preferential retention of phosphorus on ridges (Larsen et al. 2017). Other studies have focused on the extent to which surface water flow connects Everglades ecosystems longitudinally. Through examination of paleosalinity records in Florida Bay and establishment of correlations between modern salinity variations and flow, Marshall et al. (2009) concluded that historic Everglades discharge to the bay was four times greater than modern flows. Within the compartmentalized WCAs, low water surface slope is the primary factor responsible for low water flow velocities (Harvey et al. 2009). Downstream of the impoundments, in ENP, surface water flow velocities have been correlated with water level, and at high velocities and water levels particle transport increases, especially in areas with minimal vegetation (Bazante et al. 2006; Leonard et al. 2006).

Longitudinal connectivity via water flow does not necessarily coincide with biogeochemical connectivity, though the two attributes are related. Organic biomarkers and optical geochemistry are often used to understand biogeochemical connectivity related to carbon cycling (Dittmar and Paeng 2009; see Chapter 6). In the Everglades, seasonal changes in connectivity as a result of the annual dry down/rewetting cycle have been observed in long-term monitoring of optical properties of dissolved organic matter (DOM; Chen et al. 2013). Such variability is also spatially dependent throughout the greater Everglades landscape (Chen et al. 2013; Yamashita et al. 2010). A clear example of how DOM may be useful in tracing spatial connections involved a distinctive fluorescent component of the DOM found throughout the system but with highest concentrations in the canals draining the EAA (Yamashita et al. 2010). We have observed a "plume" of this DOM component, which is believed to be quite resistant to photodegradation and biodegradation (Chen et al. 2010; Chen and Jaffé 2014), downstream of canal inputs into Everglades marshes (Yamashita et al. 2010; see Fig. 4.2b) and have found that it mixes conservatively through the southwestern Everglades mangrove estuaries (Cawley et al. 2014). This suggests a clear connection in DOM transport between the northern section of the system (i.e. the EAA) and the coastal Everglades.

That said, the majority of DOM delivered to the estuarine zone is not derived from canal inputs but from the marsh itself (Lu et al. 2003; Yamashita et al. 2010). The transport of this DOM via sheet flow throughout the system is evidenced by gradual changes in DOM composition. It transitions from high soil organic matter sources (i.e., higher molecular weight, higher humic-like content, and higher aromaticity) to plant and microbial-derived sources (lower molecular weight, higher protein-like content) along a north-to-south transect (Yamashita et al. 2010). This longitudinal connectivity, as we have mentioned, is also reflected in Everglades estuaries. The DOM, and in particular dissolved organic carbon (DOC), that is exported through the mangrove estuaries of Shark River Slough is mainly derived from the upper watershed (up to 80% freshwater marsh derived; Cawley et al. 2013), but additional contributions from the mangroves are also present (Jaffé et al. 2004). Similarly, we have also observed this longitudinal connectivity between the freshwater marshes and the estuarine zone in the southern Everglades in the seasonal variations in DOM quality that result from discharges through the C-111 canal and Taylor Slough/Taylor River into Florida Bay (Maie et al. 2012). Although the connectivity between the end members is clear, a significant portion of the DOC pool in the bay is locally derived (Maie et al. 2005; Ya et al. 2014).

Similar types of geochemical tools have been used to understand the lateral connectivity of sediment transport in Everglades marshes; namely, the fluxes of floc from sloughs to ridges that we hypothesize are essential for formation and maintenance of the ridge and slough landscape (see Fig. 4.2C; Larsen and Harvey 2010). Ridge- and slough-derived floc have different *n*-alkane distributions (Saunders et al. 2006, 2015) and different optical characteristics (Larsen et al. 2010). Under present flow conditions, only limited floc exchange takes place between ridge and slough habitats (Larsen et al. 2010; Saunders et al. 2006, 2015). However, our preliminary work using sediment traps suggested that much of the remobilized particulate organic carbon (POC) in the ridge and slough environment is slough-derived (Jaffé 2013). Although floc appears to have limited mobility (Larsen et al. 2009a, 2009b), small contributions of freshwater floc and/or periphyton-derived POC have been observed in the Shark River estuary (He et al. 2014; Jaffé et al. 2006; Pisani et al. 2013). Assessing longitudinal connectivity for floc has been a challenge because this material is strongly influenced by local biomass characteristics (Neto et al. 2006) and thus is highly variable in composition on both temporal and spatial scales (Pisani et al. 2013). More research on the connectivity of POC throughout the Everglades is needed to better understand the degree of process connectivity (or lack thereof) among periphyton, floc, and soil-derived OM, and the influence of critical environmental drivers such as hydrology and primary productivity (Koch et al. 2014).

ECOLOGICAL IMPACTS OF CONNECTIVITY CHANGES IN THE EVERGLADES

In general, anthropogenic modifications to the Everglades have diminished structural, functional, and process connectivity within the system by decreasing flow velocities and water levels, as well as by altering the location of the fresh water–salt water ecotone. Some aspects of connectivity have increased: pulsed releases of water from structures and the construction of levees have enhanced surface water–groundwater connectivity at localized scales (Harvey et al. 2004; Larsen et al. 2014), while new functional connectivity pathways have been established for anthropogenic compounds introduced into the system (e.g., aromatic carbon compounds, phosphorus, and sulfur). Changing hydrologic fluxes have shifted the size and location of ecotones. For instance, diminished longitudinal fluxes of fresh water combined with accelerating SLR have effectively shifted the brackish ecotone inland, a process known as transgression (Ross et al. 2000).

Positive feedbacks between structural connectivity changes and functional and process connectivity (see Fig. 4.2) may contribute to observed threshold-type changes in landscape patterning. As the structural connectivity of sloughs diminishes, the landscape conveys flows more slowly (arrow 2 in Fig. 4.2; Kaplan et al. 2012; Yuan et al. 2015). When velocities are too slow to entrain floc, its redistribution across the slough–ridge interface ceases; this is a disruption in process connectivity (Larsen et al. 2009). The result, according to one hypothesis of ridge–slough landscape maintenance (Larsen and Harvey 2010), would be an expansion of ridges, further diminishing structural connectivity (see Fig. 4.2C). Slower flows also lengthen water residence times in Everglades marshes, affecting nutrient spiraling lengths—the extent to which solute or suspended materials are biogeochemically or photochemically transformed. This affects the fluxes of these constituents to downstream ecosystems and to coastal Everglades estuaries, with cascading impacts on ecosystem processes. Concomitant changes in water levels likewise impact the exchange of water and solutes between the surface water and groundwater; this also affects total fluxes to the coastal environment.

FRESHWATER–COASTAL CONNECTIVITY OF ANIMALS

Annual and seasonal variation in rainfall patterns and freshwater flow affect mobile species that connect freshwater and coastal habitats through movements and trophic interactions. During the wet season, marsh habitats are inundated with water and provide habitat and food sources for aquatic taxa, including larger-bodied fishes such as Florida largemouth bass (*Micropterus salmoides floridanus*), bowfin (*Amia calva*), and Florida gar (*Lepisosteus platyrhincus*), as well as smaller sunfishes (*Lepomis punctatus, Lepomis marginatus*; Boucek and Rehage 2014). When freshwater discharge decreases during the dry season, marsh water levels recede, forcing aquatic animals to seek out deep-water refuges, such as canals, sloughs, alligator holes (Parkos et al. 2011), and oligohaline mangrove creeks (Boucek and Rehage 2013).

In response to these changes in species distributions, trophic dynamics in deep-water habitats shift as predators change their behaviors in response to this influx of prey that significantly contributes to their diets (Boucek and Rehage 2013; Matich and Heithaus 2014). In turn, mobile animals also stimulate productivity through their transport of biomass and nutrients (Vanni et al. 2002). Large predators that are highly mobile play important roles in connecting habitats and ecosystem dynamics through their movements and trophic interactions. For example, American alligators (*Alligator mississippiensis*) reside in freshwater marshes, sloughs, and canals, and move within and among these habitats for mating and foraging purposes in response to changes in freshwater discharge (Mazzotti and Brandt 1994; Rosenblatt and Heithaus 2011; Rosenblatt et al. 2013). Within the Shark River estuary, some alligators frequently commute between saltwater and freshwater areas to access marine prey during the wet season, when estuarine salinities are lower (Rosenblatt and Heithaus 2011). However, this behavior stops in the dry season, when salinities increase substantially, because alligators have a low salinity tolerance (Laurén 1985; Rosenblatt and Heithaus 2011). Our long-term animal tracking data have shown that these behaviors are stable for marked individuals across years. We have also found similar movement patterns, by some bull sharks (*Carcharhinus leucas*) in the Shark River estuary, that link coastal oceans with estuaries and upstream freshwater channels (Matich and Heithaus 2015; Rosenblatt et al. 2013). However, these movements are less tied to environmental tolerances and the location and extent of the coastal ecotone. They are more closely associated with variability in biotic factors, including predation risk and food availability (Matich and Heithaus 2015). Other large-bodied species, including bottlenose dolphins (*Tursiops pectinate*), common snook (*Centropomus undecimalis*), Atlantic tarpon (*Megalops atlanticus*), and smalltooth sawfish (*Pristis pectinate*), also travel between estuarine and marine habitats in the coastal Everglades, but the importance of these movements is still under investigation.

An important aspect of the movements of alligators and bull sharks is the considerable "personality" seen in movements of different individuals ("individual specialization"; Matich et al. 2011; Rosenblatt and Heithaus 2011; Rosenblatt et al. 2013). For example, some individuals move in ways that link habitats while others remain within restricted areas or within a single habitat.

The effects of hydrologic changes on ecological connectivity among habitats, via the movements of large-bodied animals and trophic interactions, varies among behavioral types within populations. Variation in migratory and resident bird populations that is indirectly influenced by hydroperiod, through its effects on freshwater fishes, has already been observed (Fig. 4.3). Increased freshwater discharge is expected to at least temporarily increase the home ranges of freshwater and salinity-sensitive species, such as marsh fishes

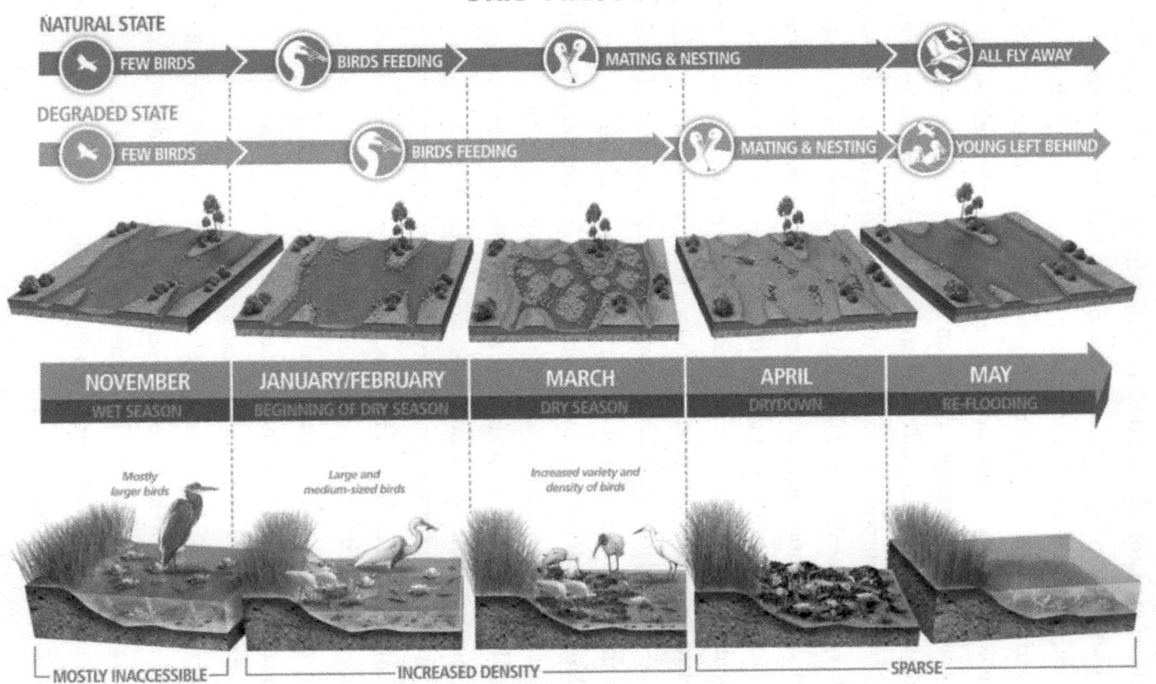

FIGURE 4.3 Altered freshwater hydrology affects the density and distribution of aquatic biota that resident and migratory birds depend upon as overwintering food resources, which are accessed by different types of birds as a function of water depth (e.g., the density of medium-to-small birds increases as water depths decrease). Freshwater food resources are critical to successful mating and offspring success of birds. Changes in the timing and magnitude of freshwater availability offset resource availability and migratory bird life histories. Figure reprinted with copyright permission from the Everglades Foundation. See page 4 of color plates.

and alligators, while reducing the home ranges of other species, such as sawfish. Increases in freshwater hydroperiod will potentially affect how various species connect disparate habitats and exert top-down or bottom-up pressures on freshwater and coastal ecosystems. Our long-term data have also shown that the magnitude and timing of migrations varies among years, based on variation in precipitation and freshwater discharge (see Fig. 4.3). In short, climate change, SLR, and restoration efforts will affect connectivity through movements and trophic interactions of small- and medium-bodied animals.

In summary, freshwater–coastal connectivity is facilitated both by physical and biologic processes, and the extent of connectivity between freshwater and coastal end members is affected by both freshwater discharge and SLR. The implications of restoration and climate change for the Florida Everglades are not clear, but the interaction of future water management with SLR will play a critical role in determining the extent and location of the coastal ecotone, which in turn will greatly influence connectivity between microhabitats within the ecosystem. This appears to be important at multiple levels of ecological organization.

CANALS BOTH PROMOTE AND DETRACT FROM ECOLOGICAL CONNECTIVITY

One of the most important aspects of human intervention in the Everglades has been the compartmentalization of the ecosystem. Over 1,600 km of canals and 1,150 km of levees today bisect Everglades marshes (Light and Dineen 1994; Sklar et al. 2005; see Chapter 1). Everglades canals were constructed to drain and reclaim wetlands, provide flood control and storage, convey water to agricultural and urban areas, and discharge water to the coast. Canals and associated structures (e.g., levees, gates, pumps) are the primary tool of water management in the south Florida landscape. Restoration efforts, as currently envisioned (see Chapter 8), will probably only remove a few canals and levees, and thus canals will continue to be dominant features of the Everglades.

Canals act to simultaneously decrease and increase hydrologic connectivity across the Everglades landscape, as well as with nearby regions, depending on the ecological process and scales of interest. The presence of canals impedes sheet flow and alters hydroperiods, resulting in wetland fragmentation and reduced connectivity (McVoy et al. 2011; see Chapter 1). Canals also reroute the path and speed of water flows. At the same time, canals promote conveyance of water among compartments, connecting areas that were previously spatially isolated. Canals foster groundwater–surface water interactions, causing an overall increase in the contribution of groundwater—relative to rainfall—to the Everglades (Harvey et al. 2004; Harvey and McCormick 2009). Canals also directly connect Everglades marshes to agricultural and urban lands, increasing the connectivity of water that may have poor quality (see Chapter 5).

Everglades canals play an important role in altering biotic connectivity across the Everglades landscape. Canals provide permanent deep-water refuges that were rare, but also less necessary, in the predrainage ecosystem (Gunderson and Loftus 1993). As a result, fish and other aquatic animals move from marshes to canals as water levels decline during the dry season (Parkos and Trexler 2014; Rehage and Trexler 2006). These movements in and out of canals can promote the mixing of seasonally isolated populations, leading to lower-than-expected genetic differentiation at regional scales (McElroy et al. 2003). Biota may also use the canal network itself to move, promoting biotic exchanges across relatively large distances, but the extent to which the presence of canals may have altered dispersal patterns is not yet known. We have found that canals break down dispersal barriers by providing an opportunity for fishes that have an affinity for deep-water habitats to expand their range (Gandy et al. 2012). Lastly, due to their depth and water permanency, canals are habitat for large numbers of large-bodied predators (Brandt et al. 2010; Gandy et al. 2012), altering species interactions and trophic connectivity across the landscape.

Perhaps one of the most important roles the canal system plays is as conduits for nutrients, other pollutants, and non-native biota. Canals connect agricultural and urban areas to Everglades wetlands, often carrying heavily impacted waters into the Everglades (McCormick et al. 1996). This results in gradients of enrichment and impact in marshes that are a function of distance from canal inflows, with pronounced consequences for ecosystem structure and function (Childers et al. 2003; Doren et al. 1997; Rehage and Trexler 2006). Along with nutrients, connectivity to agricultural and urban areas also makes canals a source of pesticides and other pollutants to Everglades marshes (Carriger et al. 2006).

Canals and levees can pose an important invasion threat, promoting connectivity and providing dispersal opportunities for unwanted biota. Levees provide disturbed upland habitat for plant and animal pests, such as fire ants, and are a source of non-native propagules to upland habitats in the Everglades, such as tree islands (Ferriter et al. 2004; Forman and Alexander 1998). Many non-native or invasive fishes and macroinvertebrates were first detected in canals and have since invaded Everglades habitats (Kline et al. 2014; Wingard et al. 2008). For example, six non-native fishes were first recorded in ENP when water was pumped into the park from a nearby canal in the 1980s (Kline et al. 2014; Light and Dineen 1994; Merritt 1996). A number of non-native species are currently present in canals but have not yet spread, and it is suspected that restoration projects that increase hydrologic connectivity between canals and marshes may promote invasions (Kline et al. 2014). Finally, non-native fish in the Everglades are tropical, and thus temperature-sensitive. Canals also provide thermal refugia for both native and non-native fishes (Kline et al. 2014; Schofield et al. 2010; Trexler et al. 2000), and the ability of canals to remain warm over extreme cold events allows non-native populations to persist.

Canals are also an important way in which people interact with the Everglades (Fig. 4.4). Canals provide access for hunters and fishers, and levees are important access points for hiking, biking, and camping. Hunters pursuing non-native Burmese pythons (*Python*

FIGURE 4.4 Fishing in canals is an important recreational activity that supports the state economy as well as provides access to food (particularly protein) for people in south Florida. Photo taken October 1959, Broward County, Florida.

molurus) use levees to gain access to Everglades marshes, but also because pythons are known to use levees (M. Bush, pers. obs.). Canals are particularly important as recreational fisheries (Edwards 2013), and these fisheries are an important ecosystem service provided by the Everglades (Fedler 2009). In fact, much of this fishery service is provided by canals. For example, the L67-A canal in the central Everglades is one of the best largemouth bass fisheries in the state of Florida (FFWCC 2011). Because canals are more accessible than other environments of the Everglades, they are a primary means by which many people access and connect with the landscape (see Fig. 4.4).

Teleconnections in the Water–Atmosphere System and Human Dimensions

The Greater Everglades landscape spans the entire south Florida peninsula. These wetlands were created and exist as a function of water availability, which today is controlled by both water management and natural variability (Moses et al. 2013; see Chapter 1). The South

Florida Water Management District (SFMWD) and U.S. Army Corps of Engineers (USACE) completely control surface water flow from Lake Okeechobee (see Box 1.1 in Chapter 1). That management is affected by changes in seasonal delivery of water via precipitation and losses via evapotranspiration. Oceanographic forcings, such as the well-documented teleconnection patterns of sea surface temperatures in the Atlantic Ocean (the AMO), have profound effects on water levels in Lake Okeechobee and the Greater Everglades (Enfield et al. 2001). Water management operations must account for short- and long-term changes in rainfall across the region and the cumulative impacts of changing water flows into and out of Lake Okeechobee (Obeysekera et al. 2011).

The neotropical Everglades is dominated by wet and dry seasons, but even this dominant climate pattern changes under certain conditions. Beckage et al. (2003) showed that during positive ENSO events (La Niña), south Florida has wetter-than-average dry seasons. Childers et al. (2006) further showed that during both La Niña and El Niño ENSO events annual total rainfall amounts tended to be average, but the timing and seasonality of rainfall shifted to wetter dry seasons and drier wet seasons. In fact, weather patterns across the south Florida peninsula impart a unique spatial heterogeneity, which is ecologically important (Moses et al. 2013). Precipitation is often extremely patchy across the landscape, with the east coast receiving the least amount of rainfall (Pielke 1974; Skinner et al. 2009). There is also a general gradient of increasing rainfall from southern Florida Bay to the northern limit of the Everglades, in Palm Beach County. Our spatial analyses of AMO, ENSO, and Pacific Decadal Oscillation (PDO) teleconnection effects on the Greater Everglades showed that both winter and summer precipitation were affected by these different teleconnections in geographically variable ways (Moses et al. 2013). We found that long-term oscillations in water depth in, and material flux to, climate-sensitive aquatic habitats in the upper Everglades watershed were highly regulated by the AMO, suggesting that water management may be muting this signal further south (Moses et al. 2013; Wachnicka et al. 2013). Such oscillatory behavior has been a part of the peninsular system since the Inter-Tropical Convergence Zone migrated northward some 5,000 years ago (Donders et al. 2005). Below-average water temperatures off the Pacific coast of North America, from Alaska to the equator, and in the North Atlantic signal the cold phases of the PDO and AMO, respectively. Computer simulations revealed that when cold PDO and AMO phases overlap, the south Florida region experiences extremely dry conditions (McCabe et al. 2004). Depending on the rainfall and temperature scenario, this significantly affects water budgets, ecosystem dynamics, and human water supply demands (Obeysekera et al. 2015).

We have documented robust evidence of these links by reconstructing salinity and water nutrient histories from sedimentary diatom assemblages in Florida Bay and Biscayne Bay. These results revealed strong links between major shifts in diatom assemblage structures and large-scale climatic oscillations (Wachnicka et al. 2013). Most of the largest shifts occurred when ENSO, AMO, and PDO were in cold phases, or when these climate patterns were oscillating between phases. The rate and magnitude of diatom assemblage restructurings, the magnitude of salinity variations, and nutrient and organic carbon concentrations in bay sediments increased during the second half of the 20th century, when Everglades drainage and water management was at its peak and when an unprecedented global climate shift began (Barnett et al. 2005; Levitus et al. 2005; Swanson and Tsonis 2009). This shift was characterized by the occurrence of more extreme climatic conditions in much of the United States, including south Florida (Easterling et al. 2000). We have linked these major climate shifts, identified in North Atlantic Oscillation (NAO), AMO, Accumulated Cyclone Energy, and North Atlantic storm activity, to corresponding changes in nutrient concentrations and ecosystem responses (e.g., primary production) in the Everglades and Florida Bay (Briceño and Boyer 2010). This has led us to conclude that water quality in Florida Bay responds to decadal or multi-decadal forcings (i.e., teleconnections) that dictate baseline conditions. Superimposed on this baseline are seasonal cycles and departures from "normality" caused by short-lived events (e.g. hurricanes). The behavior of salinity, concentrations of total nutrients, and ecosystem responses (e.g., primary production) closely follow this model. Thus, water management practices coupled with global climate oscillations have drastically altered water quality in Florida Bay and Biscayne Bay, but Everglades restoration may help reduce high coastal salinities that have characterized a human- and climate-altered south Florida for over 100 years.

Teleconnections are not just limited to climate-driven changes in the ocean–atmosphere system, which affect the rainfall and climate of the Everglades. Another important teleconnection is the flux of Saharan dust across the Atlantic that deposits nutrients, such as iron and phosphorus, across south Florida and the Caribbean (Mahowald et al. 2005; Swart et al. 2014). An increase in African dust has been attributed to increased agricultural development in the Sahel (Mulitza et al. 2010), and events and activities in this region have implications for the Everglades. It has been hypothesized that changes in tropical storm tracks throughout the late Holocene have been related to changes in the position of the Bermuda High, and this has resulted in relatively dryer conditions in south Florida over the last 3,000 years. This trend has also been related to decreased dust-derived nutrient deposition during this time (e.g., phosphorus; Glaser et al. 2013). We have correlated general trends in organic carbon accumulation, associated with increases in primary productivity, with higher periods of dustfall over the last 4,600 years (Glaser et al. 2013). Thus, the Greater Everglades landscape has been and remains truly connected to events and changes occurring across the globe.

Water management and freshwater deliveries to the Everglades are a function of weather, human demand, and actual freshwater availability. The ultimate modulating control on water quality is how much fresh water flows into the Everglades and how it counters the advance of seawater. For example, increased freshwater flows through Taylor Slough will likely decrease phosphorus concentrations in this oligotrophic and phosphorus-limited ecosystem (Childers et al. 2006). However, more studies in the mangrove ecotone are needed to understand how increased freshwater flows will affect nitrogen dynamics relative to phosphorus (Briceño et al. 2014). The responses of water quality in the coastal Everglades to changes in flow levels and salinity are not uniform across the region, posing additional uncertainties about how restoration and climate change will ultimately affect the coastal Everglades.

Future Implications for Everglades Landscape Connectivity and Resilience

Increases in hydrologic management and human demands for fresh water threaten the sustainability of freshwater and coastal ecosystems. Climate change and accelerating SLR exacerbate these threats (Zhang et al. 2011). For a century, drainage and freshwater management in the Everglades has redesigned the hydrology of this vast, flat landscape, resulting in less coastward fresh water and more landward marine water (McVoy et al. 2011; Price et al. 2006; Sklar et al. 2005; see Chapters 1 and 3). However, the largest freshwater restoration effort in the world is being planned and has been partially implemented in the Everglades (see Chapter 8). Increased ecological connectivity through the Comprehensive Everglades Restoration Plan (CERP; Fig. 4.5) should result in enhanced freshwater delivery to the coastal zone, and this should dampen the effects of SLR and coastal transgression—at least temporarily (Krauss et al. 2011; Ross et al. 2000; Saha et al. 2011; Sklar et al. 2005; see Chapter 8).

Everglades restoration offers both opportunities and challenges for rehabilitating the hydrologic regime, biologic community structure, and ecosystem function of this World Heritage Site. For example, emergent properties of the legacy of altered ecological connectivity may mean that restoring freshwater hydrology will bring about additional novel changes in structure and function (Seastedt et al. 2008). Major losses of freshwater wetlands to development and subsidence (Sklar et al. 2005; McVoy et al. 2011) and declines in many bird and mammal populations (Dorcas et al. 2012) are unlikely to be completely reversed by increases in freshwater connectivity. That said, reestablishing ecological connectivity should impart greater ecosystem resilience to coastal storms and accelerating SLR. Adaptive restoration and water management approaches will be required, though, as we continue to understand how this unique, fragile, yet resilient landscape has and will continue to respond to our designs and actions.

Ecological connectivity is multidimensional (longitudinal, vertical, lateral, and temporal). In the coastal Everglades, it is driven by the balance of freshwater and marine water supplies. Restoration efforts (CERP) should help to reconnect ecosystems that previously have been disconnected by human engineering at the same time that accelerating SLR will enhance marine connectivity. This increased bidirectional hydrologic connectivity may be most obviously manifest in a widening oligohaline ecotone, with increased freshwater influence during peak periods of the wet season and increased marine influence during peak periods of the dry season. One goal of enhanced ecological

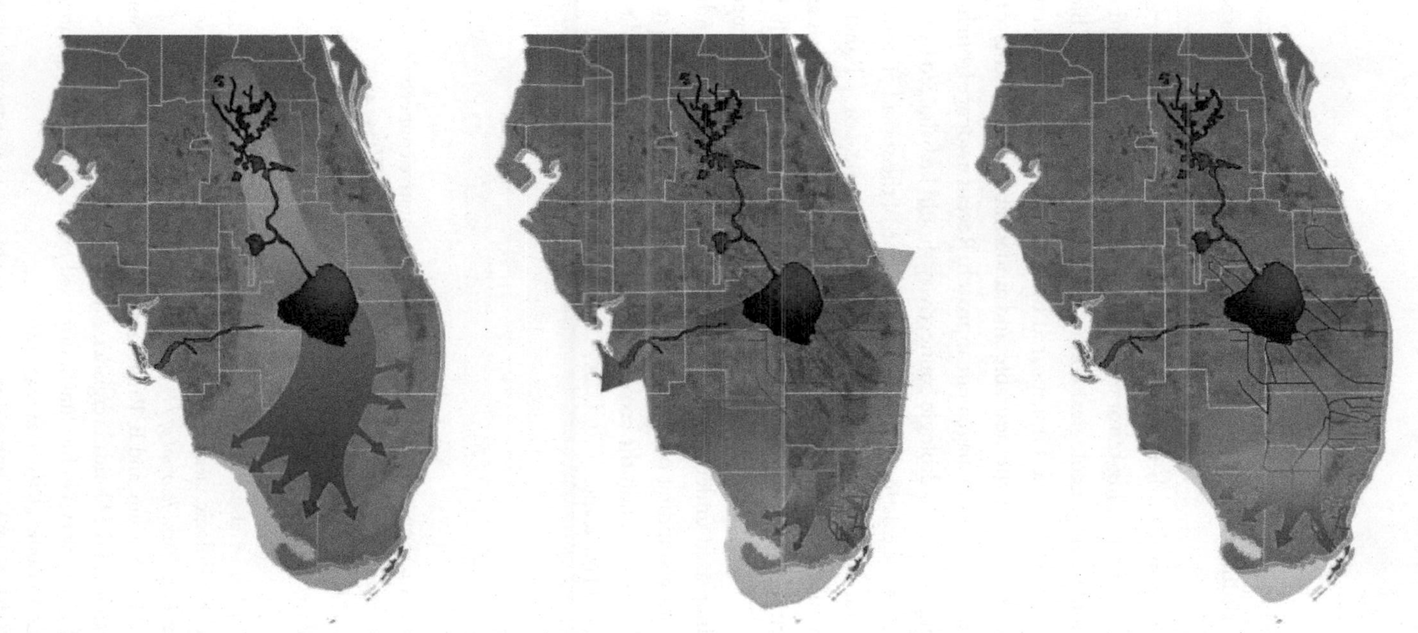

FIGURE 4.5 Conceptual diagram of changes in ecological connectivity in space and time throughout the Greater Everglades landscape. Arrows represent historic (left), current (center), and future (right) hydrologic flow paths of the Everglades and south Florida's southeastern urban boundary. Historic connectivity was manifest across the landscape as sheetflow that connected Lake Okeechobee in the North with Florida Bay to the south and Biscayne Bay to the east. The current landscape is hydrologically fragmented by an extensive system of canals and levees that drain water away from urban areas, maintaining flood control as well as a municipal water source for the urban population. The Comprehensive Everglades Restoration Plan

(CERP) is the principle mechanism to restore ecological connectivity of the Greater Everglades landscape. Reprinted with permission from evergladesplan.org. See page 3 of color plates.

connectivity should be to increase carbon storage in peat soils in order to increase soil elevations, which will counter the effects of inundation, erosion, and salinity associated with storms and SLR.

Restoring connectivity in disturbed aquatic landscapes often moves the system toward a new ecological state, making restoration targets difficult to achieve and maintain (Jackson and Pringle 2010). Ecological connectivity includes organisms (biologic connectivity) as well as material fluxes transported by water (hydrologic connectivity). Sorting out the tradeoffs among the desirable and undesirable components of both requires a multifaceted, long-term management approach. Recent evidence from human-dominated systems suggests that hydrologic connectivity should be balanced by biologic fragmentation in order to use existing fragmentation to control the spread of non-native, invasive species, pollutants, and pathogens (Jackson and Pringle 2010; Rahel 2013). Shifts in the direction, magnitude, and characteristics of ecological state changes are difficult to predict, requiring effective monitoring and adaptive management both during and after restoration. Ecological connectivity alone will not ensure the resilience of a restored Everglades. Increased understanding of effective restoration through adaptive management and an understanding of the full consequences and tradeoffs of restored connectivity are needed to maintain the ecological integrity and ecosystem resilience of a rehabilitated Everglades landscape.

References

Ali, G.A., and A.G. Roy. 2010. Shopping for hydrologically representative connectivity metrics in a humid temperate forested catchment. *Water Resources Research* 46: W12544.

Barnett, T.P., D.W. Pierce, K.M. AchutaRao, P.J. Gleckler, B.D. Santer, J.M. Gregory, and W.M. Washington. 2005. Penetration of human-induced warming into the world's oceans. *Science* 309: 284–287.

Bazante J., G. Jacobi, H.M. Solo-Gabriele, D. Reed, S. Mitchell-Bruker, D.L. Childers, L. Leonard, and M.S. Ross. 2006. Hydrologic measurements and implications for tree island formation within Everglades National Park. *Journal of Hydrology* 329: 606–619.

Beckage, B., W.J. Platt, M.G. Slocum, and B. Panko. 2003. Influence of the El Niño Southern Oscillation on fire regimes in the Florida Everglades. *Ecology* 84: 3124–3130.

Boucek, R.E., and J.S. Rehage. 2013. No free lunch: displaced marsh consumers regulate a prey subsidy to an estuarine consumer. *Oikos* 122: 1453–1464.

Boucek, R.E., and J.S. Rehage. 2014. Climate extremes drive changes in functional community structure. *Global Change Biology* 20: 1821–1831.

Bracken, L.J., J. Wainwright, G.A. Ali, D. Tetzlaff, M.W. Smith, S.M. Reaney, and A.G. Roy. 2013. Concepts of hydrological connectivity: research approaches, pathways and future agendas. *Earth-Science Reviews* 119(April): 17–34.

Brandt, L.A., M.R. Campbell, and F.J. Mazzotti. 2010. Spatial distribution of alligator holes in the central Everglades. *Southeastern Naturalist* 9: 487–496.

Briceño, H.O., and J.N. Boyer. 2010. Climatic controls on phytoplankton biomass in a subtropical estuary, Florida Bay, USA. *Estuaries and Coasts* 33: 541–553.

Briceño, H.O., G. Miller, and S.E. Davis. 2014. Relating freshwater flow with estuarine water quality in the southern Everglades mangrove ecotone. *Wetlands* 34(Suppl 1): 101–111.

Calabrese, J.M., and W.F. Fagan. 2004 A comparison-shopper's guide to connectivity metrics. *Frontiers in Ecology and the Environment* 2: 529–536.

Carriger, J.F., G.M. Rand, P.R. Gardinali, W.B. Perry, M.S. Tompkins, and A.M. Fernandez. 2006. Pesticides of potential ecological concern in sediment from south Florida canals: an ecological risk prioritization for aquatic arthropods. *Soil & Sediment Contamination* 15: 21–45.

Cawley, K.M., Y. Yamashita, N. Maie, and R. Jaffé. 2014. Using optical properties to quantify fringe mangrove inputs to the dissolved organic (DOM) pool in a subtropical estuary. *Estuaries and Coasts* 37(2): 399–410.

Chen, M., and R. Jaffé. 2014. Photo- and bio-reactivity patterns of dissolved organic matter from biomass and soil leachetes and surface waters in a subtropical wetland. *Water Research* 61(September): 181–190.

Chen, M., N. Maie, K. Parish, and R. Jaffé. 2013. Spatial and temporal variability of dissolved organic matter quantity and composition in an oligotrophic subtropical coastal wetland. *Biogeochemistry* 115(1-3): 167–183.

Chen, M., R. Price, Y. Yamashita, and R. Jaffé. 2010. Comparative study of dissolved organic matter from groundwater and surface water in the Florida Coastal Everglades using multidimensional spectrofluorometry combined with multivariate statistics. *Applied Geochemistry* 25(6): 872–880.

Childers, D.L., J.N. Boyer, S.E. Davis, C.J. Madden, D.T. Rudnick, and F.H. Sklar. 2006. Relating precipitation and water management to nutrient concentration patterns in the oligotrophic "upside down" estuaries of the Florida Everglades. *Limnology and Oceanography* 51(1): 602–616.

Childers, D.L., R.F. Doren, R. Jones, G.B. Noe, M. Rugge, and L.J. Scinto. 2003. Decadal change in vegetation and soil phosphorus pattern across the Everglades landscape. *Journal of Environmental Quality* 32: 344–362.

Dittmar, T., and Paeng, J. 2009. A heat-induced molecular signature in marine dissolved organic matter. *Nature Geoscience* 2: 175–179.

Donders, T.H., F. Wagner, and H. Visscher. 2005. Quantification strategies for human-induced and natural hydrological changes in wetland vegetation, southern Florida, USA. *Quaternary Research* 64: 333–342.

Dorcas, M.E., J.D. Willson, R.N. Reed, R.W. Snow, M.R. Rochford, M.A. Miller, W.E. Meshaka Jr., P.T. Andreadis, F.J. Mazzotti, C.M. Romagosa, and K.M Hart. 2012. Severe mammal declines coincide with proliferation of invasive Burmese pythons in Everglades National Park. *Proceedings of the National Academy of Sciences USA* 109: 2418–2422.

Doren, R.F., T.V. Armentano, L.D. Whiteaker, and R.D. Jones. 1997. Marsh vegetation patterns and soil phosphorus gradients in the Everglades ecosystem. *Aquatic Botany* 56: 145–163.

Easterling, D.R., G.A. Meehl, C. Parmesan, S.A. Changnon, T.R. Karl, and L.O. Mearns. 2000. Climate extremes: observations, modeling, and impacts. *Science* 289: 2068–2074.

Edwards, C. 2013. Recreational Angler Perspectives of Nonnative Fish Species and Mercury Advisories. Master's thesis, Florida International University.

Enfield, D.B., A.M. Mestas-Nuñez, and P.J. Trimble. 2001. The Atlantic multidecadal oscillation and its relation to rainfall and river flows in the continental US. *Geophysical Research Letters* 28: 2077–2080.

Fedler, T. 2009. The economic impact of recreational fishing in the Everglades region. Prepared for the Everglades Foundation, Miami, Florida, December.

Ferriter, A., K. Serbesoff-King, M. Bodle, C. Goodyear, B. Doren, and K. Langeland. 2004. E: Exotic species in the Everglades Protection Area. *2004 Everglades Consolidated Report*, 11–15.

Florida Fish and Wildlife Conservation Commission. 2011. *Black Bass Management Plan: 2010–2030*. Naples, FL: Fish and Wildlife Research Institute & Division of Habitat and Species Conservation.

Forman, R.T., and L.E. Alexander. 1998. Roads and their major ecological effects. *Annual Review of Ecology and Systematics* 29: 207–231+C2.

Fullerton, A.H., K.M. Burnett, E.A. Steel, R.L. Flitcroft, G.R. Pess, B.E. Feist, C.E. Torgersen, D.J. Miller, and B.L. Sanderson. 2010. Hydrological connectivity for riverine fish: measurement challenges and research opportunities. *Freshwater Biology* 55(11): 2215–2237.

Gaiser, E.E., J. Trexler, and P. Wetzel. 2012. The Everglades. In *Wetland Habitats of North America: Ecology and Conservation Concerns*, edited by D. Batzer and A. Baldwin, 231–252. Berkeley: University of California Press.

Gandy, D.A., J.S. Rehage, J.W. Munyon, K.B. Gestring, and J.I. Galvez. 2012. Canals as vectors for fish movement: potential southward range expansion of *Lepisosteus osseus* L. (longnose gar) in South Florida. *Southeastern Naturalist* 11: 253–262.

Givnish, T.J., J.C. Volin, V.D. Owen, V.C. Volin, J.D. Muss, and P.H. Glaser. 2008. Vegetation differentiation in the patterned landscape of the central Everglades: importance of local and landscape drivers. *Global Ecology and Biogeography* 17: 384–402.

Glaser, P.H., B.C. Hansen, J.J. Donovan, T.J. Givnish, C.A. Stricker, and J.C. Volin. 2013. Holocene dynamics of the Florida Everglades with respect to climate, dustfall, and tropical storms. *Proceedings of the National Academy of Sciences USA* 110: 17211–17216.

Gunderson, L.H., and W.F. Loftus. 1993. The Everglades. In *Biodiversity of the Southeastern United States/Lowland Terrestrial Communities*, edited by W.H. Martin, S.G. Boyce, and A.C. Echternact, 199–255. New York: John Wiley and Sons.

Harvey, J.W., S.L. Krupa, C. Gefvert, R.H. Mooney, J. Choi, S.A. King, and J.B. Giddings. 2002. Interactions between surface water and ground water and effects on mercury transport in the north-central Everglades. Water-Resources Investigations Report 02-4050. Reston, VA: U.S. Department of the Interior, U.S. Geological Survey.

Harvey, J.W., S.L. Krupa, and J.M. Krest. 2004. Ground water recharge and discharge in the central Everglades. *Groundwater* 42: 1090–1102.

Harvey, J.W., and P.V. McCormick. 2009. Groundwater's significance to changing hydrology, water chemistry, and biological communities of a floodplain ecosystem, Everglades, south Florida, USA. *Hydrogeology Journal* 17: 185–201.

Harvey, J.W., J.T. Newlin, and S.L. Krupa. 2006. Modeling decadal timescale interactions between surface water and ground water in the central Everglades, Florida, USA. *Journal of Hydrology* 320: 400–420.

Harvey, J.W., R.W. Schaffranek, G.B. Noe, L.G. Larsen, D.J. Nowacki, and B.L. O'Connor. 2009. Hydroecological factors governing surface water flow on a low-gradient floodplain. *Water Resources Research* 45: W03421.

He, D., R.N. Mead, L. Belicka, O. Pisani, and R. Jaffé. 2014. Assessing source contributions to particulate organic matter in a subtropical estuary: a biomarker approach. *Organic Chemistry* 75(October): 129–139.

Jackson, C.R., and C.M. Pringle. 2010. Ecological benefits of reduced hydrologic connectivity in intensively developed landscapes. *BioScience* 60: 37–46.

Jaffé, R., J.N. Boyer, X. Lu, N. Maie, C. Yang, N.M. Scully, and S. Mock. 2004. Source characterization of dissolved organic matter in a subtropics mangrove-dominated estuary by fluorescence analysis. *Marine Chemistry* 84(3-4): 195–210.

Jaffé, R., V. Ding, J. Niggemann, A.V. Vähätalo, A. Stubbins, R.G.M. Spencer, J. Campbell, and Thorsten Dittmar. 2013. Global charcoal mobilization form soils via dissolution and riverine transport to the oceans. *Science* 340(6130): 345–347.

Jaffé, R., A.I. Rushdi, P.M. Medeiros, and B.R.T. Simoneit. 2006. Natural product biomarkers as indicators of sources and transport of sedimentary organic matter in a subtropical river. *Chemosphere* 64(11): 1870–1884.

Kaplan D.A., R. Paudel, M.J. Cohen, and J.W. Jawitz. 2012. Orientation matters: patch anisotropy controls discharge competence and hydroperiod in a patterned peatland. *Geophysical Research Letters* 39: L17401.

Kline, J.L., W.F. Loftus, K. Kotun, J.C. Trexler, J.S. Rehage, J.J. Lorenz, and M. Robinson. 2014. Recent fish introductions into Everglades National Park: an unforeseen consequence of water management? *Wetlands* 34: 175–187.

Koch, G., S.E. Hagerthey, D.L. Childers, and E.E. Gaiser. 2014. Examining seasonally pulsed detrital transport in the coastal Everglades using a sediment tracking technique. *Wetlands* 34: 123–133.

Krauss, K.W., A.S. From, T.W. Doyle, T.J. Doyle, and M.J. Barry. 2011. Sea-level rise and landscape change influence mangrove encroachment onto marsh in the Ten Thousand Islands region of Florida, USA. *Journal of Coastal Conservation* 15: 629–638.

Larsen, L.G., J. Choi, M.K. Nungesser, and J.W. Harvey. 2012. Directional connectivity in hydrology and ecology. *Ecological Applications* 22: 2204–2220.

Larsen, L.G., and J.W. Harvey. 2010. How vegetation and sediment transport feedbacks drive landscape change in the Everglades and wetlands worldwide. *The American Naturalist* 176(3): E66–E79.

Larsen, L.G., J.W. Harvey, and J.P. Crimaldi. 2009a. Predicting bed shear stress and its role in sediment dynamics and restoration potential of the Everglades and other vegetated flow systems. *Ecological Engineering* 35: 1773–1785.

Larsen, L.G., J.W. Harvey, and J.P. Grimaldi. 2009b. Morphologic and transport properties of natural organic floc. *Water Resources Research*. 45(1): doi:10.1029/2008WR006990.

Larsen, L.G., J.W. Harvey, and M.M. Maglio. 2014. Dynamic hyporheic exchange at intermediate timescales: testing the relative importance of evapotranspiration and flood pulses. *Water Resources Research* 50(1): 318–335.

Larsen, L., J. Ma, and D. Kaplan. 2017. How important is connectivity for surface-water fluxes? A generalized expression for flow through heterogeneous landscapes. *Geophysical Research Letters* 44(20): 10349–10358.

Laurén, D.J. 1985. The effect of chronic saline exposure on the electrolyte balance, nitrogen metabolism, and corticosterone titer in the American alligator, *Alligator mississippiensis. Comparative Biochemistry and Physiology Part A: Physiology* 81: 217–223.

Leonard, L., A. Croft, D. Childers, S. Mitchell-Bruker, H. Solo-Gabriele, and M. Ross. 2006. Characteristics of surface-water flows in the ridge and slough landscape of Everglades National Park: implications for particulate transport. *Hydrobiologia* 569: 5–22.

Levitus, S., J. Antonov, and T. Boyer. 2005. Warming of the world ocean, 1955–2003. *Geophysical Research Letters* 32: L02604.

Light, S.S., and J.W. Dineen. 1994. Water control in the Everglades: a historical perspective. In *Everglades: the Ecosystem and its Restoration*, edited by S.M. Davis and J.C. Ogden, 47–84. Delray Beach, FL: St. Lucie Press.

Lu, X.Q., N. Maie, J.V. Hanna, D.L. Childers, and R. Jaffé. 2003. Molecular characterization of dissolved organic matter in freshwater wetlands of the Florida Everglades. *Water Research* 37(11): 2599–2606.

Mahowald, N.M., A.R. Baker, G. Bergametti, N. Brooks, R.A. Duce, T.D. Jickells, N. Kubilay, J.M. Prospero, and I. Tegen. 2005. Atmospheric global dust cycle and iron inputs to the ocean. *Global Biogeochemical Cycles* 19: GB4025.

Maie, N., Y. Yamashita, R.M. Cory, J.N. Boyer, and R. Jaffé. 2012. Application of excitation emission matrix fluorescence monitoring in the assessment of spatial and seasonal drivers of dissolved organic matter composition: sources and physical disturbance controls. *Applied Geochemistry* 27(4): 917–929.

Maie, N., C. Yange, T. Miyoshi, K. Parish, and R. Jaffé. 2005. Chemical characteristics of dissolved organic matter in an oligotrophic subtropical wetland/estuarine. *Limnology and Oceanology* 50(1): 23–35.

Marshall III, F.E., G.L. Wingard, and P. Pitts. 2009. A simulation of historic hydrology and salinity in Everglades National Park: coupling paleoecologic assemblage data with regression models. *Estuaries and Coasts* 32: 37–53.

Matich, P., and M.R. Heithaus. 2014. Multi-tissue stable isotope analysis and acoustic telemetry reveal seasonal variability in the trophic interactions of juvenile bull sharks in a coastal estuary. *Journal of Animal Ecology* 83: 199–213.

Matich, P., and M.R. Heithaus. 2015. Individual variation in ontogenetic niche shifts in habitat use and movement patterns of a large estuarine predator (*Carcharhinus leucas*). *Oecologia* 178: 347–359.

Matich, P., M.R. Heithaus, and C.A. Layman. 2011. Contrasting patterns of individual specialization and trophic coupling in two marine apex predators. *Journal of Animal Ecology* 80: 294–305.

Mazzotti, F.J., and L.A. Brandt. 1994. Ecology of the American alligator in a seasonally fluctuating environment. In *Everglades: The Ecosystem and Its Restoration*, edited by S.M. Davis and J.C. Ogden, 485–505. Delray Beach, FL: St. Lucie Press.

McCabe, G.J., M.A. Palecki, and J.L. Betancourt. 2004. Pacific and Atlantic Ocean influences on multi-decadal drought frequency in the United States. *Proceedings of the National Academy of Sciences USA* 101: 4136–4141.

McCormick, P.V., P.S. Rawlik, K. Lurding, E.P. Smith, and F.H. Sklar. 1996. Periphyton–water quality relationships along a nutrient gradient in the northern Florida Everglades. *Journal of the North American Benthological Society* 15(4): 433–449.

McElroy, T.C., K.L. Kandl, J. Garcia, and J.C. Trexler. 2003. Extinction-colonization dynamics structure genetic variation of spotted sunfish (*Lepomis punctatus*) in the Florida Everglades. *Molecular Ecology* 12: 355–368.

McVoy, C., W.P. Said, J. Obeysekera, J.A. VanArman, and T.W. Dreschel. 2011. *Landscapes and Hydrology of the Predrainage Everglades*. Gainesville: University Press of Florida.

Merriam, G. 1984. Connectivity: a fundamental ecological characteristic of landscape pattern. In *Proceedings of the first international seminar on methodology in landscape ecological research and planning*, edited by J. Brandt and P. Agger, 5–15. Denmark: Roskilde University Center.

Merritt, M.L. 1996. Simulation of the water-table altitude in the Biscayne aquifer, southern Dade County, Florida, water years 1945–89. U.S. Geological Survey Water-Supply Paper, 1–148.

Miller, G.R., J.M. Cable, A.K. McDonald, B. Bond, T.E. Franz, L. Wang, S. Gou, A.P. Tyler, C.B. Zou, and R.L. Scott. 2012. Understanding ecohydrological connectivity in savannas: a system dynamics modelling approach. *Ecohydrology* 5: 200–220.

Moses, C., W.T. Anderson, C.J. Saunders, and F.H. Sklar. 2013. Regional climate gradients in precipitation and temperature in response to climate teleconnections in the Greater Everglades ecosystem of South Florida. *Journal of Paleolimnology* 49: 5–14.

Mulitza, S., D. Heslop, D. Pittauerova, H.W. Fischer, I. Meyer, J.B. Stuut, M. Zabel, G. Mollenhauer, J.A. Collins, H. Kuhnert, and M. Schulz. 2010. Increase in African dust flux at the onset of commercial agriculture in the Sahel region. *Nature* 466: 226–228.

Neto, RR., R.N. Mead, J.W. Louda, and R. Jaffé. 2006. Organic biogeochemistry of detrital flocculent material (Floc) in a subtropical, coastal wetland. *Biogeochemistry* 77(3): 283–304.

Obeysekera, J., J. Barnes, and M. Nungesser. 2015. Climate sensitivity runs and regional hydrologic modeling for predicting the response of the greater Florida Everglades ecosystem to climate change. *Environmental Management* 55(4): 749–762.

Obeysekera, J., M. Irizarry, J. Park, J. Barnes, and T. Dessalegne. 2011. Climate change and its implications for water resources management in south Florida. *Stochastic Environmental Research and Risk Assessment* 25: 495–516.

Ogden, J.C. 2005. Everglades ridge and slough conceptual ecological model. *Wetlands* 25: 810–820.

Parkos, J.J., C.R. Ruetz, and J.C. Trexler. 2011. Disturbance regime and limits on benefits of refuge use for fishes in a fluctuating hydroscape. *Oikos* 120: 1519–1530.

Parkos III, J.J., and J.C. Trexler. 2014. Origins of functional connectivity in a human-modified wetland landscape. *Canadian Journal of Fisheries and Aquatic Sciences* 71: 1418–1429.

Pielke, R.A. 1974. A three-dimensional numerical model of the sea breezes over south Florida. *Monthly Weather Review* 102: 115–139.

Pisani, O., J.W. Louda, and R. Jaffé. 2013. Biomarker assessment of spatial and temporal changes in the composition of flocculent material (floc) in the subtropical wetland of the Florida Coastal Everglades. *Environmental Chemistry* 10(5): 424–436.

Poff, N.L., J.D. Allan, M.B. Bain, J.R. Karr, K.L. Prestegaard, B.D. Richter, R.E. Sparks, and J.C. Stromberg. 1997. The natural flow regime. *BioScience* 47(11): 769–784.

Price, R.M., P.K. Swart, and J.W. Fourqurean. 2006. Coastal groundwater discharge—an additional source of phosphorus for the oligotrophic wetlands of the Everglades. *Hydrobiologia* 569: 23–36.

Price, R.M., Z. Top, J.D. Happell, and P.K. Swart. 2003. Use of tritium and helium to define groundwater flow conditions in Everglades National Park. *Water Resources Research* 39: 1267.

Pringle, C.M. 2001. Hydrologic connectivity and the management of biological reserves: a global perspective. *Ecological Issues in Conservation* 11(4): 981–998.

Pringle, C.M. 2003. What is hydrologic connectivity and why is it ecologically important? *Hydrological Processes* 17(13): 2685–2689.

Rahel, F.J. 2013. Intentional fragmentation as a management strategy in aquatic systems. *BioScience* 63: 362–372.

Rehage, J.S., and J.C. Trexler. 2006. Assessing the net effect of anthropogenic disturbance on aquatic communities in wetlands: community structure relative to distance from canals. *Hydrobiologia* 569: 359–373.

Rosenblatt, A.E., and M.R. Heithaus. 2011. Does variation in movement tactics and trophic interactions among American alligators create habitat linkages? *Journal of Animal Ecology* 80: 786–798.

Rosenblatt, A.E., M.R. Heithaus, F.J. Mazzotti, M. Cherkiss, and B.M. Jeffery. 2013. Intra-population variation in activity ranges, diel patterns, movement rates, and habitat use of American alligators in a subtropical estuary. *Estuarine, Coastal and Shelf Science* 135: 182–190.

Ross, M.S., J.F. Meeder, J.P. Sah, P.L. Ruiz, and G.J. Telesnicki. 2000. The southeast saline Everglades revisited: 50 years of coastal vegetation change. *Journal of Vegetation Science* 11: 101–112.

Ross, M.S., S. Mitchell-Bruker, J.P. Sah, S. Stothoff, P.L. Ruiz, D.R. Reed, K. Jayachandran, and C.L. Coultas. 2006. Interaction of hydrology and nutrient limitation in the ridge and slough landscape of the southern Everglades. *Hydrobiologia* 569: 37–59.

Ruddell, B.L., and P. Kumar. 2009. Ecohydrologic process networks: 1. Identification. *Water Resources Research* 45: W03419.

Saha, A.K., S. Saha, J. Sadle, J. Jiang, M.S. Ross, R.M. Price, L.S.L.O. Sternberg, and K.S. Wendelberger. 2011. Sea level rise and South Florida coastal forests. *Climatic Change* 107: 81–108.

Saunders, C.J., M. Gao, J.A. Lynch, R. Jaffé, and D.L. Childers. 2006. Using soil profiles of seeds and molecular markers as proxies for sawgrass and wet prairie slough vegetation in Shark Slough, Everglades National Park. *Hydrobiologia* 569(1): 475–492.

Saunders, C.J., M. Gao, and Rudolf Jaffé. 2015. Environmental assessment of vegetation and hydrological conditions in Everglades freshwater marshes using multiple geochemical proxies. *Aquatic Sciences* 77(2): 271–291.

Schofield, P.J. 2010. Update on geographic spread of invasive lionfishes (*Pterois volitans* [Linnaeus, 1758] and *P. miles* [Bennett, 1828]) in the Western North Atlantic Ocean, Caribbean Sea and Gulf of Mexico. *Aquatic Invasions* 5(Supp 1): S117–S122.

Seastedt, T.R., R.J. Hobbs, and K.N. Suding. 2008. Management of novel ecosystems: are novel approaches required? *Frontiers in Ecology and the Environment* 6: 547–553.

Skinner, C., F. Bloetscher, and C.S. Pathak. 2009. Comparison of NEXRAD and rain gauge precipitation measurements in South Florida. *Journal of Hydrologic Engineering* 14: 248–260.

Sklar, F.H., M.J. Chimney, S. Newman, P. McCormick, D. Gawlik, S. Miao, C. McVoy, W. Said, J. Newman, C. Coronado, G. Crozier, M. Korvela, and K. Rutchey. 2005. The ecological-societal underpinnings of Everglades restoration. *Frontiers in Ecology and the Environment* 3: 161–169.

Sokol, E.R., J.M. Hoch, E.E. Gaiser, and J.C. Trexler. 2014. Metacommunity structure along resource and disturbance gradients in Everglades wetlands. *Wetlands* 34(Suppl 1): 135–146

Sullivan, P., R.M. Price, F. Miralles-Wilhelm, M.S. Ross, L.J. Scinto, T.W. Dreschel, F.H. Sklar, and E. Cline. 2014. The role of recharge and evapotranspiration as hydraulic drivers of ion concentrations in shallow groundwater on Everglades tree islands, Florida (USA). *Hydrological Processes* 28: 293–304.

Sullivan, P., R.M. Price, M.S. Ross, L.J. Scinto, S. Stoffella, E. Cline, T.W. Dreschel, and F.H. Sklar. 2011. Hydrologic processes on tree islands in the Everglades (Florida, USA): tracking the effects of tree establishment and growth. *Hydrogeology Journal* 19: 367–378.

Swanson, K. L., and A.A. Tsonis. 2009. Has the climate recently shifted? *Geophysical Research Letters* 36: L06711.

Swart, P.K., A.M. Oehlert, G.J. Mackenzie, G.P. Eberli, and J.J.G. Reijmer. 2014. The fertilization of the Bahamas by Saharan dust: a trigger for carbonate precipitation? *Geology* 42: 671–674.

Tischendorf, L., and L. Fahrig. 2000. On the usage and measurement of landscape connectivity. *Oikos* 90: 7–19.

Trexler, J.C., W.F. Loftus, F. Jordan, J.J. Lorenz, J.H. Chick, and R.M. Kobza. 2000. Empirical assessment of fish introductions in a subtropical wetland: an evaluation of contrasting views. *Biological Invasions* 2: 265–277.

Troxler-Gann, T.G., D.L. Childers, and D.N. Rondeau. 2005. Ecosystem structure, nutrient dynamics, and hydrologic relationships in tree islands of the southern Everglades, Florida, USA. *Forest Ecology and Management* 214: 11–27.

Vanni, M.J. 2002. Nutrient cycling by animals in freshwater ecosystems. *Annual Review of Ecology and Systematics* 33: 341–370.

Wachnicka, A., E.E. Gaiser, and L. Collins. 2013. Correspondence of historic salinity fluctuations in Florida Bay, USA, to atmospheric variability and anthropogenic changes. *Journal of Paleolimnology* 49: 103–115.

Wainwright, J., L. Turnbull, T.G. Ibrahim, I. Lexartza-Artza, S.F. Thornton, and R.E. Brazier. 2011 Linking environmental régimes, space and time: interpretations of structural and functional connectivity. *Geomorphology* 126: 387–404.

Wetzel, P.R., A.G. van der Valk, S. Newman, D.E. Gawlik, T. Troxler Gann, C.A. Coronado-Molina, D.L. Childers, and F.H. Sklar. 2005. Maintaining tree islands in the Florida Everglades: nutrient redistribution is the key. *Frontiers in Ecology and the Environment* 3(7): 370–376.

Wingard, G.L., J.B. Murray, W.B. Schill, and E.C. Phillips. 2008. Red-Rimmed Melania (*Melanoides tuberculatus*)—a snail in Biscayne National Park, Florida—harmful invader or just a nuisance? U.S. Geological Survey Fact Sheet 2008–3006.

Wu, Y., N. Wang, and K. Rutchey. 2006. An analysis of spatial complexity of ridge and slough patterns in the Everglades ecosystem. *Ecological Complexity* 3(3): 183–192.

Ya, C. 2014. Sources, Fate and Transformation of Organic Matter in Wetlands and Estuaries. Doctoral thesis, Florida International University.

Yamashita, Y., L.J. Scinto, N. Maie, and R. Jaffé. 2010. Dissolved organic matter characteristics across a subtropical wetland's landscape: application of optical properties in the assessment of environmental dynamics. *Ecosystems* 13(7): 1006–1019.

Yuan, J., M.J. Cohen, D.A. Kaplan, S. Acharya, L.G. Larsen, and M.K. Nungesser. 2015. Linking metrics of landscape pattern to hydrological process in a lotic wetland. *Landscape Ecology* 30(10): 1893–1912.

Yurek, S., D.L. DeAngelis, J.C. Trexler, F. Jopp, and D.D. Donalson. 2013. Simulating mechanisms for dispersal, production and stranding of small forage fish in temporary wetland habitats. *Ecological Modelling* 250: 391–401.

Zhang, K., J. Dittmar, M. Ross, and C. Bergh. 2011. Assessment of sea level rise impacts on human population and real property in the Florida Keys. *Climatic Change* 107: 129–146.

5

The Life of P

A BIOGEOCHEMICAL AND SOCIOPOLITICAL CHALLENGE IN THE EVERGLADES

Victor Rivera-Monroy, Jessica Cattelino, Jeffrey R. Wozniak, Katrina Schwartz, Gregory B. Noe, Edward Castañeda-Moya, and Gregory R. KochwithJoseph N. Boyer and Stephen E. Davis III

In a Nutshell

- Phosphorus (P) is an essential element for all life forms, yet to understand its life cycle and impact we need to grasp not only its biogeochemical life but also how it intersects with human activities and values.
- P is the limiting nutrient in the oligotrophic Everglades ecosystem. Thus, the anthropogenic addition of P to the landscape and its subsequent transport, transformation, and persistence throughout the Everglades are critical to both the visual appearance and the ecological integrity of the Everglades ecosystem.
- Legal and political attention to P in the Everglades has created a powerful social-cultural legacy. This legacy includes the continuing influence of P in Everglades restoration, the dissemination of knowledge about P cycling and mitigation among rural and urban Everglades residents, and social and political realignment of management priorities.
- Management decisions based on legal mandates require not only robust scientific data about P-related processes but also information on the behavior, well-being, and political decisions of tribal nations, agricultural enterprises, anglers, and others who use the south Florida landscape.

Introduction

Phosphorus (P), an essential element for all life forms, is a main character in the unfolding story of Everglades restoration, which is one of the most ambitious and expensive ecosystem restoration projects in the world (Rivera-Monroy et al. 2011; Sklar et al. 2005). Understanding the "Life of P" in the Everglades means grasping the details of how P lives through biogeochemical transformations and how such transformations interact with human actions—from everyday relationships to the environment to the construction of levees and canals for flood protection, agriculture, and water supply. Everglades ecosystems are famously oligotrophic and P-limited (Amador and Jones 1993; Boyer et al. 2006; Castaneda-Moya et al. 2011; Childers et al. 2006; Craft et al. 1995; Daoust and Childers 1999; Davis et al. 2006; Gaiser et al. 2004, 2005; Noe et al. 2001; Ross et al. 2006; Rudnick et al. 1999; Scott et al. 2005; Sharma et al. 2005; Sokol et al. 2014; Sutula et al. 2003; Troxler-Gann et al. 2005; Vaithiyanathan and Richardson 1997). As a result, P loading to this landscape triggers rapid and drastic changes in water quality attributes (i.e., eutrophication), ecosystem structure (e.g., species diversity and density), and ecosystem function (e.g., primary productivity and nutrient retention). Many of Florida's signature industries—from agriculture to real estate development—have produced P-rich wastes that have left "legacy" P in Everglades soils and increased P loads in water flowing through the remnant landscape. To understand why and how this occurs requires an interdisciplinary knowledge of P cycling, as well as attention to its social-cultural sources and legacies. As we will show in this chapter, the impacts of P are as much social-cultural and political as they are biogeochemical (Fig. 5.1).

We examine the consequences of a conundrum in south Florida and the Everglades: P generates great benefits for society while also endangering the ecosystem in ways that harm both humans and other living things. One of the best-known ecosystem services optimized by P is food provisioning by agro-ecosystems. P in fertilizers, along with nitrogen (N), revolutionized agricultural production after World War II by dramatically increasing yield (i.e., the "Green Revolution"). Agriculture is now a multi-billion-dollar industry that has fed and enabled dramatic increases in the human population over the last 75 years, not only in the United States but around the world (Childers et al. 2011; Cordell et al. 2009). At the same time, the use of P in agricultural practices has had significant negative impacts on the environment. Excess P loading from agricultural runoff into Everglades wetlands triggers dramatic ecosystem state changes that are extremely difficult to reverse (Davis et al. 2004; Gaiser et al. 2004, 2005; Koch et al. 2012; Noe et al. 2002, 2003; Noe and Childers 2007). Algal blooms, which can be triggered by nutrient inputs from a variety of anthropogenic and non-human sources, have been associated with seagrass mortality events in downstream estuaries. Land use changes associated with agricultural expansion and urbanization have imperiled the ecological integrity of the Everglades and the delivery of ecosystem services such as fisheries and tourism (Childers et al. 2003; DeBusk et al. 2001; Entry and Gottlieb 2014; Gu et al. 2001; Izuno et al.

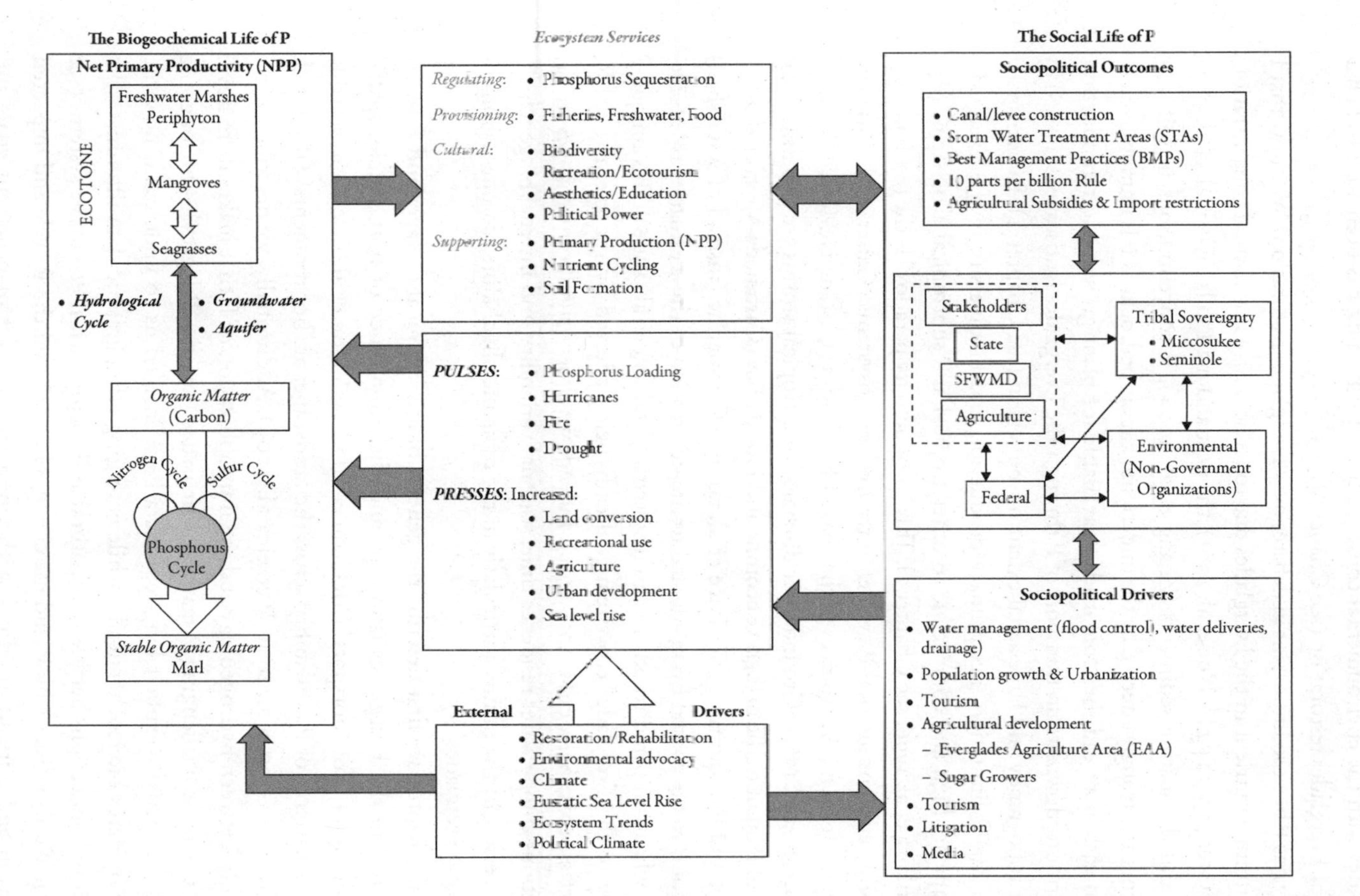

FIGURE 5.1 Social-ecological interactions among different components controlling the "Life of P". The pulsing effect of phosphorus loading strongly influences the Everglades ecosystem dynamics at different spatiotemporal scales as defined by sudden (pulses) or incremental (presses) changes in environmental or social mechanisms as discussed in the text. Adjacent boxes in a cluster (The Social Life of P) indicate that while each agency or interest group has its own approach, which has varied over time, there are commonalities within a given cluster (modified from Collins et al. 2011 and Reddy and Delaune 2008).

1999; Janardhanan and Daroub 2010; Reddy et al. 2011; Smith and McCormick 2001; Waters et al. 2013). Reconciling the social and economic benefits of agriculture in the Everglades with the environmental consequences of P loading has been a central challenge for Everglades restoration (see Chapter 8).

Why has P been so important in debates over Everglades restoration? We know that P is the limiting nutrient in the Everglades due to its pervasive low natural concentrations in the water (<5–10 $\mu g\ l^{-1}$; Noe et al. 2001), the peat and marl soils (Wozniak et al. 2008, 2012), and the marine sediments (Zhang et al. 2004). P concentrations in soils and sediments are strongly affected by the underlying karstic bedrock, as P is immobilized by calcium carbonate and therefore made unavailable for plant growth and reproduction. Furthermore, downstream transport of P through the Everglades landscape is limited by the flat topography and large aerial extent of the region (see Chapter 4). Paradoxically, despite this P limitation, net primary production and biodiversity in the Everglades landscape are high and comparable to other tropical and subtropical ecosystems (e.g., Rejmankova et al. 1996; see Chapter 2). This "productivity paradox" is the result of tight and highly efficient internal P cycling in freshwater ecosystems (Gaiser et al. 2011) and of P inputs from marine sources in the coastal Everglades (Castaneda-Moya et al. 2010; Fourqurean et al. 1992). Groundwater discharge and tropical storms are also important drivers of P input to the mangrove ecotone and Florida Bay (Castaneda-Moya et al. 2010; Herbert and Fourqurean 2009; Price et al. 2006). This marine (instead of watershed) supply of P to the coastal Everglades led to the description of these estuaries as "upside-down" when compared to other coastal systems (Childers et al. 2006). This unique P delivery to the Everglades ecosystem is already being affected by sea level rise (SLR; Koch et al. 2012; Saha et al. 2011, 2012). Indeed, we expect significant changes in the interactions among water residence time, seasonal variability, and internal biogeochemical processing in the future, particularly in the dynamic oligohaline ecotone regions of the coastal Everglades.

In this chapter, we first describe the general biogeochemical linkages and P transformations in the Everglades landscape and the importance of both patchy spatial distributions of P, or "hotspots" (McClain et al. 2003; Vidon et al. 2010), and variable temporal patterns of P availability across the landscape, or "hot moments" (Ross et al. 2009; Troxler-Gann et al. 2005; Wetzel et al. 2009). Additionally, we describe the relationship between hydrogeology and the immobilization of P that makes its removal one of the most challenging P management issues (i.e., the P legacy). Our ongoing analyses of spatially explicit long-term data have enabled us to understand how the system operates biogeochemically, while emerging ethnographic data illuminate the major interconnections between the natural and sociopolitical-economic systems (see Fig. 5.1). We also discuss how land use changes have altered the P cycle and impacted restoration and mitigation efforts, with emphasis on water management goals and long-running litigation that have shaped the water quality debate in the Everglades. We demonstrate that P is a thoroughly social-cultural element, not only because P use and management are critical drivers of ecosystem change in the Everglades, but also because P management has become part of the everyday life of some south Florida residents (see Fig. 5.1). Because urban and agricultural land uses generate P, any debate about P impacts is, ultimately, a debate about the regulation of human activity. We end the chapter by acknowledging current debates over how to deliver more fresh water to Everglades National Park (ENP) without negatively impacting the oligotrophic nutrient regime (see Fig. 5.1).

The Biogeochemical Life of P: Ecological Linkages and Transformations

OVERVIEW OF P CYCLING

The Everglades is an ecosystem that has evolved to function at low P concentrations (Davis 1994; Gaiser et al. 2004; McCormick and O'Dell 1996; Noe et al. 2001). Indeed, in much of the unimpacted Greater Everglades Ecosystem, P is found in far lower concentrations than its stoichiometric demand (Sterner et al. 2008). Despite the low availability of limiting P, the Everglades ecosystem supports comparable and at times higher annual net primary productivity than other tropical and subtropical sites (Ewe et al. 2006; Rejmankova et al. 1996). P loading and subsequent eutrophication in other aquatic systems typically generates algal blooms, which are a major environmental concern globally. However, in the Everglades the anthropogenic addition of P to the system and the subsequent transformation and transport of P throughout the landscape alter ecosystem structure and function in more fundamental ways (Noe et al. 2001; Fig. 5.2). To grasp the importance of P in the Everglades, one must first understand the unique nature of P biogeochemical processing in this system. P cycling is "simple" in comparison to other elements such as carbon, nitrogen, or sulfur (see Fig. 5.2a) because P does not form stable gaseous compounds (see Fig. 5.2b). By contrast, N can be removed from the system through the process of denitrification (conversion to N_2), and carbon (C) can be exported via ecosystem respiration (as CO_2) and methane (CH_4) production. P can be found in both particulate and dissolved forms as well as in organic and inorganic forms (see Fig. 5.2a). The bioavailable form, dissolved inorganic P (PO_4^{3-}), is found in very low concentrations in Everglades surface water (see Fig. 5.2a). Organic forms of P, both dissolved and particulate, generally must undergo biogeochemical transformation into inorganic form before they are available for biologic uptake (see Fig. 5.2b). However, in the high-alkalinity wetlands of the Everglades, inorganic P is quickly immobilized by binding to calcium carbonate, and this legacy P is difficult to remove from soils once it has been introduced.

Why is legacy P a problem? In the Everglades, even small increases in P lead to profound shifts in ecological structure and function in recipient marshes. Elevated P in the Everglades has led to hysteretic state changes from an oligotrophic to a more eutrophic environment (Figs. 5.3 and 5.4). During this transition, blue-green algae–based and diatom-based periphyton mats that precipitate calcium–phosphorus crystals are quickly replaced

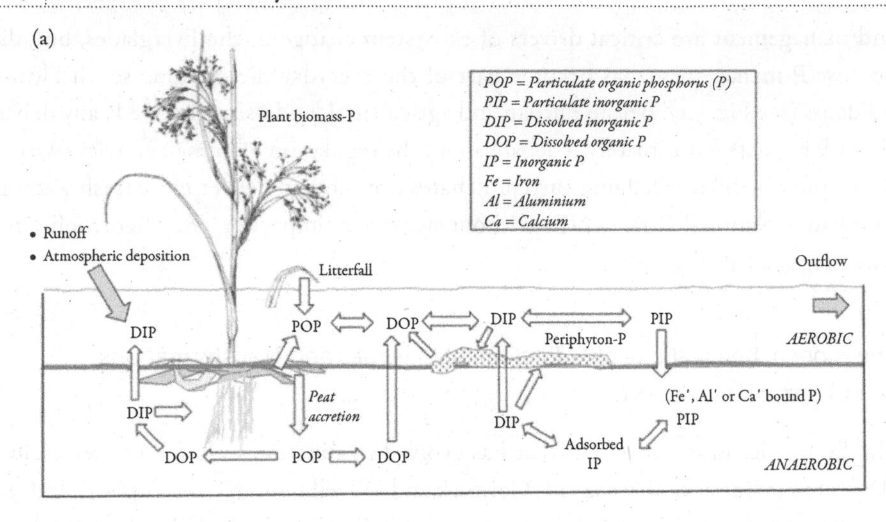

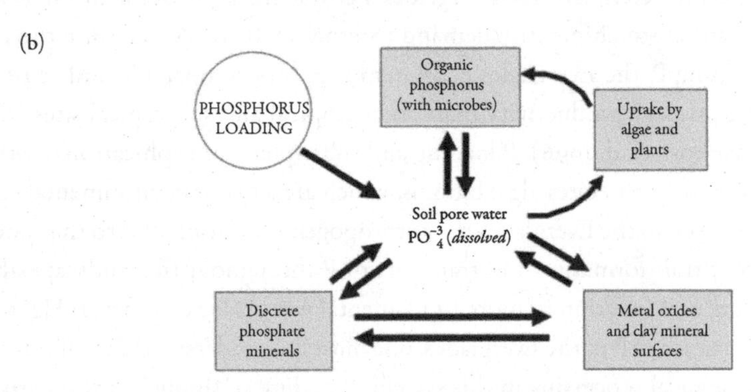

FIGURE 5.2 A) Phosphorus cycle in Everglades wetlands; B) Influence of biotic and abiotic processes and external loading on soil pore water phosphorus (modified from Reddy and Delaune 2008).

by communities of filamentous green algae that do not form mats (Gaiser et al. 2004; Fig. 5.5). Subsequently, sawgrass marshes are replaced by monotypic cattail marshes with dramatically different and less desirable ecosystem functions and services when compared with the original vegetation structure (Childers et al. 2003; Doren et al. 1997; Gaiser et al. 2005; Noe et al. 2001, 2002; see Fig. 5.5). Once the oligotrophic periphyton mats and sawgrass communities are lost, energy transfer from lower to higher trophic levels in the impacted wetlands is very different than under oligotrophic conditions. The lower-nutrient mats rich in calcium carbonate are relatively inedible, and aquatic consumers must operate under less efficient microbial pathways (Sargeant et al. 2009, 2010). These ecologically significant shifts impact all levels of the food web and are largely irreversible due to the difficulty of removing legacy P from the ecosystem (and hence hysteretic; Neto et al. 2006; Ruehl and Trexler 2013).

The limiting and recalcitrant nature of P in Everglades wetlands results in the localized stimulation of primary productivity (as observed in tree islands), site-specific

FIGURE 5.3 The hysteretic state change that results from the addition of P to Everglades wetlands. A) a traditional open water Everglades wet prairie; B) inset image shows an close-up of the marsh surface and extensive calcareous periphyton mats; C) illustrates an Everglades marsh, impacted by P loading, that has been taken over by a dense cover of cattails; D) close-up of a cattail-dominated marsh. [Photo credits: A) Photographer: Franco A. Tobias, Location: Shark River Slough/ Date: September 27, 2012; B) Photographer: Virginia Fernandez, Location: Everglades National Park, Date: December 16, 2011; C) http://sofia.usgs.gov/virtual_tour/images/photos/controlling/lox_cattailclose.jpg; D) http://digir.fiu.edu/photos/fall%2Fplants%2FCrew_1%2F12_08_05%2FP226PT120805MFTSSW.jpg]. See page 4 of color insert.

shifts in ecosystem state (loss of mat-forming periphyton communities and shifts to cattail-dominated marsh), and spatial variability of P in soil (see Fig. 5.5). These processes of ecological assimilation and adsorption effectively sequester P and thereby keep concentrations low in the water column despite high P loading rates (Gaiser et al. 2005; Troxler et al. 2014). Thus, increased inputs of P in the upstream portions of the Everglades (i.e., in water entering Lake Okeechobee as well as in and near the Everglades Agriculture Area [EAA]) result in long-term soil storage of this immobilized P, along with slow, low-level, and long-term bleeding of P downstream to less P-enriched areas of the Everglades (Fig. 5.6). Further, we know that vegetation along the upstream–downstream hydrological gradient will redistribute this legacy P that is trapped in soils as plants uptake P to grow and reproduce, and then release it back into the water column as they die and decompose. Thus, P continues entering the system even after P inputs into the water column have been reduced (Noe and Childers 2007; see Fig. 5.6). This mechanism causes complex interactions between plant production and P sequestration in soils that are often difficult for scientists to study due to the varying temporal and spatial scales at which these interactions occur. It is also often difficult for laypeople to understand how P can be so critical to ecosystem structure and function in the Everglades and yet be in such low concentration, and how even relatively minor increases in P loading can cause such profound ecological impacts. This apparent amplification of

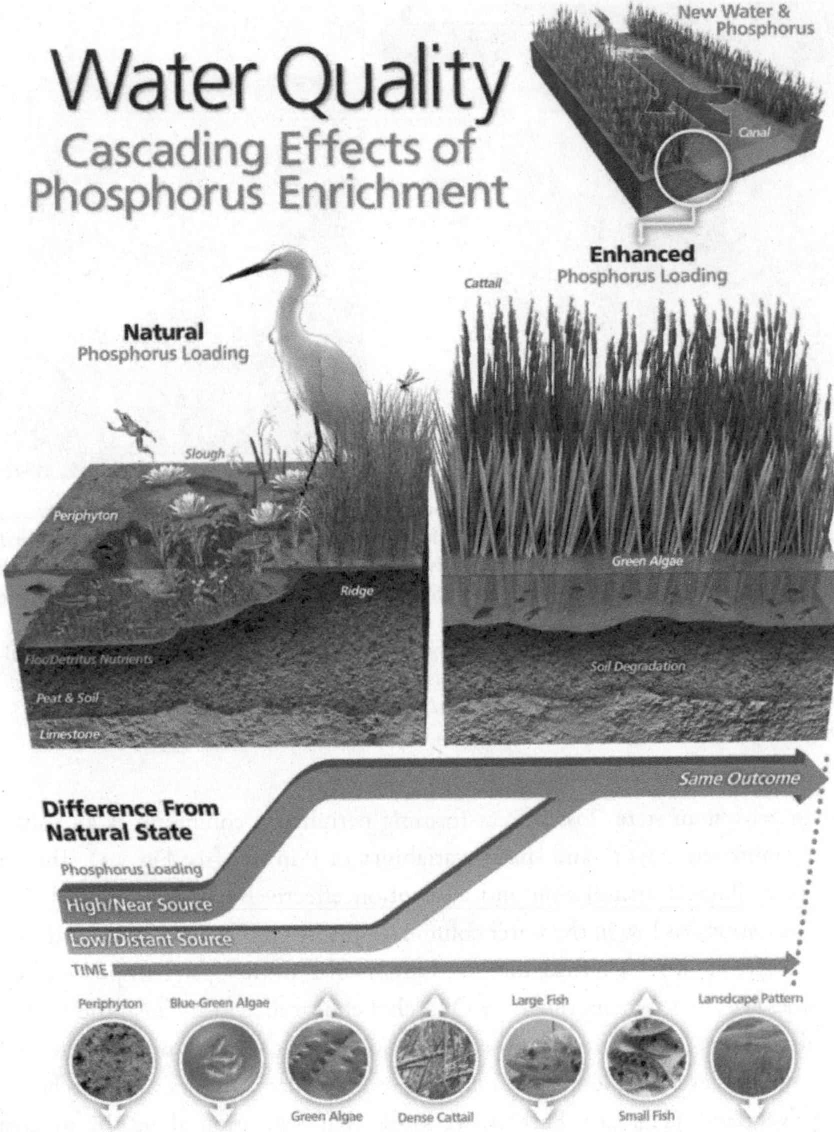

FIGURE 5.4 Cascading effects of phosphorus enrichment at different spatial scales. The main panel progresses from unimpacted on the left to impacted on the right with either rate of P loading or distance from P source (upper panel); ecological indicators of ecosystem state change are shown along the bottom (Image credit: H2H Graphics and SERES Project). See page 8 of color insert.

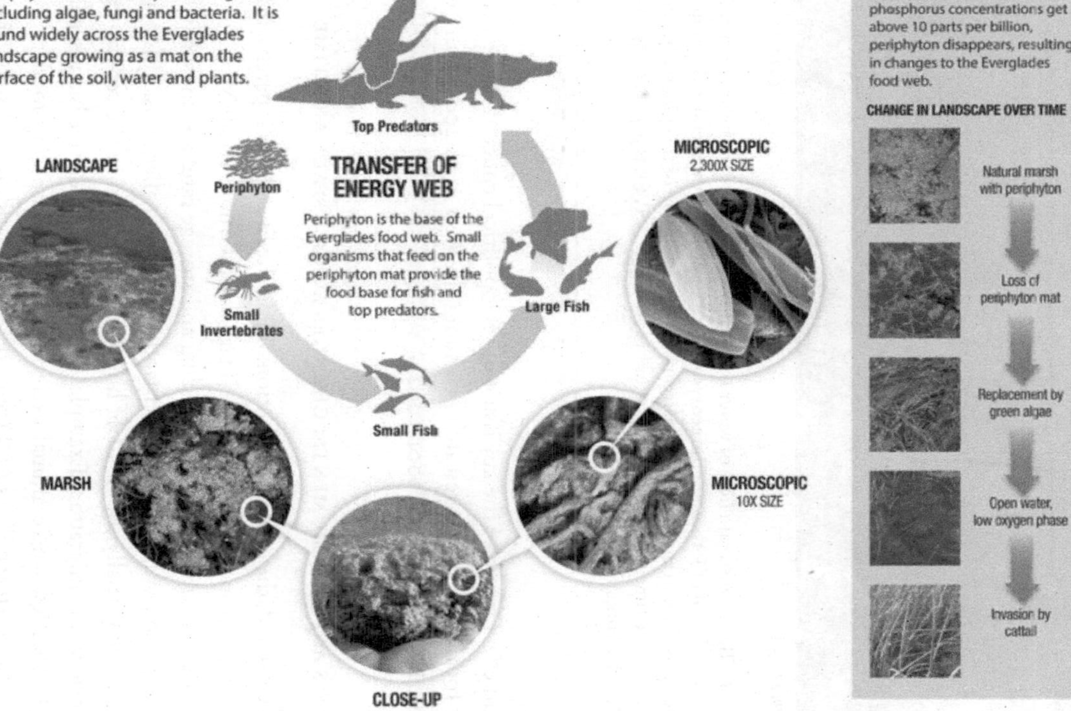

PERIPHYTON
THE BASE OF THE EVERGLADES FOOD WEB
Periphyton is created by microorganisms including algae, fungi and bacteria. It is found widely across the Everglades landscape growing as a mat on the surface of the soil, water and plants.
Top Predators
TRANSFER OF ENERGY WEB
Periphyton is the base of the Everglades food web. Small organisms that feed on the periphyton mat provide the food base for fish and top predators.
Periphyton
Small Invertebrates
Small Fish
Large Fish
LANDSCAPE
MARSH
CLOSE-UP
MICROSCOPIC
10X SIZE
MICROSCOPIC
2,300X SIZE
PERIPHYTON AND PHOSPHORUS POLLUTION
Periphyton is extremely sensitive to changes in water quality. Scientists have shown that when phosphorus concentrations get above 10 parts per billion, periphyton disappears, resulting in changes to the Everglades food web.
CHANGE IN LANDSCAPE OVER TIME
Natural marsh with periphyton
Loss of periphyton mat
Replacement by green algae
Open water, low oxygen phase
Invasion by cattail

FIGURE 5.5 Periphyton composition and ecological role in transferring energy throughout the Everglades ecosystem food chain. Phosphorus enrichment causes the collapse of calcareous periphyton communities and their replacement with non-mat-producing green filamentous algae (Image credit: H2H Graphics and SERES Project). See page 5 of color insert.

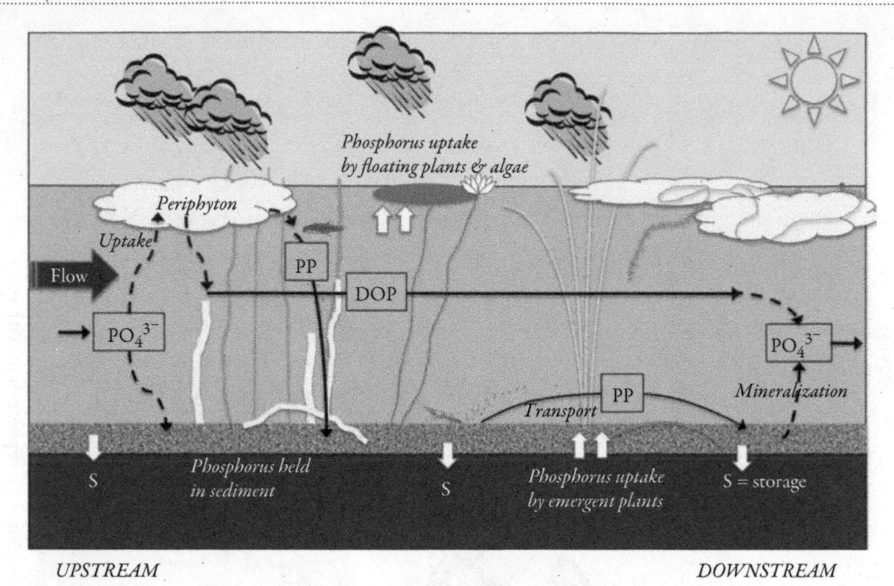

FIGURE 5.6 The downstream flow of water through the P-enriched to the unenriched Everglades results in a slow spiral of P through the ecosystem as it is repeatedly mineralized, assimilated, and thus recycled as the nutrient travels downstream. During each step of this biogeochemical spiral, P is immobilized (in sediments and biota) and thus becomes legacy P. PP is particulate phosphorus, and DOP is dissolved organic phosphorus.

cause and effect is one of the key issues of sociopolitical concern in the Everglades and a major focal point of our research program (Childers et al. 2003; Doren et al. 1997; Gaiser et al. 2005; Noe et al. 2001, 2002).

HOTSPOTS AND HOT MOMENTS: CHANGES ACROSS SPACE AND TIME

As we have noted, the spatial distribution of P in the Everglades is patchy, featuring both natural and anthropogenic "hotspots" (Harms and Grimm 2008; McClain et al. 2003). Natural hotspots are areas where P cycling and storage rates are unusually high, due to the concentration of P from the wider landscape into smaller areas. This higher concentration may be partially the result of hurricane deposition in the coastal regions or animal translocation/migration. Examples of the latter include tree islands, alligator holes, and bird rookeries, where animal excretions and decomposing food/prey accumulate (Wetzel et al. 2011; see Fig. 5.6). Due to declines in wading bird populations since the 1940s (Frederick et al. 2009), avian redistribution of P is less important today than it has been historically (Frederick and Powell 1994; Noe and Childers 2007). Anthropogenic hotspots, in contrast, are areas where P concentrations are elevated due to anthropogenic inputs of P. These include canals, marshes downstream of canal inflows, and constructed treatment wetlands (see Figs. 5.1 and 5.4).

Biogeochemical "hot moments" are defined as short periods of time when P cycling rates exceed long-term averages for a particular location (Harms and Grimm 2008; McClain et al. 2003). Events such as tropical storms and changes in the timing and distribution of freshwater deliveries can produce biogeochemical hot moments (Noe et al. 2010; see Chapter 7). The Everglades ecosystem can tolerate events of low frequency and magnitude (e.g., seasonal wet/dry cycles) without experiencing changes in structure and function. However, if hot moments occur at sufficiently high frequencies and across sufficiently large spatial scales, they can lead to major changes in ecosystem structure and loss of complexity (Hagerthey et al. 2011, 2012, 2014). This has occurred, for example, in water conservation area (WCA) 2A, where the native sawgrass marsh has been displaced by a cattail monoculture as a result of recurrent changes in hydrology and high P inputs from canals draining the EAA (see Figs. 5.1, 5.3, and 5.4). Other examples of hot moments in the Everglades include pulsing events, such as flooding immediately after a fire, or P-rich sediment deposition during hurricane storm surges. For instance, the hot moment associated with Hurricane Wilma significantly increased the total P density in mangrove soils across the southwestern region of the Florida Costal Everglades Long Term Ecological Research (FCE LTER) study area (Castañeda-Moya et al. 2010; see Chapter 7). P accumulation in soils as a consequence of Wilma's storm surge ranged from 20% to 54% of the pre-storm soil content in these mangrove forests, confirming previous hypotheses about how pulsed deposition at large spatial scales contributes to high mangrove forest primary productivity in this region (Castañeda-Moya et al. 2010, 2013; Chen and Twilley 1998, 1999; see Chapter 7).

P SPIRALING IN THE EVERGLADES

The patchy distribution of hot spots and the pulsed timing of hot moments result in the heterogeneous transport of P throughout the landscape and the mosaic of habitat types that characterizes the Everglades. P is transported from the northern Everglades (or Kissimmee River/Chain of Lakes) through Lake Okeechobee to downstream marshes and tree islands and, ultimately, through the mangrove ecotone and into Florida Bay or the Gulf of Mexico. Moments of rapid P mobilization and hydrologic transport can result in the dramatic downstream transfer of P among these habitats (e.g., Walker and Kadlec 2011). The slow downstream flow of water from P-enriched areas to unimpacted areas is the driver of a slow spiraling of P through the ecosystem (see Fig. 5.6). This spiraling concept involves measurements of both P cycling and downstream transport along longitudinal scales (Ensign and Doyle 2006; Koch et al. 2014; Newbold 1992). Thus, P is assimilated from the water column into benthic biomass, temporally retained, and mineralized as organic matter is decomposed and its nutrients, including P, return to the water column. Understanding P cycling is important not only because the biota may affect nutrient concentrations, but also because P concentrations may affect the biota. Thus, flocculent organic particles ("floc"), periphyton, and suspended organic and inorganic particles quickly incorporate phosphate from the water column, while organic P is cycled through the food web by aquatic invertebrates and fish (Noe et al. 2003; see Fig. 5.5). P transport downstream is dominated by loads of both dissolved inorganic and organic P; the latter is typically in higher in concentration (Noe et al. 2007; see Fig. 5.2a) and some of this organic P (~25%) is associated with suspended particles (Noe et al. 2010; see Chapter 4). Episodic flooding events triggering high water flow velocities can entrain significant amounts of P-rich organic particles from floc and periphyton, causing an increase of up to an order of magnitude in the transport of particulate P downstream (Harvey et al. 2011). This particulate P is trapped in periphyton or deposited in the floc downstream, eventually to be mineralized as the spiral continues (see Fig. 5.6). Seasonally higher water velocities also entrain and move floc downstream within the coastal mangrove ecotone (Koch et al. 2014). This downstream spiraling of P is difficult to predict, particularly in situations when its slow transport via sheet flow is interrupted by episodic mobilization of P-rich organic particles during periods of high water flow. This spiraling pattern can explain the lack of immediate changes in surface water P concentrations despite evidence of water column enrichment by P derived from entrained periphyton, soil, floc, and vegetation material in the water column (Childers et al. 2003; Gaiser et al. 2005, 2011; Larsen et al. 2009; Reddy et al. 2011).

POTENTIAL EFFECTS OF SLR

SLR affects P inputs and chemical transformations in the coastal Everglades through changes in hydrodynamics, water levels, and salinity (see Fig. 5.1). P inputs to coastal wetlands are expected to increase with accelerating SLR because, as we have noted, the main sources of P to Everglades estuaries are via groundwater discharge during the dry season and via tidal flows from the Gulf of Mexico and Florida Bay throughout the year. Episodic pulses of sedimentary P during large storms add to this source, and these inputs will increase if climate change increases the frequency and/or strength of these storms (Castañeda-Moya et al. 2010; see Chapter 7). Research has already demonstrated that as estuaries transgress in response to SLR throughout the southeastern United States, the conversion of tidal freshwater marshes to oligohaline wetlands causes increases in soil P mineralization rates, and thus in P availability (Noe et al. 2013). More research is needed to elucidate the effects of this transgressional ecosystem change in coastal Everglades wetlands (Chambers et al. 2014).

The near-shore marine portions of the Greater Everglades Ecosystem, including Florida Bay, are influenced by large regional current systems including the Florida Current-Gulf Stream and the Loop Current (Kourafalou and Kang 2012). Any change in these currents, such as a greater incursion of flow into the coastal zone of peninsular Florida due to SLR, may have profound effects on coastal geomorphology and associated water chemistry. A key morphologic feature of Florida Bay is the wide marl mudbanks along its boundary with the Gulf of Mexico and associated extensive carbonate sediment deposits throughout. These mudbanks restrict the circulation of water not only between the ocean and the bay, but also within the bay, partitioning it into over 50 basins (Fourqurean et al. 1992a; Nuttle et al. 2000; Wanless and Taggett 1989). Liu et al. (2014) showed that these carbonate sediments have a high capacity for rapid P uptake. The combined influence of these features contributes to long P residence times, reduced exchange among basins, and a distinct gradient in P availability (especially relative to N) across Florida Bay. A key uncertainty is the extent to which the elevation of the bay's mudbanks will change as a result of SLR (Koch et al. 2015). Such elevation changes will strongly influence hydrologic connectivity and the exchange rate of nutrients and other materials between the bay and the ocean, and water residence times within the bay. Such changes may ultimately impact the net exchange and availability of P along the coastline.

A direct effect of SLR is increased water depths. For example, a 0.5-m increase in sea level would increase the average depth of Florida Bay by approximately 30% to 50%. Another direct effect would be to decrease light availability to the benthos, limiting benthic primary production and potentially altering community structure. This decreased light penetration will likely modify the structural and functional properties of benthic macrophyte communities (seagrass and macroalgae). However, if the rate of SLR is sufficiently rapid (e.g., at least 9.5 mm per year), more significant changes would be caused indirectly as coastal wetlands erode and these eroded sediments and nutrients are transported to Florida Bay (Koch et al. 2015). This climate-induced nutrient enrichment of Everglades estuaries and coastal zones will likely have cascading impacts on fisheries and other wildlife that are dependent on benthic seagrass communities, as has been occurring globally with nutrient enrichment from cultural eutrophication (Boesch 2002; Nixon 1995) and other stressors (Orth et al. 2006). While patterns of freshwater flow into Florida Bay are expected to be modified by Everglades restoration projects, more drastic changes may be caused by accelerating SLR and associated changes in the elevation of adjacent mangrove wetlands, as well as by changing precipitation patterns. We expect that the quantity of flow will change roughly in proportion to the balance of precipitation and evaporation in the Everglades watershed (see Chapter 3), thereby affecting P cycling in coastal wetlands and associated estuaries and bays.

The Social and Political Life of P

LITIGATING P

Because P is such a critical driver of ecosystem structure and function in the Everglades, and because humans have introduced large amounts into the landscape during the last century, P has been the focus of sustained conflicts over Everglades restoration and water management. Over the past quarter-century, the Social Life of P in the Everglades has been shaped by litigation and its legacies (see Fig. 5.1). Numerous parties have filed lawsuits and administrative appeals related to water quality issues, but one suit in particular has had the greatest impact: the "USA lawsuit," which was brought by the federal government in 1988 and remains active (*U.S. v. South Florida Water Management District [SFWMD]*). Key controversies and divisions among stakeholders have persisted from the late 1980s to the present day, as revealed through extensive ethnographic and archival research conducted since 2009. This research included observations in policy making and advocacy settings, over 140 hours of open-ended interviews with 120 individuals (federal, state, local, and tribal agency personnel; academic researchers; environmental, agricultural, and recreational advocates), and review of court documents and government reports (Schwartz 2014).

The target of the USA lawsuit was P-enriched runoff from the EAA, and both the course of the litigation and perceptions about it have been informed by the political power of EAA agricultural interests. The fertile muck soils south of Lake Okeechobee have supported the country's largest production of sugarcane. This production was made possible by the flood protection and drainage provided by the Central & South Florida Project (see Chapters 1 and 3) and by the Cuban Revolution of 1959 (Blake 1980; McCally 1999). Two vertically integrated corporations—U.S. Sugar Corporation (USSC) and Florida Crystals—collectively own over half of the land in the EAA and generate over half of total production today. USSC is the country's largest sugarcane grower and produces 10% of our domestic sugar supply, while Florida Crystals and its multinational subsidiaries (including Domino) is the world's largest sugar refiner (Green 2012). The profitability of sugar production in the EAA continues to be propped up by direct and indirect governmental support, including import restrictions and price supports at the federal level (Hollander 2008), and by subsidized water supply and flood control at the state level (Diamond et al. 2012). Sugar growers exert considerable political influence at the local, state, and federal levels and have lobbied successfully, with numerous other agricultural interests, against efforts to reform trade and commodity policies (e.g., Green 2012; Roberts 1999).

As early as the mid-1970s, SFWMD scientists warned of possible ecological impacts of P on the Everglades, but these reports were not widely circulated (John 1994). In 1988, Dexter Lehtinen, acting U.S. attorney in Miami, sued the state of Florida for discharging nutrient-laden water from the EAA onto federal lands, namely ENP and the Arthur R. Marshall Loxahatchee National Wildlife Refuge (LNWR). Although "it was highly unusual for the Department of Justice to litigate questions of state law" (John 1994), limits on available legal instruments led the District Attorney's office to sue the state for violating water quality regulations, including a "narrative standard" forbidding alteration of nutrient concentrations so as to "cause an imbalance in natural populations of aquatic flora or fauna." In 1990, Lawton Chiles was elected governor after campaigning on a pledge to end exorbitant legal fees by quickly settling the lawsuit. In a legendary move, Chiles appeared in the courtroom to personally represent the state of Florida, where he "surrendered his sword" and "stipulated that the water was dirty" (Rizzardi 2001).

This paved the way for the 1991 Settlement Agreement, which set deadlines for achieving specific levels of P reduction through construction of treatment wetlands (stormwater treatment areas [STAs]) and adoption of mandatory best management practices (BMPs) by EAA growers. It also called for research to determine a numeric standard for P concentrations in the Everglades to replace the narrative standard. The sugar industry filed numerous administrative appeals and denounced the Settlement Agreement for requiring farm runoff to be "cleaner than rain" (Box 5.1) and for excluding them from the negotiations (John 1994). In 1994 the Florida legislature implemented the terms of the Settlement Agreement by passing the Everglades Forever Act (EFA), which assessed an "agricultural privilege tax" of $25 per acre in the EAA to help fund research and cleanup. It also required the Florida Department of Environmental Protection (FDEP) to establish the numeric total phosphorus (TP) concentration standard, or default to a standard of 10 parts per billion (ppb), by the end of 2006. In 2003, the FDEP promulgated, and the U.S. Environmental Protection Agency (EPA) approved, a TP concentration standard of 10 ppb (see Fig. 5.1).

On the same year, the legislature enacted controversial amendments to the EFA that delayed and weakened implementation of the TP standard. In 2004 the Miccosukee Tribe (represented by Dexter Lehtinen) sued the EPA for failing to determine that those amendments changed the state's water quality standard for the Everglades. In his strongly worded 2008 and 2010 rulings, federal judge Alan Gold castigated the EPA and state agencies for prolonged delays in achieving cleanup goals and ordered the EPA to produce a compliance plan. The SFWMD rejected the EPA's plan as unaffordable and released its own alternative, which was approved by the legislature in 2013. The state plan, known as the Restoration Strategies Regional Water Quality Plan, included $880 million in additional STAs and other structures that are scheduled for completion in 2024.

In the quarter-century since the USA lawsuit was settled, the combination of STAs and mandatory and voluntary BMPs has succeeded in reducing P loads (Daroub et al. 2011). The SFWMD's 57,000 acres of STAs have removed a cumulative total of 1,874 metric tons of P since operations began (SFWMD 2015). In Water Year 2014, TP concentrations in EAA outflows were 63% lower than the pre-Settlement baseline, making it the 19th consecutive year in which reductions surpassed the required 25% (SFWMD 2015). The total reduction in P runoff from the EAA from 1996 through 2014 was 2,853 metric tons, or 55% of the baseline (SFWMD 2015). Some landowners have suggested that P concentrations are lower in water draining from their land than in water they are using for irrigation, yet the interpretation of water quality data remains disputed (Faridmarandi and Naja 2014). Moreover, state–federal intergovernmental relations appear to have become significantly more amicable in recent years (Schwartz 2014). Nevertheless, P concentrations exceeded the limit in water entering the LNWR in 2008–2009 and 2014, and in water entering ENP in 2012 and 2014, so the USA lawsuit remains active. The Department of the Interior has expressed support of restoration strategies if implemented as planned, but in the meantime, the state remains under the scrutiny

BOX 5.1
"CLEANER THAN RAIN"

This has become one of the most politically charged phrases in the long and litigious history of P in the Everglades. The slogan emerged during the contentious years between the filing of Lehtinen's lawsuit in 1988 and the agreement to settle the case in 1991. Some scientists had been proposing 10 ppb as a reasonably protective standard since the mid-1980s, well before the Florida legislature adopted it as the "default" numeric standard in the 1994 EFA. But critics in agriculture and other sectors, drawing on some reports by SFWMD and by ecologist Curtis J. Richardson, whose Everglades research at the time was funded by the sugar industry, claimed that a 10-ppb standard would require TP concentrations in EAA discharges to be lower than rainfall concentrations in the Everglades (Fairbanks 1990; Richardson 1989). At meetings of the SFWMD's governing board in 1990 and 1991, opponents passed out stickers declaring "Cleaner than rain is insane."

An expert witness hired by the U.S. Department of Justice, environmental engineering consultant William W. Walker Jr., questioned the SFWMD's figures, sampling methods, and methods of analysis. Walker noted that in order to accurately estimate TP concentrations, measurements needed to be weighted by rainfall volume, a method now considered standard practice (e.g., Eklund et al. 1997). Walker (1989) analyzed the same SFWMD data set and found that once rainfall volume was accounted for, the range of TP concentrations was actually 5 to 9 ppb (Walker 1989). Another weakness of the early SFWMD and Richardson analyses was that they failed to account for contamination of samples by bird droppings, insects, and dust—a problem that was not well understood at the time. In a subsequent peer-reviewed study, researchers using more rigorous sampling methods to analyze data from 1992 to 1996 concluded that the statewide average TP concentration in rain was 5 ppb (Pollman et al. 2002).

In short, scientists now concur that the 10-ppb P standard is roughly double that of rain, but the correct figures for rain concentrations are not well known beyond the scientific community. In the meantime, the "cleaner than rain" slogan has taken on a life of its own. During the "penny-a-pound" sugar tax battle in 1996, opponents of the tax used the phrase in letters to the editor (Aromandi 1996; Powell 1996). In 1999, a USSC executive told the *New York Times*: "We are the only people in the country who have to clean water cleaner than the water that falls on us" (McKinley 1999). As late as 2015, the executive editor of a widely read Florida political blog wrote that if the 10-ppb standard is achieved through implementation of restoration strategies, "the water flowing into the Everglades from the [EAA] . . . [will be] cleaner than the rain water that falls there" (Schorsch 2015). This widely used distortion of facts includes variations that substitute the word "rain" for brands of bottled water (e.g., Evian) to emphasize "cleanness" in claims that have no basis in actual P concentration data. Even acclaimed environmental journalist Michael Grunwald (2015), author of the popular Everglades history

The Swamp and a critic of the Florida sugar industry, recently referred to restoration as a "deal to get runoff from sugar farms even cleaner than Evian."

of two federal judges. The term "restoration strategies" includes a Science Plan to guide investigation of the critical factors that collectively influence the performance of STAs, particularly at relatively low TP concentrations (<20 ppb; SFWMD 2013). STAs retain P through several mechanisms, including plant nutrient uptake, organic soil accumulation, sedimentation and sorption, co-precipitation with minerals, and microbial uptake (see Fig. 5.2). Additional research being conducted by scientists from federal and state agencies and universities will continue to inform the design and operations of STAs and other treatment projects, toward the goal of achieving compliance with the P standard.

ONGOING CONTROVERSIES

Many issues in the sociopolitical life of P remain highly contentious (Fig. 5.7). The question of who should pay to reduce P inflows has been debated since the 1980s. Many environmental advocates argue that growers should pay more and that no public lands should be used for treatment. The agricultural industry maintains that this would drive growers out of business, damaging the Florida economy, destroying jobs, and reducing the food supply. In 1996, environmentalists succeeded in placing two constitutional amendment initiatives on the Florida state ballot: a "penny-per-pound" tax on EAA sugar growers and a provision stating that EAA polluters "shall be primarily responsible for paying the costs of the abatement of that pollution" (Kleindienst 1997) (see Fig. 5.7). In what was at the time Florida's costliest political campaign, the tax was defeated but the "polluter pays" measure passed with nearly 70% of the vote. However, the legislature has yet to implement it. This 1996 referendum solidified an "agriculture versus environment" narrative in Florida and focused criticism on "Big Sugar" (Hollander 2008). This narrative continues to shape management practices, public debates, and personal identities across the region.

A second debate, still active in some circles, concerns whether the 10-ppb TP standard was "the right number" (see Figs. 5.1 and 5.7 and Box 5.1). The initial determination of 10 ppb as a default criterion was made during two months of intense interagency negotiations leading up to the 1991 Settlement Agreement, but it was reaffirmed by FDEP in 2003, drawing upon some $70 million worth of research conducted over 12 years by agency and university scientists. Agricultural spokespeople still maintain that the standard is too low, that it fails to reflect the variability of P across the ecosystem, and that the science underpinning it was rushed and flawed. Officials at state agencies sometimes voice similar views, whereas EPA and Department of the Interior officials, as well as state scientists and other state officials, reject these views. Some growers continue to suggest that the goal of environmentalists is to take EAA land out of production, and their view is not completely unfounded (see Fig. 5.7). In 2008, for example, when the state announced a deal to buy all of USSC's land and assets in the interest of Everglades restoration, David Guest of EarthJustice said that taking the sugar land out of production "relieves an immediate stress" on the ecosystem (as quoted in Cave 2008).

These debates feed into a larger question of whether P litigation has been more helpful or harmful to Everglades restoration overall. Environmentalists and some federal officials herald it as the only mechanism capable of compelling reluctant state agencies and lawmakers to take meaningful action. But many others contend that litigation has forced an unnecessary tradeoff between improved water quality and increased water flows to the Everglades, all in the name of an overly protective P standard. While those in the former camp counter that the P standard is not to blame for delays in implementing storage and flow restoration projects, none would dispute that the state's concerns about triggering water quality violations continue to influence intergovernmental relations and the implementation of key projects (Schwartz 2014).

The new Central Everglades Planning Project (CEPP), which bundles together several critical Comprehensive Everglades Restoration Plan (CERP) projects that had faced prolonged implementation delays, was widely considered to be a major breakthrough in reconciling the goals of P removal and hydrologic restoration (see Fig. 5.7). CERP did not include water quality projects, given that P levels were expected to be brought into compliance in a timely manner through completion of ongoing STA construction that was funded separately. CEPP includes a water quality component—construction of "flow-equalization basins" intended to improve the functioning of STAs—and every CEPP project is premised on the constraint that increased water flow to the Everglades will not violate water quality standards (see overview of the CEPP plan in Davis et al. 2014). As this book goes to press, CEPP was still awaiting congressional authorization, but protecting the Everglades from further P impacts remains both a top priority and a key challenge (see Fig. 5.7).

THE EVERYDAY LIFE OF P

In the wake of legal, political, and administrative attention, P pervades the Everglades as both biogeochemical and sociocultural legacies (see Figs. 5.1 and 5.7). Our ethnographic research has shown that rural residents in the agricultural areas along Lake Okeechobee's southern shore (known as "the Glades") talk about P, debate over P, and otherwise use P to help make sense of their social and political worlds. The sociocultural legacy of P includes the continuing symbolic power of P in Everglades restoration, the diffusion of detailed and differentiated knowledge about P cycling and mitigation among rural Everglades residents, and the effects of P on social fracturing and coalescence. In the Glades, farmers, anglers, and public officials exhibited extensive knowledge of the history, politics, and science of P. In particular, farmers were quick to discuss on-farm

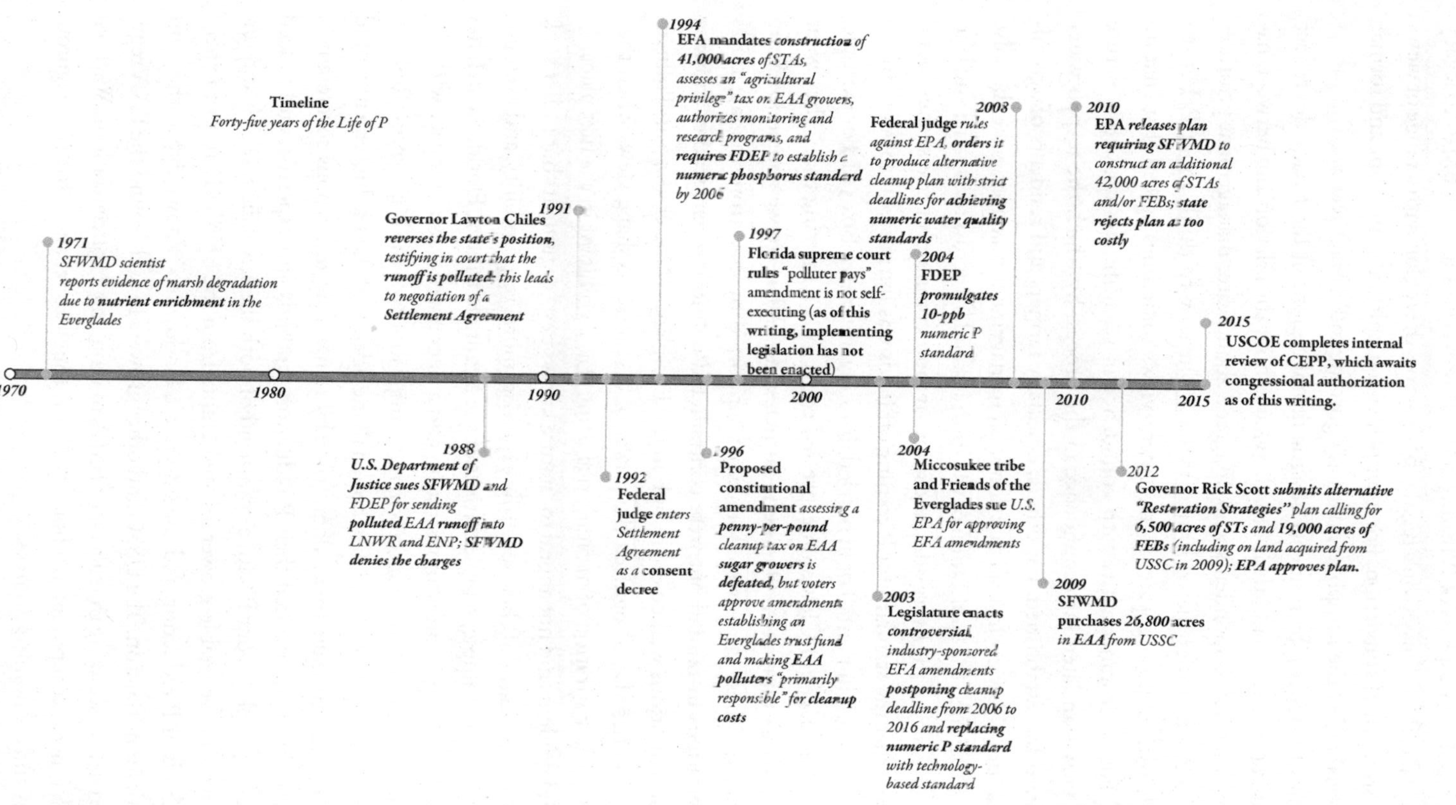

Timeline
Forty-five years of the Life of P
1970
1980
1990
2000
2010
2015
1971
SFWMD scientist reports evidence of marsh degradation due to nutrient enrichment in the Everglades
1988
U.S. Department of Justice sues SFWMD and FDEP for sending polluted EAA runoff into LNWR and ENP; SFWMD denies the charges
1991
Governor Lawton Chiles reverses the state's position, testifying in court that the runoff is polluted; this leads to negotiation of a Settlement Agreement
1992
Federal judge enters Settlement Agreement as a consent decree
1994
EFA mandates construction of 41,000 acres of STAs, assesses an "agricultural privilege" tax on EAA growers, authorizes monitoring and research programs, and requires FDEP to establish a numeric phosphorus standard by 2006
1996
Proposed constitutional amendment assessing a penny-per-pound cleanup tax on EAA sugar growers is defeated, but voters approve amendments establishing an Everglades trust fund and making EAA polluters "primarily responsible" for cleanup costs
1997
Florida supreme court rules "polluter pays" amendment is not self-executing (as of this writing, implementing legislation has not been enacted)
2003
Legislature enacts controversial industry-sponsored EFA amendments postponing cleanup deadline from 2006 to 2016 and replacing numeric P standard with technology-based standard
2004
FDEP promulgates 10-ppb numeric P standard
2004
Miccosukee tribe and Friends of the Everglades sue U.S. EPA for approving EFA amendments
2008
Federal judge rules against EPA, orders it to produce alternative cleanup plan with strict deadlines for achieving numeric water quality standards
2009
SFWMD purchases 26,800 acres in EAA from USSC
2010
EPA releases plan requiring SFWMD to construct an additional 42,000 acres of STAs and/or FEBs; state rejects plan as too costly
2012
Governor Rick Scott submits alternative "Restoration Strategies" plan calling for 6,500 acres of STs and 19,000 acres of FEBs (including on land acquired from USSC in 2009); EPA approves plan.
2015
USCOE completes internal review of CEPP, which awaits congressional authorization as of this writing.

FIGURE 5.7 Timeline of key litigation issues, decisions, and controversies in the sociopolitical life of P. SFWMD: South Florida Water Management District; EAA: Everglades Agricultural Area; FDEP: Florida Department of Environmental Protection; LNWR: Loxahatchee National Wildlife Refuge; ENP: Everglades National Park; EFA: Everglades Forever Act; STA: Stormwater Treatment Areas; FDEP: Florida Department of Environmental Protection; USEPA: U.S. Environmental Protection Agency: FEBs: flow-equalization basins; USCOE: U.S. Army Corps of Engineers; CEPP: Central Everglades Planning Project.

practical and regulatory aspects of P, readily addressing everything from BMPs to legacy P and plant uptake. P is discussed on a regular basis across the region with a range of views and with the strength of conviction. For example, an agricultural consultant and former SFWMD official called the 10-ppb standard a "lightning rod" (MacVicar 2005).

Assessments of P science by residents of agricultural regions of the Everglades reflect both their interest positions and their broader views about the relationship between nature and human action. Respondents who disagreed about water management priorities with regard to P sometimes shared similar views about how to think about P. For example, people with opposing water management views sometimes agreed that "nature knows best." For environmentalists with confidence in scientific findings, this "nature knows best" view translated into confidence in the process by which the 10-ppb water quality standard was established. For others, especially farmers and residents of agricultural communities, this view instead cast doubt on management practices, either because respondents questioned the application of the 10-ppb standard across the spatial and biogeochemical diversity of the Everglades landscape, or more commonly because they found it unreasonable that the standard for P coming off fields was purportedly lower than the perceived amount of P in rainfall or in bottled drinking water (see Box 5.1). For example, the former city manager of Clewiston expressed concern at a meeting that "They want us to clean up God's rainwater." Or, as one farmer said: "What are we gonna do? Shoo away the clouds?" These views were somewhat supported by data from the late 1980s on P concentrations in rainfall but were contradicted by more recent studies showing rainfall P concentrations well below 10 ppb in the Everglades (Noe et al. 2001; Pollman et al. 2002). Skeptics also recognized the power of discourse to shape views when they questioned the categorization of naturally occurring P as a pollutant in the first place.

Other rural residents did not appeal to "nature knows best" arguments but rather took P levels to be an indicator of their scientific expertise, moral responsibility, and even political authority. For example, representatives of the Seminole tribe of Florida, which has a complex water management administration and a negotiated water compact with the state of Florida, have touted their success in reducing P inputs to the Big Cypress Reserve. They focused on P in part because they hoped that evidence of their achievement would encourage outsiders to leave them alone (thereby reinforcing indigenous sovereignty). Some farmers also demonstrated their P reduction methods and expertise, and indeed farmers regularly spoke about P with a command of both the scientific research and on-farm expertise. One vegetable grower secured grants from the SFWMD to test experimental methods of P reduction and touted private landowners' knowledge as the most effective path for mitigation. The USSC and other growers in and around the EAA regularly released figures showing the success of BMPs in reducing P concentrations. Without a doubt, P has become a potent symbol, a political touchstone, and a focus of vigorous negotiations and economic resources.

Efforts to reduce P concentrations have altered both the physical and the sociocultural landscapes of interior south Florida. As we noted, 57,000 acres of STAs have been built in the EAA, and more are planned. These STAs increasingly have become destinations for waterfowl hunters, birders, and alligator hunters. The nearly 17,000-acre STA5, for example, is regularly named as a top birding destination in Florida. The U.S. National Audubon Society leads tours there that bring birders and their money to Clewiston and nearby communities, and the SFWMD has responded by expanding user access times and infrastructure. P mitigation practices also have altered the landscape by interrupting connectivity among agricultural lands where compliance with state-mandated and voluntary BMPs required on-property water retention to prevent P runoff. And farmers have lost productive acreage to this water retention requirement. For example, the McDaniels ranch, located in southern Hendry County, has converted over 3,000 acres of farmland for water and P retention.

Most dramatically, P reduction was one motivation for a project that would have dramatically altered the EAA landscape. In 2008, the state agreed to purchase all of the USSC's land and assets in order to gain 180,000 acres of farmland for restoration. The proposed deal, which was reduced to 26,800 acres in part because of the Great Recession, reorganized longstanding political loyalties and affected the political economy of the agricultural communities along the southern shore of Lake Okeechobee. Real estate agents reported that property values in Clewiston and surrounding areas declined at far greater rates than in nearby south Florida counties after the deal was announced, and at least one major employer pulled out of a plan to relocate to Clewiston in the wake of the deal's announcement. Local leaders worked with the SFWMD and some environmentalists to mitigate the economic impact of the proposed deal, and new, albeit short-lived, alliances formed. Despite long-established narratives that pitted agriculture against the environment, the Sociocultural Life of P does not simply reflect preexisting interest positions; rather, it also brings interests and stakes into being.

P has a dual geography in the Everglades. From one perspective, variations in naturally occurring P levels differentiate the Everglades landscape (see "Hot Spots" section earlier in the chapter). From another perspective, P generates a geography of social and legal responsibility whereby certain areas of the region are deemed more vulnerable, culpable, or measurable—one could call these "sociopolitical hotspots" (e.g., Parker and Hackett 2012). Lake Okeechobee, which has been a P battleground since the 1970s, is one example of a sociopolitical hotspot. Florida Audubon has devoted significant staff resources to the lake, and federal and state agencies have restored a portion of the Kissimmee River that flows into the lake. But coastal communities continue to decry the effects of releases of high-P water from the lake on ecologically sensitive and economically significant estuaries on both of Florida's coasts (see Chapter 1). Historically, a much-debated issue was "backpumping," a flood control and agricultural drainage practice whereby excess water was pumped from the EAA back into the lake (Izuno and Bottcher 1994). This practice was limited by P litigation beginning in the 1980s, and "backpumping" remains a fraught term today. As P loading to the lake continues to increase despite various restoration initiatives and regulatory programs, debates continue over the possibility of allowing "water augmentation" (i.e., backpumping by a less politicized name) into Lake Okeechobee. Residents and growers south of the lake felt unfairly targeted as polluters, given the large P loads entering the lake from the north (Cattelino 2015). Restoration initiatives include programs that offer payments to ranchers on lands north of the lake for on-ranch water retention and filtration services. Meanwhile, water quality concerns continue to complicate efforts to move more water from the lake south into the WCAs and ENP.

In rural south Florida, P influences how people understand nature, their shifting interest positions, the political economy, and changes in the landscape. Drivers of P loads are not generalizable to forces such as "agriculture" but rather reflect the sociocultural diversity of a region that includes large-scale growers, field laborers, extension agents, smaller independent farmers, agricultural contractors, civic leaders in agricultural towns, and a wide range of others (Ogden 2008). P remains both a water management priority and a lived experience for a range of social actors, and its legacies endure.

Summary: A Complicated, Enduring Life

The social-ecological approach employed here to learn about the life and legacy of P in South Florida highlights the complex ways human and nonhuman aspects of social-ecological systems are co-constituted (Collins et al. 2011). This approach helped us not only elucidate how social, political, and economic factors influenced P cycling, but also to understand how ecological change has affected the South Florida social landscape over the last century. However, it is politically challenging to agree on and implement cost-effective and ecologically sustainable strategies to restore and/or rehabilitate biogeochemical cycles and hydrologic patterns in such a diverse and extensive landscape.

Long-term hydrologic and water quality data collected since the mid-1990s by FCE-LTER collaborators and other academic researchers—as well as by state, federal, and local government agencies—are now providing valuable information to help assess the operational realities of restoring the Everglades ecosystem. The question of whether efforts to further reduce P concentrations must be in place and successful before freshwater flows to the Everglades can be increased continues to be debated. Water management decisions require not only knowledge about P fluxes and loading, residence times, and immobilization rates, but also information about the behavior, well-being, and political commitments of local and regional end users—such as tribal nations, agricultural enterprises, and anglers. The dual geography of P creates ecosystem management and restoration challenges along the heavily modified hydrologic gradient from upstream of Lake Okeechobee to the coastal Everglades and the Gulf of Mexico. Indeed, P pervades the iconic Everglades as both a sociocultural and a biogeochemical legacy (see Chapter 2). This legacy includes the continuing power of P in Everglades restoration, the differentiated knowledge about P cycling and mitigation among rural and urban residents, and the socially fracturing and consolidating effects that P has had and continues to have. These effects reverberated in litigation and were encapsulated in diverse visions of how, and how much, to eliminate excess P in waters entering the Everglades.

One promising approach to restoring the health of south Florida ecosystems involves a monitoring and adaptive management strategy that incorporates iterative learning processes whereby new data on ever-changing regional, social, and economic conditions are incorporated into management strategies (see Chapter 8; Christensen et al. 1996; Sklar et al. 2005). However, implementation of such adaptive management approaches requires trust among a diverse constellation of public and interagency institutions and players. After more than 30 years of contentious litigation and legislation, this trust may be a tall order in some cases. These approaches also require improved communication among scientists and managers. Still, we are optimistic that our research and findings will provide the scientific basis to continue anchoring and justifying Everglades restoration efforts aimed at reducing ecological and economic threats to both human well-being and ecosystem sustainability.

References

Amador, J.A., and R.D. Jones. 1993. Nutrient limitations on microbial respiration in peat soils with different total phosphorus-content. *Soil Biology & Biochemistry* 25: 793–801.

Aromandi, C. 1996. Letter to the editor. *Palm Beach Post*, October 27.

Blake, N.M. 1980. *Land Into Water—Water Into Land. A History of Water Management in Florida*. Tallahassee. University Presses of Florida.

Boesch, D.F. 2002. Challenges and opportunities for science in reducing nutrient over-enrichment of coastal ecosystems. *Estuaries* 25(4): 886–900.

Boyer, J.N., S.K. Dailey, P.J. Gibson, M.T. Rogers, and D. Mir-Gonzalez. 2006. The role of dissolved organic matter bioavailability in promoting phytoplankton blooms in Florida Bay. *Hydrobiologia* 569: 71–85.

Castañeda-Moya, E., R.R. Twilley, and V.H. Rivera-Monroy. 2013. Allocation of biomass and net primary productivity of mangrove forests along environmental gradients in the Florida Coastal Everglades, USA. *Forest Ecology and Management* 307: 226–241.

Castañeda-Moya, E., R.R. Twilley, V.H. Rivera-Monroy, B.D. Marx, C. Coronado-Molina, and S.M.L. Ewe. 2011. Patterns of root dynamics in mangrove forests along environmental gradients in the Florida Coastal Everglades, USA. *Ecosystems* 14: 1178–1195.

Castañeda-Moya, E., R.R. Twilley, V.H. Rivera-Monroy, K.Q. Zhang, S.E. Davis, and M. Ross. 2010. Sediment and nutrient deposition associated with Hurricane Wilma in mangroves of the Florida Coastal Everglades. *Estuaries and Coasts* 33: 45–58.

Cattelino, J.R. 2015. The cultural politics of water in the Everglades and beyond: transcript of the Lewis Henry Morgan Lecture given on October 14, 2015. *HAU: Journal of Ethnographic Theory* 5(3): 235–250.

Cave, D. 2008. Florida to buy sugar maker in bid to restore Everglades. *The New York Times*, June 25, 2008.

Chambers, L.G., S.E. Davis, T. Troxler, J.N. Boyer, A. Downey-Wall, and L.J. Scinto. 2014. Biogeochemical effects of simulated sea level rise on carbon loss in an Everglades peat soil. *Hydrobiologia* 726(1): 195–211.

Chen, R., and R.R. Twilley. 1998. A gap dynamic model of mangrove forest development along gradients of soil salinity and nutrient resources. *Journal of Ecology* 86: 1–12.

Chen, R.H., and R.R. Twilley. 1999. Patterns of mangrove forest structure and soil nutrient dynamics along the Shark River estuary, Florida. *Estuaries* 22: 955–970.

Childers, D.L., J.N. Boyer, S.E. Davis, C.J. Madden, D.T. Rudnick, and F.H. Sklar. 2006. Relating precipitation and water management to nutrient concentrations in the oligotrophic "upside-down" estuaries of the Florida Everglades. *Limnology and Oceanography* 51: 602–616.

Childers, D.L., J. Corman, M. Edwards, and J.J. Elser. 2011. Sustainability challenges of phosphorus and food: solutions from closing the human phosphorus cycle. *Bioscience* 61: 117–124.

Childers, D.L., R.F. Doren, R. Jones, G.B. Noe, M. Rugge, and L.J. Scinto. 2003. Decadal change in vegetation and soil phosphorus pattern across the everglades landscape. *Journal of Environmental Quality* 32: 344–362.

Christensen, N.L., A.M. Bartuska, J.H. Brown, S. Carpenter, C. D'Antonio, R. Francis, J.F. Franklin, J.A. MacMahon, R.F. Noss, D.J. Parsons, C.H. Peterson, M.G. Turner, and R.G. Woodmansee. 1996. The report of the Ecological Society of America committee on the scientific basis for ecosystem management. *Ecological Applications* 6: 665–691.

Collins, S.L., S.R. Carpenter, S.M. Swinton, D.E. Orenstein, D.L. Childers, T.L. Gragson, N.B. Grimm, M. Grove, S.L. Harlan, J.P. Kaye, A.K. Knapp, G.P. Kofinas, J.J. Magnuson, W.H. McDowell, J.M. Melack, L.A. Ogden, G.P. Robertson, M.D. Smith, and A.C. Whitmer. 2011. An integrated conceptual framework for long-term social-ecological research. *Frontiers in Ecology and the Environment* 9: 351–357.

Cordell, D., J.-O. Drangert, and S. White. 2009. The story of phosphorus: global food security and food for thought. *Global Environmental Change: Human and Policy Dimensions* 19: 292–305.

Craft, C.B., J. Vymazal, and C.J. Richardson. 1995. Response of Everglades plant-communities to nitrogen and phosphorus additions. *Wetlands* 15: 258–271.

Daoust, R.J., and D.L. Childers. 1999. Controls on emergent macrophyte composition, abundance, and productivity in freshwater everglades wetland communities. *Wetlands* 19: 262–275.

Daroub, S.H., S. Van Horn, T.A. Lang, and O.A. Diaz. 2011. Best management practices and long-term water quality trends in the Everglades Agricultural Area. *Critical Reviews in Environmental Science and Technology* 41: 608–632.

Davis, S.E., III, J.E. Cable, D.L. Childers, C. Coronado-Molina, J.W. Day, C.D. Hittle, C.J. Madden, E. Reyes, D. Rudnick, and F. Sklar. 2004. Importance of storm events in controlling ecosystem structure and function in a Florida Gulf Coast estuary. *Journal of Coastal Research* 20: 1198–1208.

Davis, S.E., III, D.L. Childers, and G.B. Noe. 2006. The contribution of leaching to the rapid release of nutrients and carbon in the early decay of wetland vegetation. *Hydrobiologia* 569: 87–97.

Davis, S.E., III, G.M. Naja, and A. Arik. 2014. Restoring the heart of the Everglades: the challenges and benefits. *National Wetlands Newsletter* 36: 5–9.

Davis, S.M. 1994. Phosphorus inputs and vegetation sensitivity in the Everglades. In *Everglades: The Ecosystem and Its Restoration*, edited by S. Davis and J.C. Ogden, 357–378. Boca Raton, Florida: St. Lucie Press.

DeBusk, W.F., S. Newman, and K.R. Reddy. 2001. Spatio-temporal patterns of soil phosphorus enrichment in Everglades Water Conservation Area 2A. *Journal of Environmental Quality* 30: 1438–1446.

Doren, R.F., T.V. Armentano, L.D. Whiteaker, and R.D. Jones. 1997. Marsh vegetation patterns and soil phosphorus gradients in the Everglades ecosystem. *Aquatic Botany* 56: 145–163.

Eklund, T.J., W.H. McDowell, and C.M. Pringle. 1997. Seasonal variation of tropical precipitation chemistry: La Selva, Costa Rica. *Atmospheric Environment* 31: 3903–3910.

Ensign, S.H., and M.W. Doyle. 2006. Nutrient spiraling in streams and river networks. *Journal of Geophysical Research-Biogeosciences* 111: G04009.

Entry, J.A., and A. Gottlieb. 2014. The impact of stormwater treatment areas and agricultural best management practices on water quality in the Everglades Protection Area. *Environmental Monitoring and Assessment* 186: 1023–1037.

Ewe, S.M.L., E.E. Gaiser, D.L. Childers, D. Iwaniec, V.H. Rivera-Monroy, and R.R. Twilley. 2006. Spatial and temporal patterns of aboveground net primary productivity (ANPP) along two freshwater-estuarine transects in the Florida Coastal Everglades. *Hydrobiologia* 569: 459–474.

Fairbanks, J.N. 1990. In saving the lake, we must save farms, too. *Palm Beach Post*, July 15.

Faridmarandi, S., and G.M. Naja. 2014. Phosphorus and water budgets in an agricultural basin. *Environmental Science & Technology* 48: 8481–8490.

Fourqurean, J.W., J.C. Zieman, and G.V.N. Powell. 1992. Phosphorus limitation of primary production in Florida Bay—Evidence from C-N-P ratios of the dominant seagrass *Thalassia testudinum*. *Limnology and Oceanography* 37: 162–171.

Frederick, P., D.E. Gawlik, J.C. Ogden, M.I. Cook, and M. Lusk. 2009. The White Ibis and Wood Stork as indicators for restoration of the Everglades ecosystem. *Ecological Indicators* 9: S83–S95.

Frederick, P.C., and G.V.N. Powell. 1994. Nutrient transport by wading birds in the Everglades. In *Everglades: The Ecosystem and Its Restoration*, edited by S.M. Davis and J.C. Ogden, 571–584. Delray Beach, FL: St. Lucie Press.

Gaiser, E.E., P.V. McCormick, S.E. Hagerthey, and A.D. Gottlieb. 2011. Landscape patterns of periphyton in the Florida Everglades. *Critical Reviews in Environmental Science and Technology* 41: 92–120.

Gaiser, E.E., L.J. Scinto, J.H. Richards, K. Jayachandran, D.L. Childers, J.C. Trexler, and R.D. Jones. 2004. Phosphorus in periphyton mats provides the best metric for detecting low-level P enrichment in an oligotrophic wetland. *Water Research* 38: 507–516.

Gaiser, E.E., J.C. Trexler, J.H. Richards, D.L. Childers, D. Lee, A.L. Edwards, L.J. Scinto, K. Jayachandran, G.B. Noe, and R.D. Jones. 2005. Cascading ecological effects of low-level phosphorus enrichment in the Florida Everglades. *Journal of Environmental Quality* 34: 717–723.

Grunwald, M. 2015. Could Obama's Everglades stop hurt the Everglades? A rare bipartisan environmental success turns political. *Politico Magazine*, April 22, 2015. https://www.politico.com/magazine/story/2015/04/obama-everglades-trip-117219

Gu, B., T.A. DeBusk, F.E. Dierberg, M.J. Chimney, K.C. Pietro, and T. Aziz. 2001. Phosphorus removal from Everglades agricultural area runoff by submerged aquatic vegetation/limerock treatment technology: an overview of research. *Water Science and Technology* 44: 101–108.

Hagerthey, S.E., B.J. Bellinger, K. Wheeler, M. Gantar, and E. Gaiser. 2011. Everglades periphyton: a biogeochemical perspective. *Critical Reviews in Environmental Science and Technology* 41: 309–343.

Hagerthey, S.E., M.I. Cook, R. Mac Kobza, S. Newman, and B.J. Bellinger. 2014. Aquatic faunal responses to an induced regime shift in the phosphorus-impacted Everglades. *Freshwater Biology* 59: 1389–1405.

Hagerthey, S.E., S. Newman, and S. Xue. 2012. Periphyton-based transfer functions to assess ecological imbalance and management of a subtropical ombrotrophic peatland. *Freshwater Biology* 57: 1947–1965.

Harvey, J.W., G.B. Noe, L.G. Larsen, D.J. Nowacki, and L.E. McPhillips. 2011. Field flume reveals aquatic vegetation's role in sediment and particulate phosphorus transport in a shallow aquatic ecosystem. *Geomorphology* 126: 297–313.

Harms, T.K., and N.B. Grimm. 2008. Hot spots and hot moments of carbon and nitrogen dynamics in a semiarid riparian zone. *JGR Biogeosciences* 113(G1): 148–227.

Herbert, D.A., and J.W. Fourqurean. 2009. Phosphorus availability and salinity control productivity and demography of the seagrass *Thalassia testudinum* in Florida Bay. *Estuaries and Coasts* 32: 188–201.

Hollander, G.M. 2008. *Raising Cane in the 'Glades: The Global Sugar Trade and the Transformation of Florida*. Chicago: University of Chicago Press.

Izuno, F.T., and A.B. Bottcher. 1994. Introduction. In *Everglades Agricultural Area (EAA): Water, Soil, Crop, and Environmental Management*, edited by A.B. Bottcher and F.T. Izuno, 1–12. Gainesville: University Press of Florida.

Izuno, F.T., R.W. Rice, and L.T. Capone. 1999. Best management practices to enable the coexistence of agriculture and the Everglades environment. *Hortscience* 34: 27–33.

Janardhanan, L., and S.H. Daroub. 2010. Phosphorus sorption in organic soils in south Florida. *Soil Science Society of America Journal* 74: 1597–1606.

John, D. 1994. *Civic Environmentalism: Alternatives to Regulation in States and Communities*. Washington, DC: CQ Press.

Kleindienst, L. 1997. Everglades group: make polluters pay for cleanup. *Sun-Sentinel*, May 6, 1997.

Koch, G.R., D.L. Childers, P.A. Staehr, R.M. Price, S.E. Davis, and E.E. Gaiser. 2012. Hydrological conditions control P loading and aquatic metabolism in an oligotrophic, subtropical estuary. *Estuaries and Coasts* 35: 292–307.

Koch, G.R., S. Hagerthey, D.L. Childers, and E. Gaiser. 2014. Examining seasonally pulsed detrital transport in the coastal Everglades using a sediment tracing technique. *Wetlands* 34: S123–S133.

Koch, M.S., C. Coronado, M.W. Miller, D.T. Rudnick, E. Stabenau, R.B. Halley, and F.H. Sklar. 2015. Climate change projected effects on coastal foundation communities of the Greater Everglades using a 2060 scenario: need for a new management paradigm. *Environmental Management* 55: 857–875.

Kourafalou, V.H., and H. Kang. 2012. Florida current meandering and evolution of cyclonic eddies along the Florida Keys Reef Tract: are they interconnected? *Journal of Geophysical Research* 117(C05028): doi:10.1029/2011JC007383.

Larsen, L.G., J.W. Harvey, G.B. Noe, and J.P. Crimaldi. 2009. Predicting organic floc transport dynamics in shallow aquatic ecosystems: insights from the field, the laboratory, and numerical modeling. *Water Resources Research* 45: W01411.

Liu, K.J., H.P. Li, and S.E. Davis, III. 2014. Benthic exchange of C, N, and P along the estuarine ecotone of Lower Taylor Slough, Florida (USA): effect of seasonal flows and phosphorus availability. *Wetlands* 34: S113–S122.

MacVicar, T. 2005. Oral history interview with Brian Gridley, May 20, 2001, in West Palm Beach, Florida, transcript, Samuel Proctor Oral History Program Collection, P.K. Yonge Library of Florida History, University of Florida. http://ufdc.ufl.edu/UF00005375/00001

McCally, D. 1999. *The Everglades: An Environmental History*. Gainesville: The University Press of Florida.

McClain, M.E., E.W. Boyer, C.L. Dent, S.E. Gergel, N.B. Grimm, P.M. Groffman, S.C. Hart, J.W. Harvey, C.A. Johnston, E. Mayorga, W.H. McDowell, and G. Pinay. 2003. Biogeochemical hot spots and hot moments at the interface of terrestrial and aquatic ecosystems. *Ecosystems* 6: 301–312.

McCormick, P.V., and M.V. O'Dell. 1996. Quantifying periphyton responses to phosphorus in the Florida Everglades: a synoptic-experimental approach. *Journal of the North American Benthological Society* 15(4): 450–468.

McKinley, J.C.J. 1999. Sugar companies play a pivotal role in effort to restore Everglades. *New York Times*, April 16, 1999.

Mock-Roos & Associates. 2003. Lake Istokpoga and Upper Chain of Lakes phosphorus source control: Task 4 final report. South Florida Water Management District Contract No. C-13413, West Palm Beach, Florida.

Neto, R.R., R.N. Mead, J.W. Louda, and R. Jaffe. 2006. Organic biogeochemistry of detrital flocculent material (floc) in a subtropical, coastal wetland. *Biogeochemistry* 77: 283–304.

Newbold, J.D. 1992. Cycles and spirals of nutrients. In *Rivers Handbook*, edited by P. Calow and G. Petts, 379–408. Oxford, UK: Blackwell Scientific Publications.

Nixon, S.W. 1995. Coastal marine eutrophication: a definition, social causes, and future concerns. *Ophelia* 41(1): 199–219.

Noe, G.B., and D.L. Childers. 2007. Phosphorus budgets in Everglades wetland ecosystems: the effects of hydrology and nutrient enrichment. *Wetlands Ecology and Management* 15: 189–205.

Noe, G.B., D.L. Childers, A.L. Edwards, E. Gaiser, K. Jayachandran, D. Lee, J. Meeder, J. Richards, L.J. Scinto, J.C. Trexler, and R.D. Jones. 2002. Short-term changes in phosphorus storage in an oligotrophic Everglades wetland ecosystem receiving experimental nutrient enrichment. *Biogeochemistry* 59: 239–267.

Noe, G.B., D.L. Childers, and R.D. Jones. 2001. Phosphorus biogeochemistry and the impact of phosphorus enrichment: why is the Everglades so unique? *Ecosystems* 4: 603–624.

Noe, G.B., K.W. Krauss, B.G. Lockaby, W.H. Conner, and C.R. Hupp. 2013. The effect of increasing salinity and forest mortality on soil nitrogen and phosphorus mineralization in tidal freshwater forested wetlands. *Biogeochemistry* 114: 225–244.

Noe, G.B., L.J. Scinto, J. Taylor, D.L. Childers, and R.D. Jones. 2003. Phosphorus cycling and partitioning in an oligotrophic Everglades wetland ecosystem: a radioisotope tracing study. *Freshwater Biology* 48: 1993–2008.

Nuttle, W.K., J.W. Fourqurean, B.J. Cosby, J.C. Zieman, and M.B. Robblee. 2000. Influence of net freshwater supply on salinity in Florida Bay. *Water Resources Research* 36(7): 1805–1822.

Ogden, L. 2008. The Everglades ecosystem and the politics of nature. *American Anthropologist* 110: 21–32.

Orth, R.J., T.J.B. Carruthers, W.C. Dennison, C.M. Duarte, J.W. Fourqurean, K.L. Heck, A.R. Hughes, G.A. Kendrick, W.J. Kenworthy, S. Olyarnik, F.T. Short, M. Waycott, and S.L. Williams. 2006. A global crisis for seagrass ecosystems. *BioScience* 56(12): 987–996.

Parker, J.N., and E.J. Hackett. 2012. Hot spots and hot moments in scientific collaborations and social movements. *American Sociological Review* 77: 21–44.

Pollman, C.D., W.M. Landing, J.J. Perry, and T. Fitzpatrick. 2002. Wet deposition of phosphorus in Florida. *Atmospheric Environment* 36: 2309–2318.

Powell, D. 1996. Letter to the editor. *Sarasota Herald-Tribune*, November 3, 1996.

Price, R.M., P.K. Swart, and J.W. Fourqurean. 2006. Coastal groundwater discharge—an additional source of phosphorus for the oligotrophic wetlands of the Everglades. *Hydrobiologia* 569: 23–36.

Reddy, K.R., S. Newman, T.Z. Osborne, J.R. White, and H.C. Fitz. 2011. Phosphorus cycling in the Greater Everglades Ecosystem: legacy phosphorus implications for management and restoration. *Critical Reviews in Environmental Science and Technology* 41: 149–186.

Rejmankova, E., K.O. Pope, R. Post, and E. Maltby. 1996. Herbaceous wetlands of the Yucatan Peninsula: communities at extreme ends of environmental gradients. *Internationale Revue Der Gesamten Hydrobiologie* 81: 223–252.

Richardson, C.J. 1989. Ecological Analysis of the Water Conservation Areas: A Preliminary Analysis. Workshop, West Palm Beach, Florida.

Rivera-Monroy, V.H., R.R. Twilley, S.E. Davis, D.L. Childers, M. Simard, R. Chambers, R. Jaffe, J.N. Boyer, D.T. Rudnick, K. Zhang, E. Castaneda-Moya, S.M.L. Ewe, R.M. Price, C. Coronado-Molina, M. Ross, T.J. Smith, B. Michot, E. Meselhe, W. Nuttle, T.G. Troxler, and G.B. Noe. 2011. The role of the Everglades Mangrove Ecotone Region (EMER) in regulating nutrient cycling and wetland productivity in south Florida. *Critical Reviews in Environmental Science and Technology* 41: 633–669.

Rizzardi, K. 2001. Translating science into law: phosphorus standards in the Everglades. *Journal of Land Use and Environmental Litigation* 17: 149–168.

Roberts, D.H. 1999. A Framework for Analyzing Technical Trade Barriers in Agricultural Markets. USDA technical bulletin no. 1876, National Agricultural Library.

Ross, M.S., S. Mitchell-Bruker, J.P. Sah, S. Stothoff, P.L. Ruiz, D.L. Reed, K. Jayachandran, and C.L. Coultas. 2006. Interaction of hydrology and nutrient limitation in the ridge and slough landscape of the southern Everglades. *Hydrobiologia* 569: 37–59.

Ross, M.S., J.J. O'Brien, R.G. Ford, K.Q. Zhang, and A. Morkill. 2009. Disturbance and the rising tide: the challenge of biodiversity management on low-island ecosystems. *Frontiers in Ecology and the Environment* 7: 471–478.

Rudnick, D.T., Z. Chen, D.L. Childers, and T.D. Fontaine. 1999. Phosphorus and nitrogen inputs to Florida Bay: the importance of the Everglades watershed. *Estuaries* 22: 398–416.

Ruehl, C.B., and J.C. Trexler. 2013. A suite of prey traits determine predator and nutrient enrichment effects in a tri-trophic food chain. *Ecosphere* 4: 75.

Saha, A.K., C.S. Moses, R.M. Price, V. Engel, T.J. Smith, III, and G. Anderson. 2012. A hydrological budget (2002–2008) for a large subtropical wetland ecosystem indicates marine groundwater discharge accompanies diminished freshwater flow. *Estuaries and Coasts* 35: 459–474.

Saha, A.K., S. Saha, J. Sadle, J. Jiang, M.S. Ross, R.M. Price, L.S.L.O. Sternberg, and K.S. Wendelberger. 2011. Sea level rise and South Florida coastal forests. *Climatic Change* 107: 81–108.

Sargeant, B.L., E.E. Gaiser, and J.C. Trexler. 2010. Biotic and abiotic determinants of intermediate-consumer trophic diversity in the Florida everglades. *Marine and Freshwater Research* 61: 11–22.

Schorsch, P. 2015. With new TV ad, Everglades supporters break the peace with the sugar industry. *Florida Politics*, February 23, 2015. http://floridapolitics.com/archives/8360-with-new-tv-ad-everglades-supporters-break-the-peace-with-the-sugar-industry-2

Schwartz, K.Z.S. 2014. The anti-politics of biopolitical disaster on Florida's coasts. Paper presented at the Western Political Science Association conference, Seattle, Washington.

Scott, J.T., R.D. Doyle, and C.T. Filstrup. 2005. Periphyton nutrient limitation and nitrogen fixation potential along a wetland nutrient-depletion gradient. *Wetlands* 25: 439–448.

Sharma, K., P.W. Inglett, K.R. Reddy, and A.V. Ogram. 2005. Microscopic examination of photoautotrophic and phosphatase-producing organisms in phosphorus-limited Everglades periphyton mats. *Limnology and Oceanography* 50: 2057–2062.

Sklar, F.H., M.J. Chimney, S. Newman, P. McCormick, D. Gawlik, S.L. Miao, C. McVoy, W. Said, J. Newman, C. Coronado, G. Crozier, M. Korvela, and K. Rutchey. 2005. The ecological-societal underpinnings of Everglades restoration. *Frontiers in Ecology and the Environment* 3: 161–169.

Smith, E.P., and P.V. McCormick. 2001. Long-term relationship between phosphorus inputs and wetland phosphorus concentrations in a northern Everglades marsh. *Environmental Monitoring and Assessment* 68: 153–176.

Sokol, E.R., J.M. Hoch, E. Gaiser, and J.C. Trexler. 2014. Metacommunity structure along resource and disturbance gradients in Everglades wetlands. *Wetlands* 34: S135–S146.

Sutula, M.A., B.C. Perez, E. Reyes, D.L. Childers, S. Davis, J.W. Day, D. Rudnick, and F. Sklar. 2003. Factors affecting spatial and temporal variability in material exchange between the southern Everglades wetlands and Florida Bay (USA). *Estuarine Coastal and Shelf Science* 57: 757–781.

Sterner, R.W., T. Andersen, J.J. Elser, D.O. Hessen, J.M. Hood, E. McCauley, and J. Urabe. 2008. Scale-dependent carbon:nitrogen:phosphorus seston stoichiometry in marine and freshwaters. *Limnology and Oceanography* 53(3): 1169–1180.

Troxler-Gann, T., D.L. Childers, and D.N. Rondeau. 2005. Ecosystem structure, nutrient dynamics, and hydrologic relationships in tree islands of the southern Everglades, Florida, USA. *Forest Ecology and Management* 214: 11–27.

Troxler-Gann, T., C. Coronado-Molina, D.N. Rondeau, S. Krupa, S. Newman, M. Manna, R.M. Price, and F.H. Sklar. 2014. Interactions of local climatic, biotic and hydrogeochemical processes facilitate phosphorus dynamics along an Everglades forest-marsh gradient. *Biogeosciences* 11: 899–914.

Vaithiyanathan, P., and C.J. Richardson. 1997. Nutrient profiles in the Everglades: examination along the eutrophication gradient. *Science of the Total Environment* 205: 81–95.

Vidon, P., C. Allan, D. Burns, T.P. Duval, N. Gurwick, S Inamdar, R. Lowrance, J. Okay, D. Scott, and S. Sebestyen. 2010. Hot spots for hot moments in riparian zones: potential for improved water quality management. *Journal of the American Resources Association* 46(2): 278–298.

Walker, W.W., and K.E. Havens. 1995. Relating algal bloom frequencies to phosphorus concentrations in Lake Okeechobee. *Lake and Reservoir Management* 11: 77–83.

Walker, W.W., Jr., and R.H. Kadlec. 2011. Modeling phosphorus dynamics in Everglades wetlands and stormwater treatment areas. *Critical Reviews in Environmental Science and Technology* 41: 430–446.

Wanless, H.R. and M.G. Tagett. 1989. Origin, growth and evolution of carbonate mudbanks in Florida Bay. *Bulletin of Marine Science* 44(1): 454–489.

Waters, M.N., J.M. Smoak, and C.J. Saunders. 2013. Historic primary producer communities linked to water quality and hydrologic changes in the northern Everglades. *Journal of Paleolimnology* 49: 67–81.

Wetzel, P.R., F.H. Sklar, C.A. Coronado, T.G. Troxler, SL. Krupa, P.L. Sullivan, S. Ewe, R.M. Price, S. Newman, and W.H. Orem. 2011. Biogeochemical processes on tree islands in the greater Everglades initiating a new paradigm. *Critical Reviews in Environmental Science and Technology* 41(sup1): 670–701.

Wetzel, P.R., A.G. van der Valk, S. Newman, C.A. Coronado, T.G. Troxler-Gann, D.L. Childers, W.H. Orem, and F.H. Sklar. 2009. Heterogeneity of phosphorus distribution in a patterned landscape, the Florida Everglades. *Plant Ecology* 200: 83–90.

Wozniak, J.R., W.T. Anderson, D.L. Childers, E.E. Gaiser, C.J. Madden, and D.T. Rudnick. 2012. Potential N processing by southern Everglades freshwater marshes: are Everglades marshes passive conduits for nitrogen? *Estuarine Coastal and Shelf Science* 96: 60–68.

Wozniak, J.R., D.L. Childers, W.T. Anderson, D.T. Rudnick, and C.J. Madden. 2008. An in situ mesocosm method for quantifying nitrogen cycling rates in oligotrophic wetlands using (15)N tracer techniques. *Wetlands* 28: 502–512.

Zhang, J.Z., C.J. Fischer, and P.B. Ortner. 2004. Potential availability of sedimentary phosphorus to sediment resuspension in Florida Bay. *Global Biogeochemical Cycles* 18: GB4008.

6

Carbon Cycles in the Florida Coastal Everglades Social-Ecological System Across Scales

Tiffany Troxler, Greg Starr, Joseph N. Boyer, Jose D. Fuentes, and Rudolf Jaffé with Sparkle Malone, Jordan Barr, Stephen E. Davis III, Ligia Collado-Vides, Josh Breithaupt, Amartya Saha, Randolph Chambers, Christopher Madden, Joseph Donny Smoak, James Fourqurean, Gregory R. Koch, John Kominoski, Len Scinto, Steve Oberbauer, Victor Rivera-Monroy, Edward Castañeda-Moya, Nick Schulte, Sean Charles, Jennifer Richards, Dave Rudnick, and Kevin Whelan

In a Nutshell

- Extreme weather events such as tropical storms and cold fronts temporarily perturb the carbon sequestration capacity of Everglades coastal wetland ecosystems, as both marshes and mangroves switch from being carbon sinks to being carbon sources.
- The mangrove forests of the coastal Everglades are an effective sink for atmospheric carbon dioxide, with an unusually high annual net ecosystem productivity of about 1,200 g carbon m^{-2} that is a result of the year-round growing season and uncommonly low ecosystem respiration rates.
- Salinity, tidal flushing, and inundation modulate the magnitude and seasonal variability of this carbon sink.
- Everglades mangrove carbon stocks are estimated to approximate 45 million metric tons.

- Florida Bay is an interesting case study in microbial carbon dynamics, with sources, sinks, age, fate, transport, air–water exchange of carbon dioxide, and phytoplankton–bacteria productivity coupling.

Introduction

Wetlands are among the most productive ecosystems in the world. Although wetlands occupy just 2% to 4% of the Earth's surface, they account for about 6.3% of terrestrial net primary production (Schlesinger 1997). As wetlands are one of the largest components of the terrestrial carbon (C) pool (Whiting and Chanton 1993), an understanding of wetland C dynamics is important because of their influence on the global climate system via greenhouse gas fluxes. Much of the net productivity in wetlands is retained as peat, which accumulates when litter decomposition rates under anoxic soil conditions are slower than deposition rates (Gorham 1991). However, wetlands represent a double-edged sword for greenhouse forcing—they sequester significant amounts of atmospheric carbon dioxide (CO_2) but also release the more potent greenhouse gases, including methane (CH_4) and nitrous oxide (N_2O). Further complicating the situation, the C stored in wetland soils is highly vulnerable to rapid release when wetlands are degraded through direct human activities and the impacts of climate change.

The future of many wetland ecosystems is uncertain due to a combination of climate change impacts (sea level rise [SLR], changes in storm activity, altered freshwater availability, soil warming) and human activities (population growth, changes in resource and land use, wetland drainage and filling). The consequences of coastal wetland loss are not limited to the regional services they provide (e.g., storm mitigation, aquifer recharge, fisheries) but also extend globally through impacts on biodiversity, biogeochemical cycling, and atmospheric interactions. For instance, recent studies have shown that vegetated coastal systems store up to 50 times more C than tropical forests (1,000 Mg organic C ha^{-1}) as a result of high productivity and low respiration rates (Bouillon 2011; Fourqurean et al. 2012; McLeod et al. 2011). Studies that examine controls on long-term spatial and temporal patterns of coastal wetland C budgets should reduce uncertainty about their persistence, influence on global climate, and the future of ecosystem services and benefits provided to coastal communities and to society at large.

More than a century of water management and hydrologic modification in the Everglades has dramatically altered wetlands throughout the system (Davis et al. 2005; Light and Dineen 1994; Myers and Ewel 1990). The construction of roads, canals, levees, and flow control structures has modified the quantity, quality, timing, and location of water delivery to the Everglades (see Chapters 1 and 3). These changes have altered hydroperiods, salinity and nutrient levels, community assemblages, fire regimes, and ultimately C cycles and storage (Gaiser et al. 2006; McCormick and Laing 2003; Snyder and Davidson 1994; see Chapter 7). Particularly striking changes include the loss of enormous quantities of C from peat soils in the northern reaches of the Everglades, in response to drainage for agriculture (Aich et al. 2013; Gleason and Stone 1994), and the inland expansion, or transgression, of mangroves in the coastal regions of Everglades National Park (ENP) in response to reduced freshwater flows and SLR (Davis et al. 2005). Increasing attention is being paid to the critical ecosystem services provided by the coastal wetlands of the Everglades. In south Florida, coastal wetlands support vulnerable, rare, and endemic species; reduce risks of storm impacts; protect vulnerable groundwater supplies; and contribute to significant C storage (FOCC 2010; Hopkinson et al. 2012; Pearlstine et al. 2010). Our Florida Coastal Everglades (FCE) research has highlighted these critically important resources, how human modifications have degraded them, and how they may be sustained with changes in how they are managed.

In this chapter, we synthesize our C-related research in which we have sought to determine (1) how, in the coastal Everglades landscape, population- and ecosystem-level dynamics are controlled by the relative importance of water source, water residence time, and local biotic processes and (2) how climate change and resource management decisions interact to influence freshwater availability, ecosystem dynamics, and the value and utilization of ecosystem services associated with C dynamics. Our synthesis incorporates C dynamics from large-scale, landscape-level processes, such as synoptic weather patterns and their influences on C fluxes, to "life in the small," which includes microbial and small material contributions.

Our framework for understanding Everglades C cycling begins with a conceptualization of interacting ecosystems positioned across the coastal ecotone (Fig. 6.1). The Everglades may be thought of as a series of linked habitats, including short- and

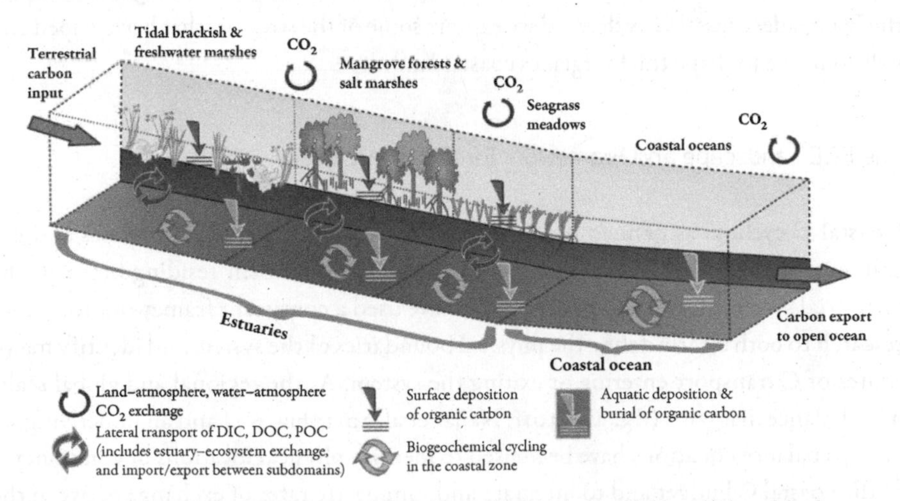

FIGURE 6.1 Conceptual model of coastal C cycling, which includes tidal brackish and freshwater marshes, mangrove forests, saltwater marshes, seagrass meadows, and comprises estuaries and the coastal ocean (created by Jordan Barr). See page 10 of color insert.

long-hydroperiod marshes, mangroves, and subtidal areas where seagrasses are the dominant producer. Freshwater marshes are characterized as short hydroperiod (flooded for less than 6 months of the year) or long hydroperiod (flooded for more than 6 months of the year) and are dominated by sawgrass, spikerush, and expansive mats of periphyton (an assemblage of microorganisms including algae, bacteria, and fungi; Gaiser et al. 2012). Everglades periphyton assemblages form mats that are dominated by calcium carbonate (30–50% of dry mass); the remaining C is primarily organic and detrital (Donar et al. 2004), with autotrophic and heterotrophic microbes making up the small remainder of biomass. These freshwater marshes transition into mangrove forests through a coastal ecotone, which is characterized by marked variability in salinity and nutrient availability depending on freshwater flows and marine exposure (Rivera-Monroy et al. 2011; Troxler et al. 2013). Mangrove forests vary from short stature to tall stature and, in Shark River Slough, they are influenced by semi-diurnal tides (Chen and Twilley 1999; Ewe et al. 2006). Shallow seagrass ecosystems, with associated macro- and micro-algal assemblages, dominate Florida Bay, where the distribution of seagrass species and macroalgal assemblages is a product of salinity and nutrient gradients (Zieman et al. 1989).

In this chapter, we will take a journey through the "biogeochemical reactor" we know as our Everglades coastal ecotone (see Fig. 6.1; Barr et al. 2014), beginning with C cycling research in central Everglades freshwater marshes and progressing downstream through the mangrove estuaries and Florida Bay ecosystems. C enters the biogeochemical reactor in a variety of forms, including dissolved and particulate forms of organic and inorganic C. C that is not stored via burial in soils and sediments may leave through CO_2 outgassing or export to the coastal ocean, but C inflows are also possible across these interfaces. We provide context for the flows by considering some of the aquatic fluxes and C transformations that have guided our understanding of the interlinkages among ecosystems of the Everglades coastal C cycle. And we explore some of the stressors that have shaped and will continue to shape the Everglades coastal landscape.

The FCE Landscape and Framework for C Research

Coastal C cycling, as defined here, is the set of all biogeochemical processes and lateral aquatic fluxes of C that occur within the coastal domain residing between the terrestrial system and the open ocean. We have used a consistent framework for our C research to both clearly define the physical boundaries of the system and identify major routes of C transport entering or exiting the system. At the regional and global scale, mass balance diagrams (e.g., Cai 2011; Najjar et al. 2012; Fig. 6.2) and an underlying set of mass balance equations have been used to identify physical locations or components of the coastal C budget and to integrate and summarize rates of exchange between the components. These mass balance equations are subject to mass balance closure such that inputs − outputs + generation = accumulation. These budgets can be estimated

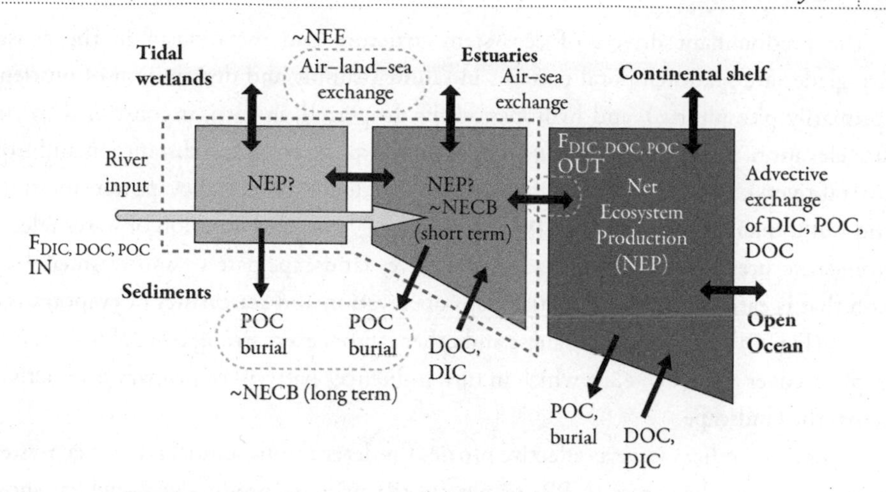

FIGURE 6.2 Coupling coastal carbon cycle science at the regional scale in order to integrate coastal and shelf carbon science with carbon processes occurring in tidal wetlands and estuaries.

separately for inorganic, organic, and total C, as well as for individual subdomains (e.g., estuarine open water). Our interdisciplinary collaborations (Najjar et al. 2010, 2012) have endeavored to constrain and quantify regional coastal C budgets that include riverine inputs, estuarine fluxes, tidal wetland fluxes, air–sea CO_2 exchange, benthic fluxes, advective exchange between the continental shelf and the open ocean, primary productivity, and respiration (see Fig. 6.2). Linking these approaches in space and time provides a fundamental framework whereby our research, and that of other coastal Long Term Ecological Research (LTER) programs, is used to constrain spatial and temporal variability in national- and global-scale C research and simulation models.

The quality and integrity of landscape C budgets rest on mass balance approaches developed at much finer scales, such as individual rivers and habitats. At these scales, the net ecosystem C balance (NECB) represents a key concept for understanding C cycling in coastal ecosystems, where NECB is defined as the rate of C accumulation in an ecosystem (Chapin et al. 2006). The NECB for a system is quantified by summing the change in all of the organic C pools in that system from one year to the next (Lovett et al. 2006). In the coastal environment, the NECB equals the net CO_2 fixed from the atmosphere (i.e., net ecosystem exchange, or –NEE) plus the net lateral exchanges of all forms of carbon (F_{tot}), such that $NECB = -NEE + F_{tot}$ (Chapin et al. 2006). The F_{tot} includes exchange of dissolved inorganic carbon (DIC), dissolved organic carbon (DOC), particulate inorganic carbon (PIC), and particulate organic carbon (POC) and has a positive sign convention for C entering the ecosystem. This approach has provided an important entry point for developing C budgets in our marshes and mangrove forests (Troxler et al. 2013). Dynamic spatial and temporal integration and forecasting are key areas of further development.

The predominant drivers of ecosystem structure and metabolism in the coastal Everglades are spatiotemporal patterns in salinity, supply and distribution of nutrients (primarily phosphorus), and hydroperiod. Hydroperiod, in turn, is controlled by surface elevation, physiography (including proximity to the coast and distance inland from coastal rivers), climate variability (including rainfall and potential evapotranspiration), and water management practices (including timing, rate, and duration of water releases from structures). Variation in these factors drives landscape pattern and seasonal variation that is easily observed in spatial maps of elevation and seasonality of evapotranspiration (Fig. 6.3). Consequently, these and other factors exert strong control on variation in plant cover type (Fig. 6.4), which in turn influences patterns of primary productivity across the landscape.

Vegetation indices serve as effective proxies for determining annual rates of ecosystem productivity—either gross (GPP) or net (NPP) primary productivity—which shows significant spatial heterogeneity across the Everglades (Fig. 6.5). Mangrove forests are some of the most productive areas in the Everglades. These forested regions are interspersed with lower-productivity dwarf mangroves and freshwater and brackish marshes of moderate productivity (Fig. 6.6). Notably, our long-term research sites may not necessarily capture the range of variability to sufficiently or appropriately scale C and energy exchanges, especially across the spatially heterogeneous coastal ecotone. In this context, remotely sensed data have been an important tool to understand this spatial heterogeneity, as well as a means to provide some of the first estimates of Everglades C cycling and C budgets. For example, relationships between vegetation class and satellite-derived estimates of GPP and NPP have provided the first spatially articulate estimates of gross C production and net C uptake across the coastal Everglades (see Fig. 6.6). This is an important area of developing research. We will now individually explore some of the drivers and patterns of spatial and temporal variability for freshwater marsh, mangrove, and Florida Bay ecosystems.

Freshwater Marsh Dynamics

We measure aboveground biomass and estimate NPP using (1) allometric models that relate change in aboveground biomass to nondestructive measurements of the dominant, "charismatic" macrophyte, sawgrass (*Cladium jamaicense*; Childers et al. 2006) and (2) gas exchange methods that measure gross ecosystem exchange (GEE) and ecosystem respiration (ER), allowing us to estimate NEE (Malone et al. 2013). We have shown that the largest sources of variation in long-term records of sawgrass aboveground net primary production (ANPP) are hydrology and salinity, with P also known to be an important driver (Childers et al. 2006). Long-term trends in sawgrass productivity at FCE sites that include freshwater peat marshes, brackish peat marshes, and marl marshes have confirmed that salinity is important driver of sawgrass productivity in the coastal ecotone.

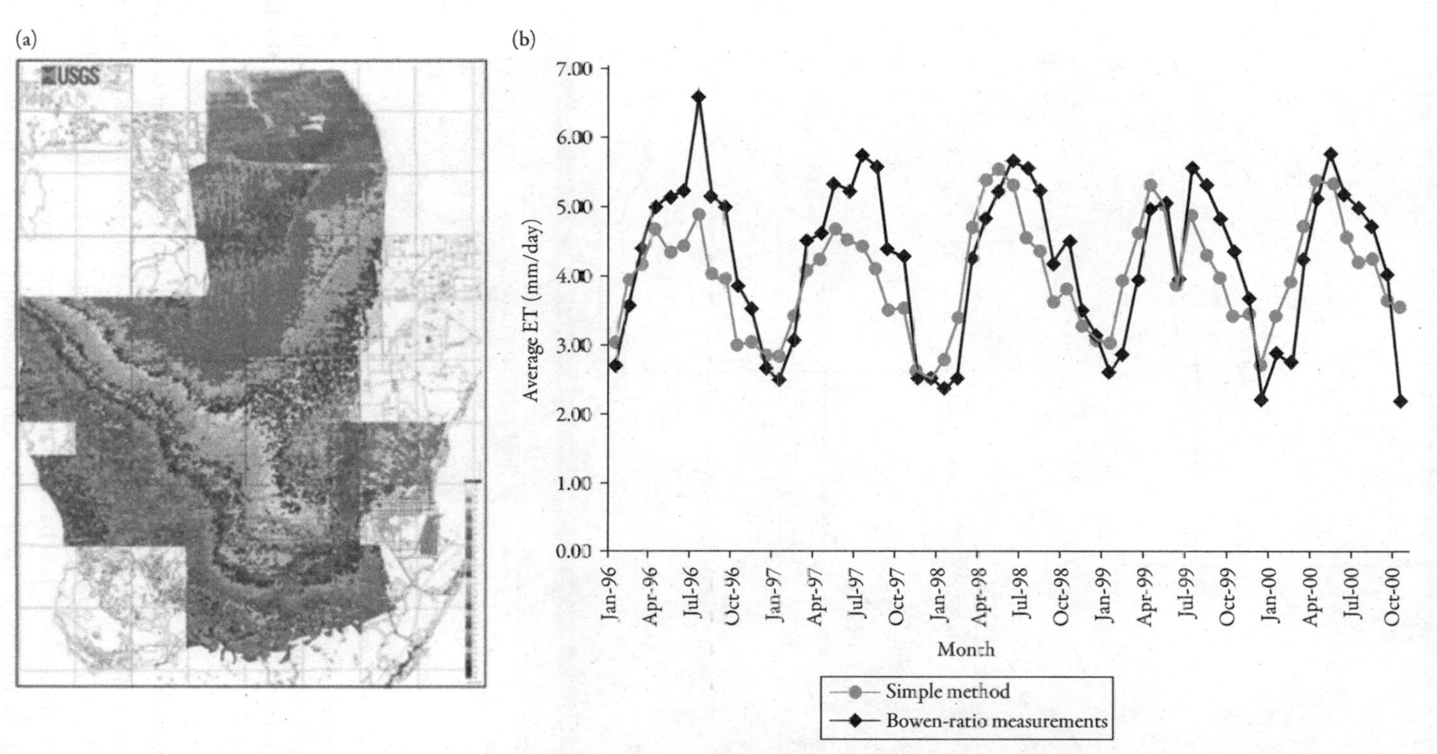

FIGURE 6.3 Spatial and temporal drivers of variation in carbon cycling and budgets, through influences on primary production, respiration, and transport. A) hypsographic map generated from the GPS-derived elevation data points detailing the topography of the Everglades (Desmond 2003). B) average monthly evapotranspiration (ET) over a 2-year period calculated using the Simple method and Bowen-ratio measurements.

FIGURE 6.4 Vegetation classes within Everglades National Park, as developed by the Center for Remote Sensing and Mapping Science, Department of Geography at The University of Georgia, and the South Florida Natural Resources Center at Everglades National Park (Welch, R. and M. Madden. 1999). See page 9 of color insert.

Expanding on work by Childers et al. (2006), Troxler et al. (2013) found that water level and length of inundation controlled aboveground sawgrass biomass as well as the density of spikerush (*Eleocharis cellulosa*), another macrophyte responsive to shifts in hydrologic conditions in freshwater marshes. Based on these long-term trajectories in both *Cladium* and *Eleocharis*, we anticipate that shifts in plant community structure may occur within 5 to 10 years of sustained changes in water management because of the important role

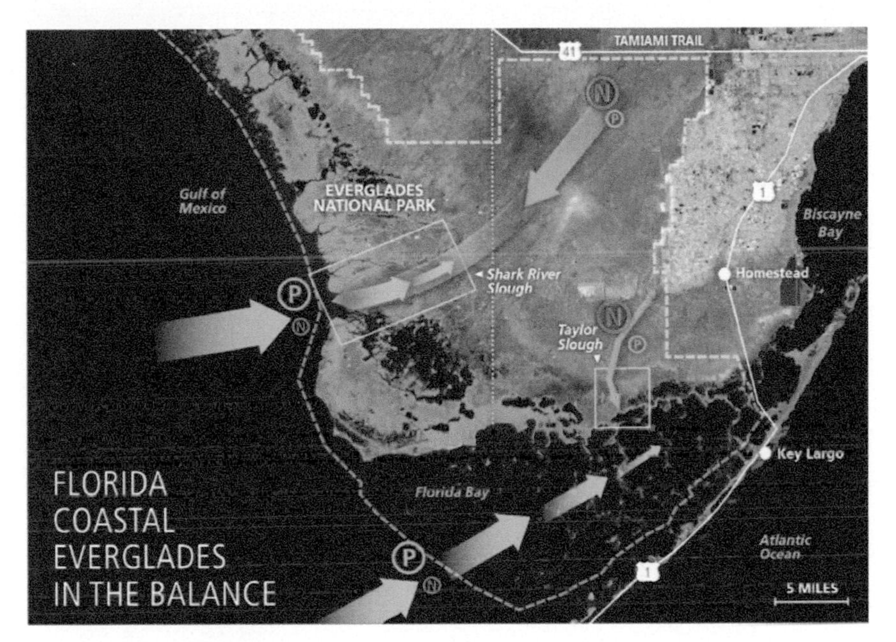

The upside-down estuary with productivity gradients shaped by the supply of phosphorus (P) from the coast, rather than from upstream marshes.

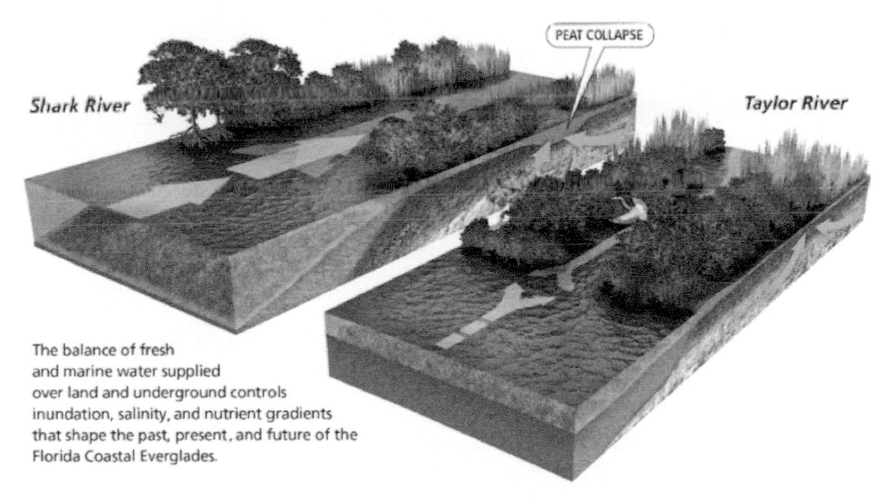

PEAT COLLAPSE
Shark River
Taylor River
The balance of fresh
and marine water supplied
over land and underground controls
inundation, salinity, and nutrient gradients
that shape the past, present, and future of the
Florida Coastal Everglades.

The balance of fresh and marine water supply, therefore, shapes the past, present, and future function of the Everglades ecosystem through its effect on P supply, inundation, and salinity. Infographic: Hiram Henriquez.

Sawgrass marsh builds peat soil on top of the limestone only in freshwater areas. Mangroves develop peat soil in saline and brackish conditions.

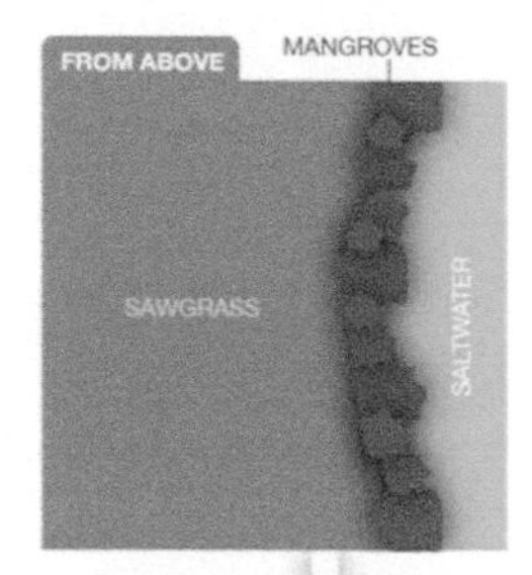

2 Saltwater Intrusion

Intrusion of saltwater causes sawgrass dieback and mangrove expansion. Freshwater peat soil begins to degrade with exposure to saltwater.

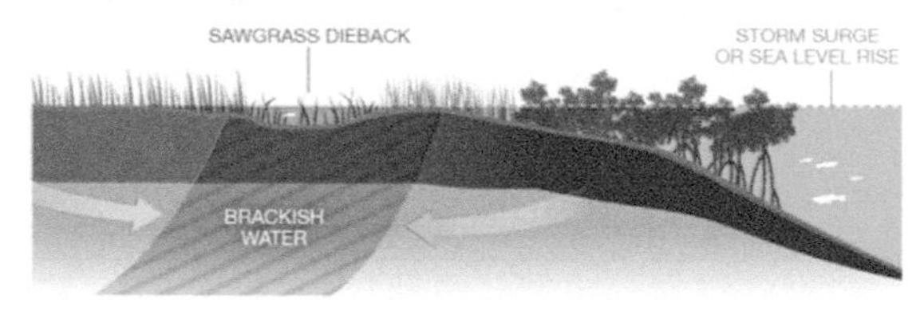

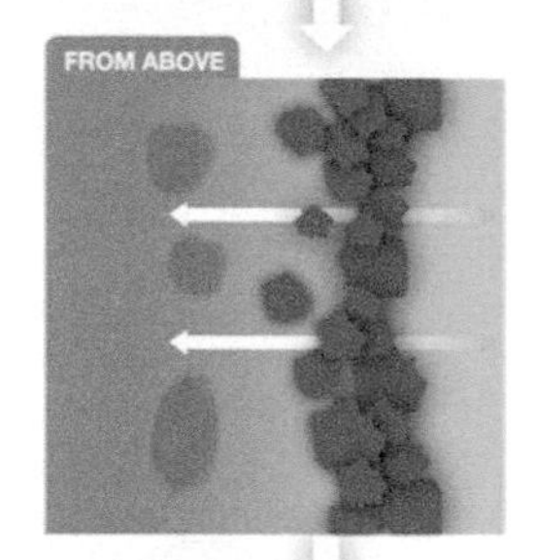

3 Peat Collapse

Freshwater peat collapses and the water is too deep for plants to become established. Mangroves established elsewhere help to re-stabilize soil.

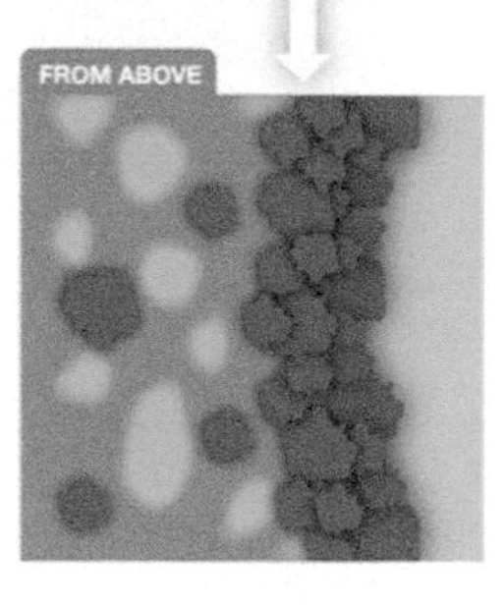

Conceptual schematic of the process of peat collapse in response to saltwater intrusion and sea level rise, which is another aspect of the marine influence.

Conceptual diagram of changes in ecological connectivity in space and time throughout the Greater Everglades landscape, from historical to current to future.

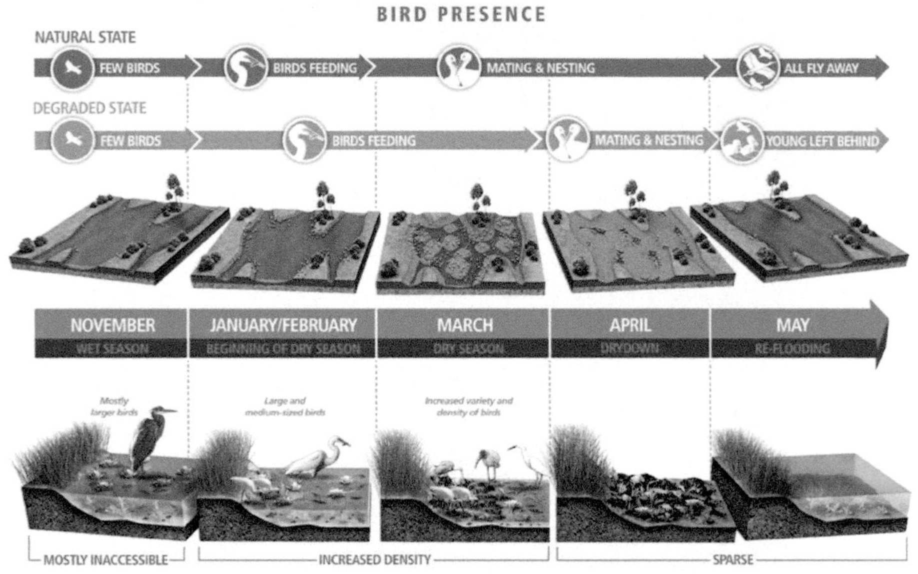

Schematic of how altered freshwater hydrology affects the density and distribution of aquatic biota that resident and migratory birds depend on as overwintering food resources.

The state change that results from the addition of P to Everglades wetlands, from an Everglades ridge and slough marsh (a) with extensive calcareous periphyton mats (b) to a dense cattail marsh (c & d).

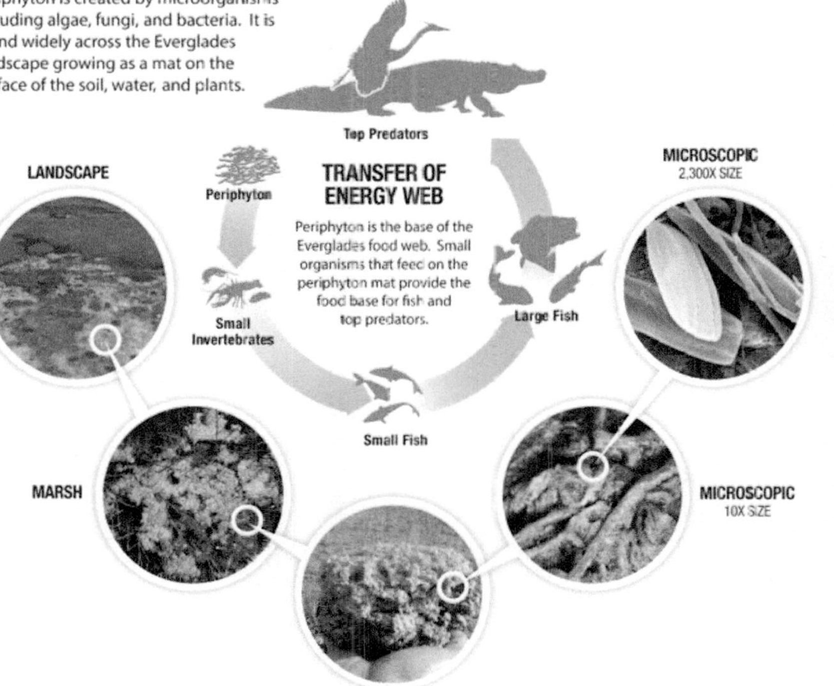
PERIPHYTON THE BASE OF THE EVERGLADES FOOD WEB
Periphyton is created by microorganisms including algae, fungi, and bacteria. It is found widely across the Everglades landscape growing as a mat on the surface of the soil, water, and plants.
Top Predators
TRANSFER OF ENERGY WEB
Periphyton is the base of the Everglades food web. Small organisms that feed on the periphyton mat provide the food base for fish and top predators.
Periphyton
Small Invertebrates
Small Fish
Large Fish
LANDSCAPE
MARSH
CLOSE-UP
MICROSCOPIC
10X SIZE
MICROSCOPIC
2,300X SIZE
PERIPHYTON AND PHOSPHORUS POLLUTION
Periphyton is extremely sensitive to changes in water quality. Scientists have shown that when phosphorus concentrations get above 10 parts per billion, periphyton disappears, resulting in changes to the Everglades food web.
CHANGE IN LANDSCAPE OVER TIME
Natural marsh with periphyton
Loss of periphyton mat
Replacement by green algae
Open water, low oxygen phase
Invasion by cattail

Periphyton composition and ecological role in transferring energy throughout the Everglades ecosystem food chain.

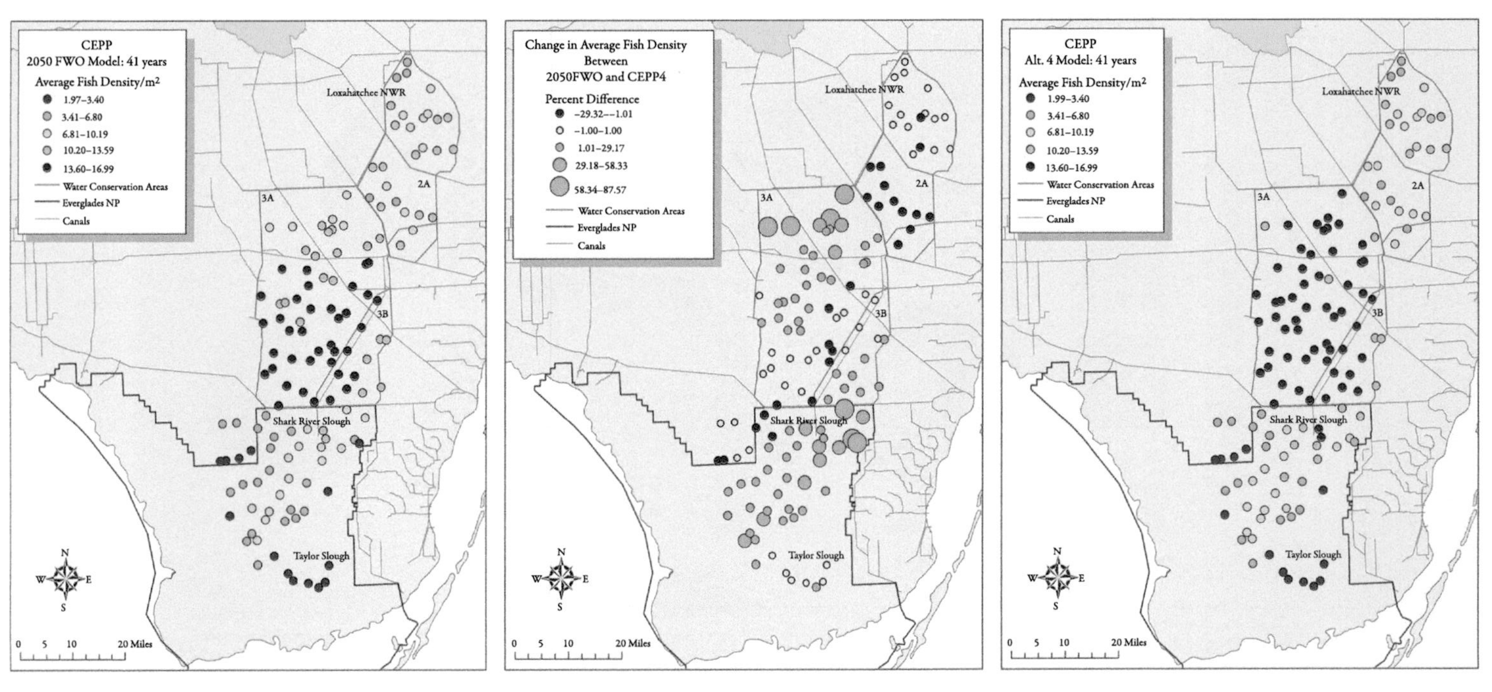

The difference between Alt4 and the future without CEPP (FWO) restoration model runs for the fish Habitat Suitability Index.

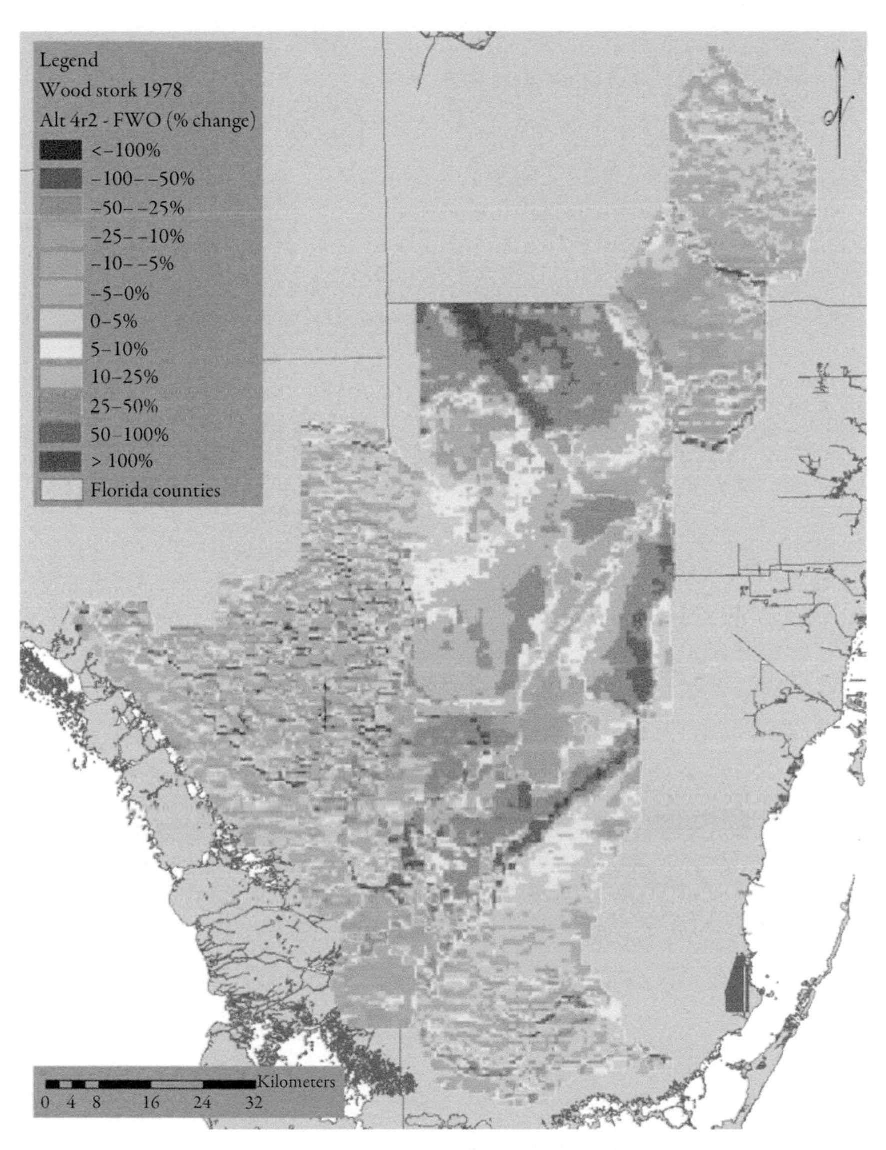

Mean percent change in Wood Stork spatial foraging conditions in an average year (1978) for the CEPP Alt 4r2 restoration model run relative to future without restoration (FWO) model runs.

The FCE program has focused on understanding the origins of productivity in estuaries, and this figure depicts four major outcomes of this research.

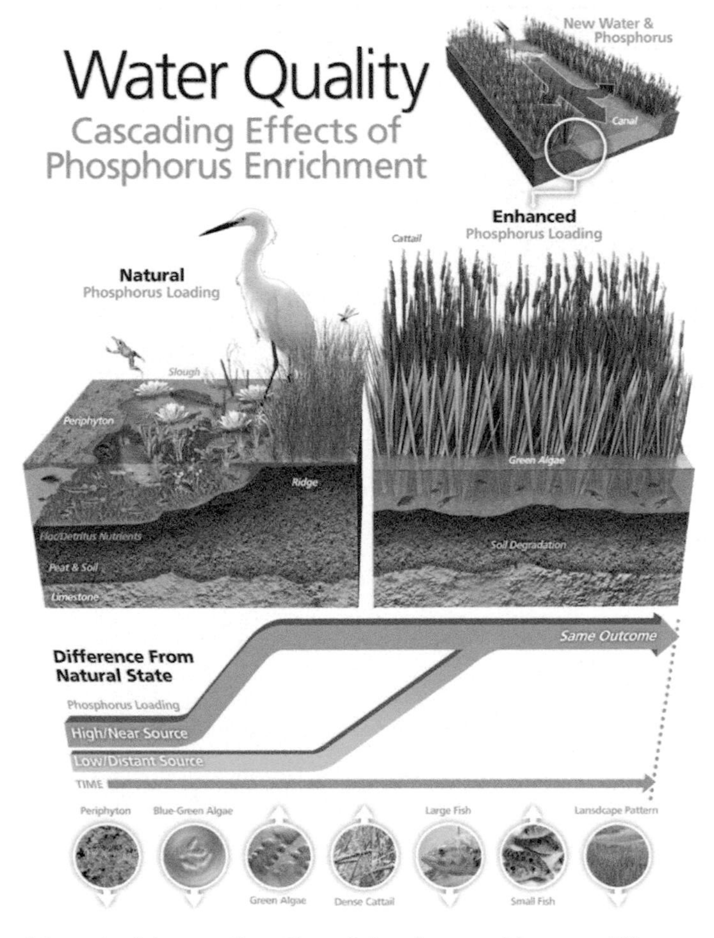

Schematic of the cascading effects of phosphorus enrichment at different spatial scales.

Vegetation classes in Everglades National Park.

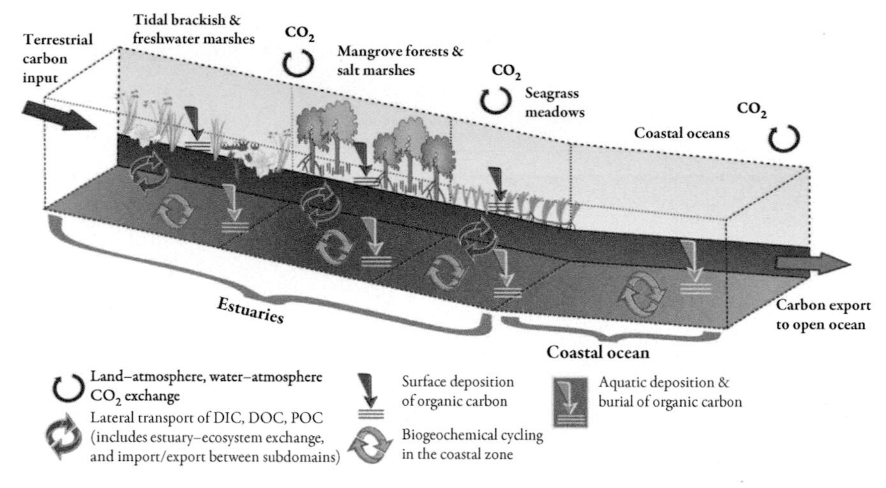

Conceptual model of coastal C cycling, from upstream freshwater marshes to the coastal ocean.

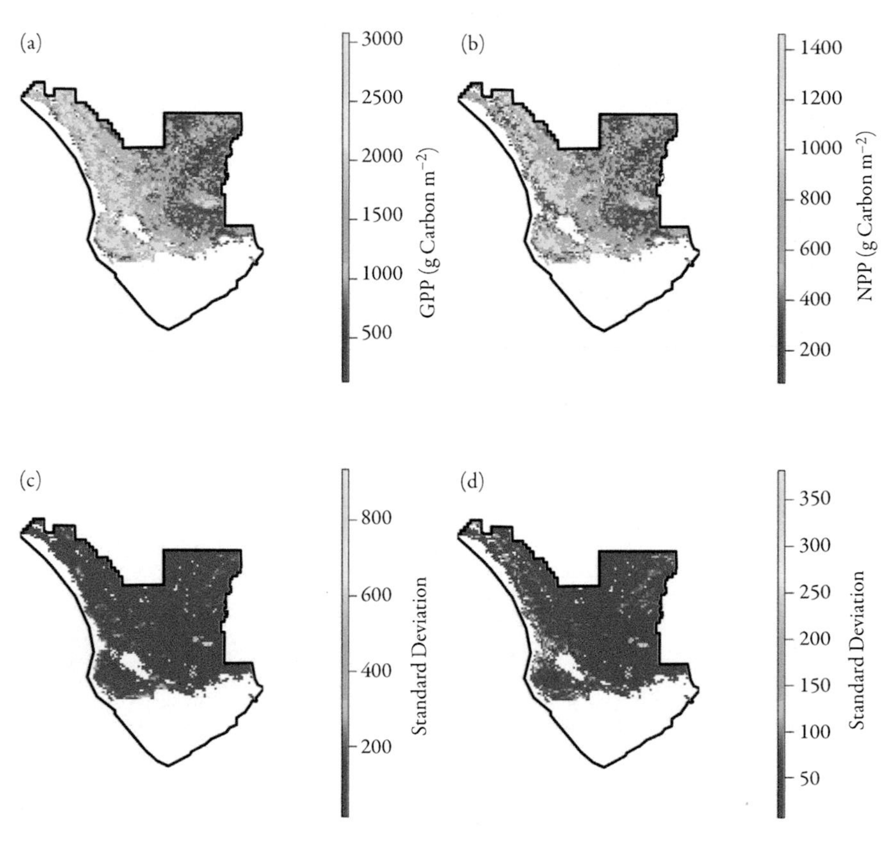

Spatial patterns in mean 2000–2013 GPP (a) + standard deviation (c) and mean NPP (b) + standard deviation (d) for Everglades National Park.

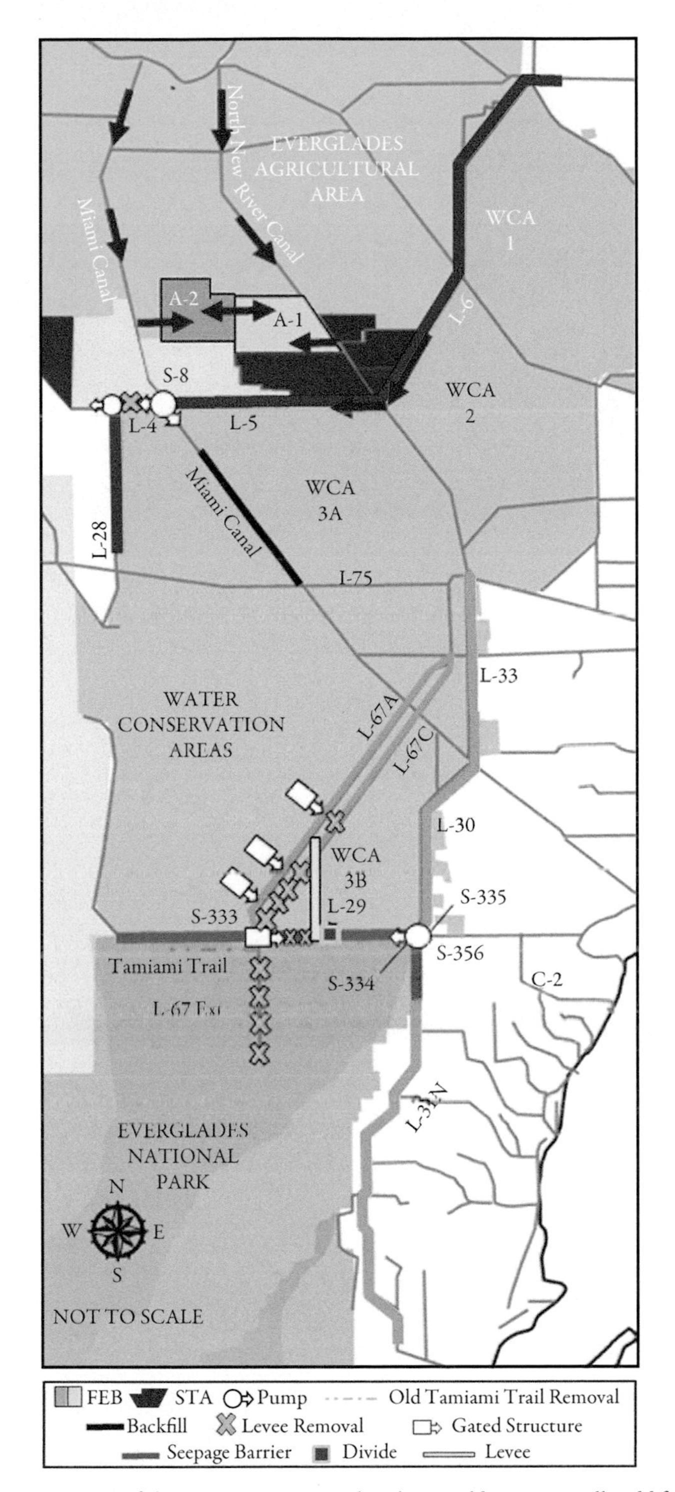

Map of main components of the CEPP restoration plan that would incrementally add features across critical boundaries and divide structures such as canals and levees (see Chapter 8 for details on the Redline, Greenline, Blueline, and Yellowline).

(b)

Storm surge water levels during Hurricane Wilma were estimated from field observations at SRS-6 after the storm, including water marks on tree trunks and sediment deposition observed on equipment at the eddy flux tower located 100 m from the shore.

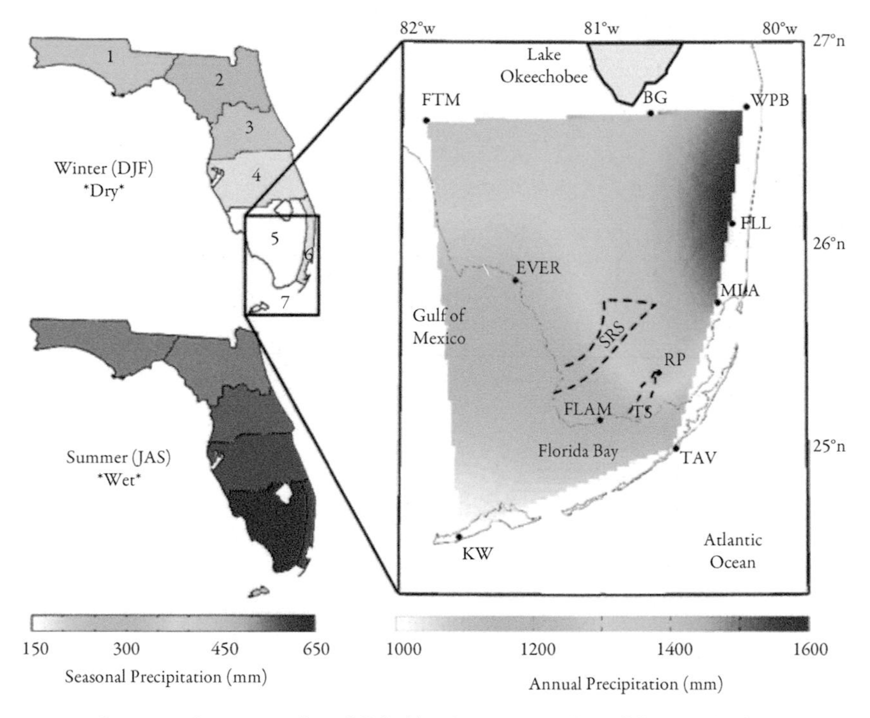

State-wide seasonal patterns of rainfall (left)and mean annual rainfall in south Florida.

Storm-derived marine sediment (gray) deposited on mangrove peats (dark brown) across mangrove forests after Hurricane Wilma.

Mangrove forests at SRS-6 in Shark River estuary before (a) and after (b and c) Hurricane Wilma. Hurricane damage to the mangrove forest at the mouth of this estuary (d).

Close-up view showing depth of collapse
View from the edge of a collapsed pool

Peat collapse in freshwater marshes of lower Shark River Slough experiencing saltwater intrusion, showing about 20 cm of soil collapse around sawgrass plants that are likely <10 years old.

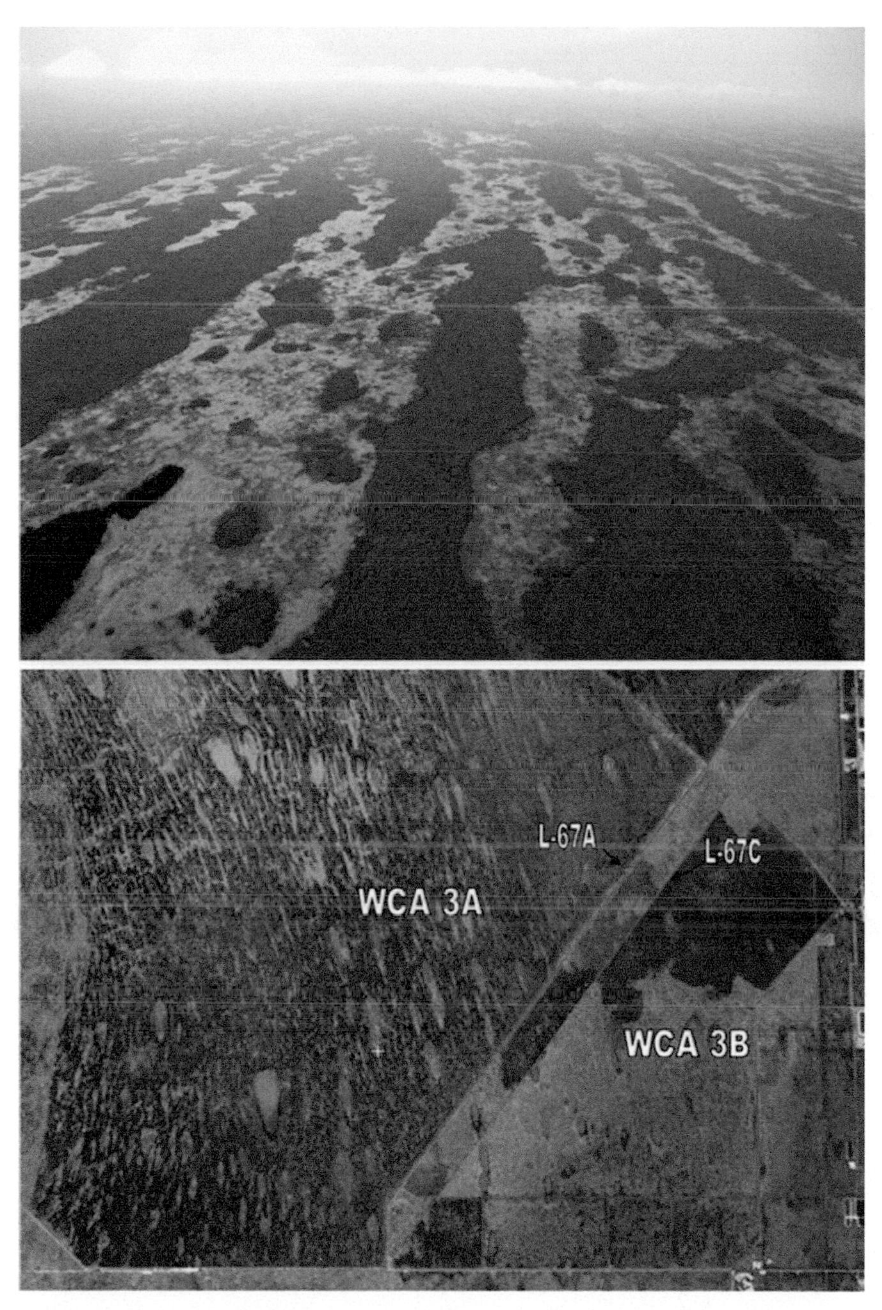

The "corrugated" ridge and slough considered typical of the historical Everglades and remaining in only a fraction of its historical range (top) and a marsh with degraded ridge and slough patterning.

Subtropical Karstic Estuary

↑N:P
↓PIM
P-limited Marsh
Mangrove forest
Ecotone
Tidal exchange
↓N:P
Fresh Groundwater discharge
Seagrass
ΔSoil = OM±PIM
Porous limestone
Brackish Groundwater discharge

Temperate Saltmarsh Estuary

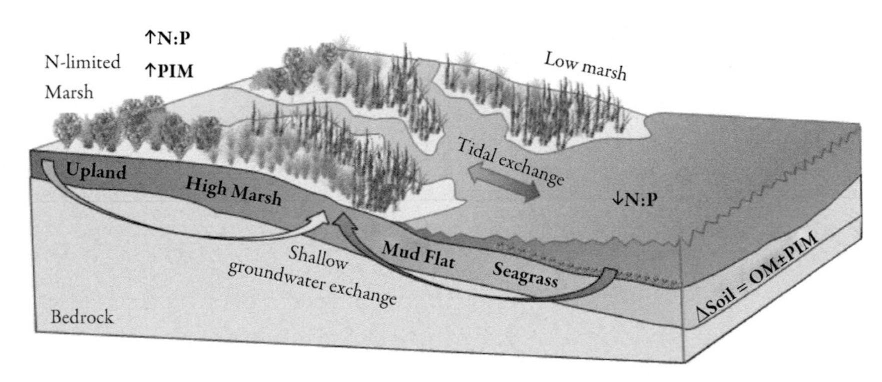

Depiction of generalized differences between subtropical karst (e.g., the Coastal Everglades) and temperate saltmarsh estuaries.

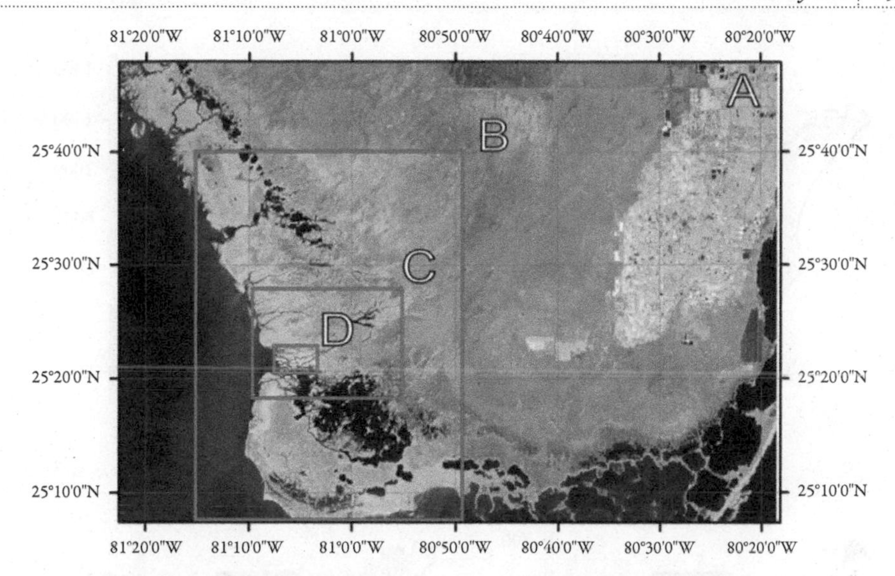

FIGURE 6.5 False color composite (RGB bands 5, 4, 3) of a LANDSAT 5 image from January 23, 2004 used to identify the spatial heterogeneity of green vegetation indices and productivity. Highly productive regions are lighter in color (Thematic Mapper data, wavelengths: 0.63-1.75). The clipped image that includes ENP (A) was divided into smaller subsets (B, C, and D) for further analysis. Subset B includes most of the mangrove forests in ENP. Subset C includes several important estuarine drainages of Shark River Slough. Subset D is centered on SRS-6, a region of peak mangrove forest productivity and standing biomass.

management plays in coastal salinity, sawgrass productivity, and C dynamics in the ecotone. Long-term data on soil pore water salinity may be more informative for landscape-scale assessments of sawgrass response to changing freshwater inflows and coastal salinity (Fig. 6.7; Troxler et al. 2017). We have demonstrated the effects of SLR and transgression over the last 50 years in the southern Everglades, as well as the influences of upstream water management (Ross et al. 2000). Coastal disturbances, including storm surge, are also important triggers of marine transgression (see Chapter 7). These and other long-term studies that link spatial and temporal patterns of biomass and productivity with hydrology, salinity, and P availability will improve our ability to predict macrophyte responses, landscape-scale vegetation patterns, and ultimately changes in coastal landscape marsh C stocks. Given the importance of sawgrass to freshwater marsh community structure, controls on sawgrass productivity likely drive freshwater marsh NEE.

We use gas exchange approaches to measure the net exchange of CO_2 between the marsh and the atmosphere (=NEE). These data illustrate the important role of hydroperiod in Everglades freshwater marsh ecosystems and show marked differences in seasonal patterns of CO_2 exchange rates. Because the waterborne movement of C in these marshes is minimal, patterns of NEE reflected the capacity of the marsh to sequester and store C (Jimenez et al. 2012; Malone et al. 2014; Schedlbauer et al. 2012). In short-hydroperiod marshes, net annual CO_2 uptake was generally greater than in

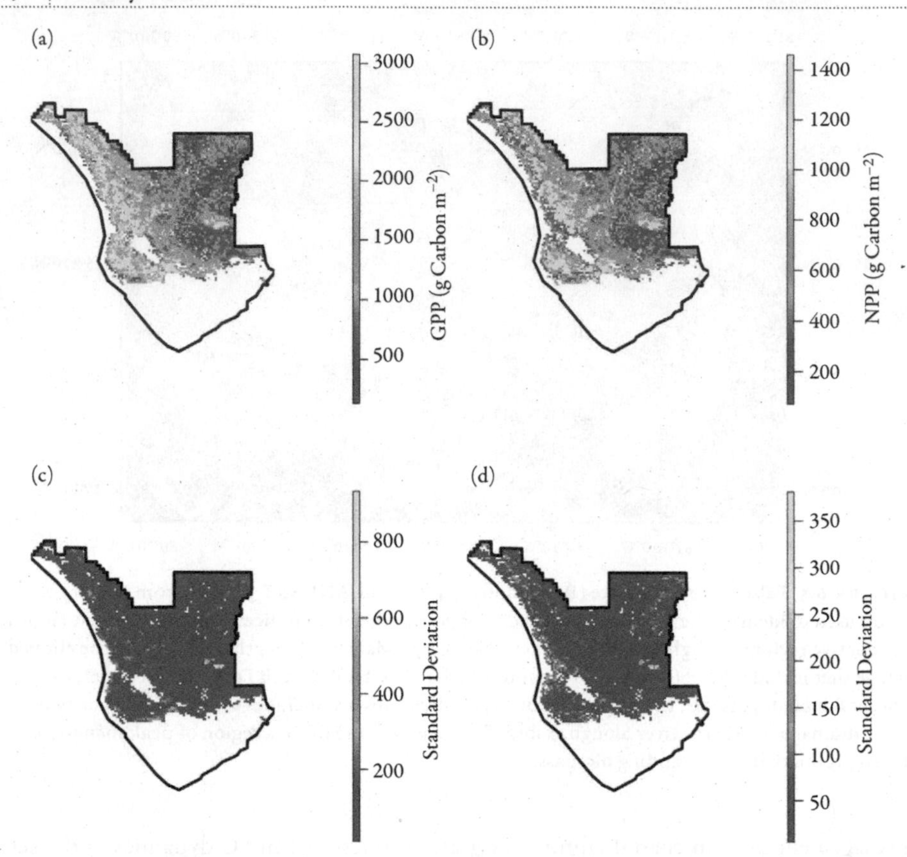

FIGURE 6.6 Spatial patterns in A) mean 2000-2013 GPP and B) mean NPP for Everglades National Park. The standard deviation of GPP (C) and NPP (D) shows variance in time and space. See page 10 of color insert.

long-hydroperiod marshes, and variability was correlated with season length (Malone et al. 2014). In short-hydroperiod marshes, net C uptake rates were highest during the dry season, while in long-hydroperiod marshes we observed greater net C uptake during the wet season (Jimenez et al. 2012; Schedlbauer et al. 2012). Hydroperiod also interacts with water depth to influence NEE. As water depths increase, less leaf area of marsh plants is exposed for atmospheric C exchange, and leaf photosynthetic uptake of CO_2 under water becomes diffusion-limited (Jimenez et al. 2012; Schedlbauer et al. 2010, 2012).

Periphyton is an important component of freshwater marsh GPP. Periphyton GPP rates are difficult to scale, though, because different methods of quantifying productivity often differ by an order of magnitude when they are scaled to annual NPP. In spite of these challenges, we have found that periphyton NPP is roughly half that of sawgrass aboveground plus belowground NPP in long-hydroperiod marshes, but it is roughly double sawgrass NPP in short-hydroperiod marshes (Gaiser et al. 2006; Troxler et al. 2013).

We have also found that ER is the dominant flux contribution to NEE (Jimenez et al. 2012). The primary components of freshwater marsh ER are plant, microbial, and

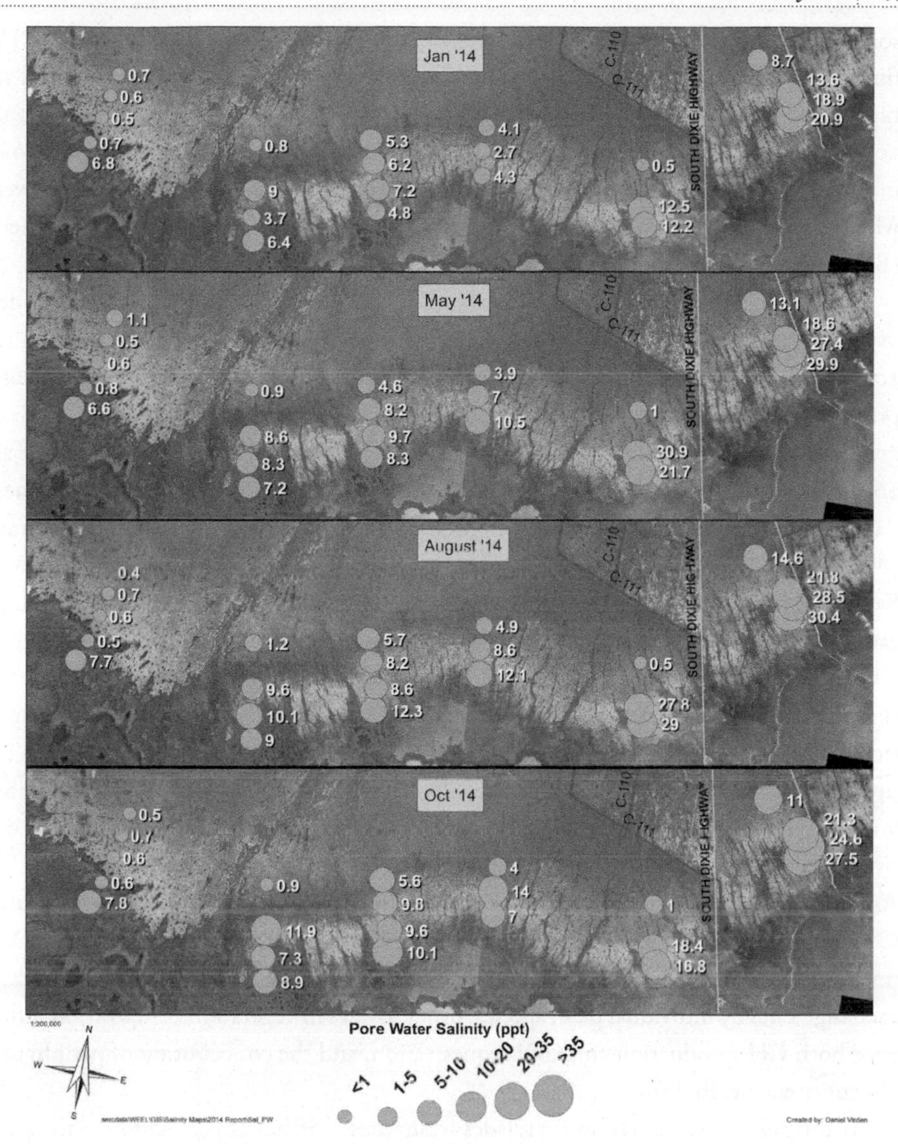

FIGURE 6.7 Quarterly soil porewater salinities in the coastal marsh ecotone of the southern Everglades in 2014 (Troxler, unpublished data).

macroinvertebrate respiration. Leaf litter decomposition contributes to ER whether it is inundated or exposed to the atmosphere. Litter that is inundated undergoes both microbial and macroinvertebrate decomposition, and the C is respired into the water column. Rubio and Childers (2006) showed that annual rates of leaf litter decomposition depended on location and the plant species of the litter. Decomposition rates were higher in long-hydroperiod marshes, for spikerush compared to sawgrass, and in the coastal ecotone than in freshwater sites. Heterotrophic respiration rates in soils are strongly controlled by oxygen availability and the depth of soil oxygenation within the soils (Kramer and Boyer 1995; Webster et al. 2013). Oxygenation also influences CH_4 fluxes. We have found that CH_4 fluxes in freshwater marshes are positively related to higher soil moisture levels (Bachoon and Jones 1992; Smith et al. 2003; Webster et al. 2013; Whelan 2005) and fluxes decrease as the depth of oxygenation increases until soils may actually become a sink for CH_4 (Bachoon and Jones 1992; Smith et al. 2003). Even when inundated, sawgrass marshes are only a weak source of CH_4 to the atmosphere (Bachoon and Jones 1992; Malone et al. 2013).

Seasonal changes in productivity, and thus GEE relative to ER, vary as different modes of CO_2 uptake and release become relatively more or less important (Jimenez et al. 2012; Schedlbauer et al. 2012). For example, as water levels rise during the wet season, photosynthetic capacity is reduced and ER dominates NEE. In the dry season, GEE is greater than ER, as declining water levels increase exposed leaf areas (Schedlbauer et al. 2012). We have found that during periods of inundation in short-hydroperiod marshes, daytime stomatal closure of dominant species, sawgrass (a C3 plant) and muhly grass (*Muhlenbergia capillaris;* a C4 plant), reduces photosynthesis (Schedlbauer et al. 2010). Sawgrass has relatively inefficient aerenchyma, and muhly grass is less flood tolerant than sawgrass (Jimenez et al. 2012). In these marshes, inundation also contributes to higher CO_2 uptake because of high periphyton biomass and enhanced $CaCO_3$ production by the mats (Schedlbauer et al. 2010), though this is not enough to account for the reduction in primary productivity by emergent plants. In a similar manner, photosynthetic uptake of C in long-hydroperiod marshes also increases during the dry season, as water levels decline. However, ER also increases with decreasing water levels, resulting in the ecosystem being a small source of C to the atmosphere on an annual basis (negative NEE). As a result, longer dry seasons in these marshes are associated with lower rates of net C uptake while longer wet seasons are correlated with greater C uptake annually. Our understanding of ecosystem C dynamics will improve with more information on CO_2 exchange rates by individual plant species, how changes in vegetation composition influence both CH_4 production and CO_2 sequestration, and the contribution of periphyton to ecosystem metabolism.

Historically, C was stored in Everglades freshwater marshes as peat and marl through processes tightly coupled with hydroperiod (Davis and Ogden 1994). Peat accumulates organic C in marshes with relatively high water levels and long hydroperiods, and areas with short hydroperiods are characterized by marl soils that store inorganic C (Myers and Ewel 1990). Prior to the last 100 years, the Everglades was almost certainly a net sink for organic C, accreting peat soils to depths exceeding 3 m in areas (McVoy et al. 2011). Two mechanisms maintain coastal peat soil elevations: net organic matter (OM) accumulation from plant production, and mineral sediment accretion via tidal transport (Smoak et al. 2013). In mangrove and marsh ecosystems that receive little sediment input, root production is the primary driver of vertical peat accretion and soil C accumulation (Baustian et al. 2012; McKee 2011; Nyman et al. 2006). However, the autotrophic and heterotrophic mechanisms that contribute to peat soil maintenance, and how these processes are affected by salinity exposure, are not well understood. We do know that salinity is an important driver of ecosystem productivity and composition in the coastal Everglades (Barr et al. 2010; Troxler et al. 2014), and we have found dry-season pore water salinity levels in some sawgrass marshes that exceeded 30 ppt (Troxler et al. 2014). Salinity is also an important driver of soil microbial processes. Enhanced sulfate reduction due to increased soil salinity increases soil respiration and sulfide concentrations while suppressing methane production (Chambers et al. 2011; Weston et al. 2006).

The effect of carbonate precipitation on whole-ecosystem C balance is unknown in Everglades marshes. However, the calcite that precipitates in short-hydroperiod marsh periphyton mats is the basis of the marl soil, which can accrete up to 1 mm per year—particularly during dry years (Gaiser et al. 2006). Using periphyton growth rates on artificial substrates as representative of natural mat accrual, we found that extant periphyton mats represent from 3 to 16 years of accumulated organic C. In contrast, biological oxygen demand (BOD) methods result in high rates of NPP when expressed annually (270–780 g m^{-2} yr^{-1}), suggesting that the C in periphyton mats may turn over three to five times per year. Neither method is perfect, but these findings do imply that autotrophic processes dominate during the wet season; the result is thick mat accumulations (Troxler et al. 2013).

Other important parts of the C budget freshwater Everglades marshes are DOC and POC. The latter is most commonly found in the form of unconsolidated flocculent organic matter ("floc"). Floc is an important component in detrital food webs (Belicka et al. 2012c), and it is also an important player in building and maintaining the "corrugated" ridge and slough landscape patterning and topography in Everglades marshes (Larsen et al. 2009, 2011; Chapter 4). Floc derives from local plant communities, both macrophytic and periphytic (Neto et al. 2006), but we have found it to be highly variable in composition depending on the spatial and temporal scales (Pisani et al. 2013). Hydroperiod influences floc OM preservation, and the floc in short-hydroperiod marshes is more oxidized because of regular drying. We have also found that seasonal differences in floc composition are linked to seasonality in contributions by the dominant primary producers. Jaffe et al. (2007), Pisani et al. (2013), and He et al. (2014) also used molecular markers to show limited hydrologic transport of floc from freshwater marshes to the estuarine ecotone. The amount of C in floc that is available for respiration is larger during the dry season compared to the wet season (Pisani et al. 2011), and the older, partially degraded floc that is present in the dry season is more photoreactive than fresher wet-season floc (Pisani et al. 2011). In summary, our research on floc dynamics suggests a complex set of biogeochemical controls on this C pool, making it a challenge to estimate its contribution to Everglades C cycling.

In contrast to floc C, DOC dynamics have been well documented, on both spatial and temporal scales. Yamashita et al. (2010) reported a decrease in DOC concentrations along a north–south gradient across the Greater Everglades landscape. They found strong evidence that the Everglades Agricultural Area (EAA) is an important source for DOC to Everglades wetlands. The composition of DOC also changes as it moves through the landscape and incorporates locally produced plant-derived materials. We have found a similar pattern in southern Everglades marshes, where canal-derived DOC gradually mixed with marsh-derived DOC (Lu et al. 2003). Both studies strongly suggest that DOM characteristics are controlled by landscape-scale hydrology, local primary production, soil type, and season (Chen et al. 2013). Long-term trends of DOC export from the Everglades to the Gulf of Mexico established that over the past decade DOC fluxes have decreased and are expected to continue to do so based on climate/restoration scenarios modeling (Regier et al. 2016).

In the coastal Everglades, where tidal influences and hurricanes shape wetland structure and function, freshwater marshes give way to coastal marshes, macro- and microtidal mangroves, estuarine creeks and rivers, and mangrove lakes with high densities of submerged aquatic vegetation (see Fig. 6.1). The Everglades mangrove region covers approximately 1445 km^2 (Simard et al. 2006) and is home to some of the most productive mangrove forests in the world (Barr et al. 2010); the macrotidal, tall mangrove forests in the western Everglades exhibit unusually high NEE. We have found that the components of the C balance (see Fig. 6.1; Barr et al. 2012) in the Everglades mangrove ecosystems are highly dynamic in response to variable salinity and inundation patterns and to the unique atmospheric conditions of south Florida that at times include large disturbances.

In mangrove forests, maximum daytime NEE typically ranges from −20 to −25 μmol CO_2 m^{-2} s^{-1}. Mangrove forest respiration is highly variable in response to changing atmospheric conditions and levels of tidal inundation. Daily average respiration amounts to 2.81 ± 2.41 mmol CO_2 m^{-2} s^{-1}, with maximum values during the summer (Barr et al. 2010, 2012). We have observed strong seasonal patterns in NEE in response to variations in atmospheric conditions. For example, during the winter dry season, forest C assimilation substantially increases in response to increases in the proportion of diffuse solar irradiance that penetrates into and through the forest canopy. During the passage of cold fronts, low air temperatures (<10°C) result in reduced photosynthetic activity and thus reduced NEE (i.e., NEE declines to nearly zero). We have used remote sensing analyses and light-use efficiency models optimized for quantifying mangrove C dynamics to show that air temperature variation is a dominant control on seasonal NEE patterns in subtropical mangrove forests (Barr et al. 2013). For example, in January 2010, when south Florida experienced 10 days with nighttime temperatures at or near freezing (<5°C), mangrove photosynthetic uptake (GEE) declined to nearly zero (see Chapter 7).

Soil pore water salinity and tidal activity are also important controls on NEE. For example, daily light-use efficiency of the trees is reduced by as much as 46% at high salinities (>34 ppt) compared with low salinities (<17 ppt; Barr et al. 2012). Also, tidal inundation reduces respiration rates by as much as 1 mmol CO_2 m^{-2} s^{-1} during the daytime and by 0.5 mmol CO_2 m^{-2} s^{-1} at night. Troxler et al. (2013) found similar results for mangrove soil respiration. Over the course of the full growing season, the mangrove forest is a net sink for atmospheric CO_2, with an annual net ecosystem productivity of about 1,200 g

C m^{-2}. We attribute this unusually high net ecosystem productivity to the year-round growing season and low ER. In addition, tidal exports of DIC derived from belowground respiration contribute to the relatively low estimates of mangrove ER, as these losses are not accounted for in C gas flux. Mangrove ER rates also depend on the species composition of the forest (e.g., the presence/absence of pneumatophores typical of black mangrove, *Avicennia germinans*) and the presence of coarse woody debris (Troxler et al. 2013). Our ongoing research on tidal exchanges of C will better elucidate the importance of tidal exports to the ecosystem C budget.

The effects of seasonal changes in water source on salinity, C, and nutrients (mainly nitrogen [N] and phosphorus [P]) in the mangrove ecotone have been well documented (Childers et al. 2006; Davis et al. 2003). Concentrations of DOC in estuarine tidal creeks and rivers are inversely related to salinity (Cawley et al. 2013; Childers et al. 2006; Davis et al. 2003; Jaffe et al. 2004; Timko et al. 2014), suggesting a mangrove source of organic C. This pattern also reflects differences in water residence time and effects of freshwater flow between the southern Everglades and Shark River Slough estuaries. The latter are flushed daily while the former are flushed seasonally. We have also found that the sources of organic C in southern Everglades estuaries vary spatially and seasonally (Chen et al. 2013), affecting the composition of DOC in these waters (Maie et al. 2005). Thus, DOC concentrations do not follow a simple conservative mixing pattern or control by a single dominant end-member source. In contrast, we have reported that, at best (wet season; Harney River), only about 20% of the DOC in Shark River Slough estuarine waters is derived from mangroves; most of this C pool is derived from freshwater end members (Cawley et al. 2013; Jaffe et al. 2004). Furthermore, complex mixing dynamics in the oligohaline ecotone, including the decoupling between DOC and chromophoric DOM, suggest a transition of DOC sources in this region from freshwater to mangrove-derived (Maie et al. 2014). Most of the humic-like components experience non-conservative mixing along the salinity gradient, suggesting additional DOC inputs from the mangrove wetlands (Cawley et al. 2013). In contrast, we found that protein-like components were non-conservatively mixed only during the dry season because of enhanced bioavailability of these labile DOC components at higher P levels that characterize the middle and lower estuaries. In short, DOC in these estuaries is not a "black box" of bulk materials that behaves as a single C pool; rather, it is composed of a variety of components from different sources, with distinct variation in transport and fate. While this complicates our budgeting and assessment of C cycling in Everglades estuaries, it illustrates the importance of other factors controlling transport and fate, including total and dissolved nutrients.

Total N (TN) in the southern Everglades mangrove zone is strongly tied to seasonal flows, reflecting that much of the TN in this more seasonally driven system is also organic in nature. Indeed, much of the TN in the greater Everglades landscape is dissolved organic nitrogen (DON; Boyer et al. 1997), and we have characterized this DON to a molecular level (Maie et al. 2006). We found that the DON composition was generally indicative of an elevated degree of "freshness" (high bioavailability index), suggesting either that production and consumption rates were similar or that bioavailability was low due to P limitation. In fact, Boyer et al. (unpublished) found that bioavailability of DON in Everglades wetlands was 10% at best but increased to roughly 25% after solar exposure and photodegradation.

Total P (TP) availability is an important control on NEE, NECB, and C cycling, and TP concentrations increase with salinity, reflecting the upside-down nature of these estuaries (Childers et al. 2006). In southern Everglades estuaries TP typically shows a fairly dramatic spike at the high-salinity peak of the dry season (Childers et al. 2006). Davis and Childers (2007) observed a similar salinity-driven accumulation of surface water TP in mangrove leaching experiments with different water sources. They hypothesized that P may be accumulating in the water column because of a lack of labile organic C during the height of the dry season, when water residence times are longest. However, Chen et al. (2013) reported that DOM in the lower Taylor River estuary was strongly influenced by seagrass sources associated with saltwater intrusions during the dry season. Koch et al. (2012) showed that water column metabolic rates were related to elevated upstream TP concentrations and salinities. They also found that pulses in aquatic metabolism were associated with P being supplied by groundwater upwelling. In the absence of changes in freshwater discharge, coupled with SLR, brackish water will continue to move inland and aquatic metabolism in the estuarine ecotone will continue to be driven by the balance of P-poor freshwater inflows and the discharge of brackish P-rich groundwater (Price et al. 2006).

Romigh et al. (2006) reported the highest DOC concentrations and DOC flux into the mangrove forests of the lower Shark River estuary in the late dry season (May) and lowest DOC concentrations and net export during the wet season (October and December). Cawley et al. (2013) also found that mangrove–creek fluxes of DOC in these estuaries were greatest during the wet season, and He et al. (2014) reported POC fluxes that were similar in magnitude to these DOC fluxes. Tidal processes are clearly an important driver of DOM flux between the mangrove wetlands and tidal creeks, and ultimately the coastal ocean. A logical next step in these C cycling studies is to expand these flux measurements across spatial scales, but this is a challenging task. However, validation of our direct flux estimates with estimates of C flux from our whole-system metabolism budgets has yielded fairly good agreement (Barr et al. 2009; Bergamaschi et al. 2012; Ho et al. 2014; Troxler et al. 2013), suggesting that we have reasonably tight estimates of the C cycle in our mangrove estuaries.

Dynamics of Florida Bay

Florida Bay is a shallow (<2 m average depth), subtropical estuary between the Florida Keys and the southernmost tip of the Florida peninsula. The bay is approximately 2,000 km^2 in area and is characterized by a collection of shallow basins separated by seagrass-covered carbonate mudbanks and mangrove islands. Freshwater inflows to Florida Bay come from the Everglades, mainly through Taylor Slough and Taylor River (a direct source of freshwater inflow). Florida Bay's "ice cube tray-like" morphology limits the exchange of water between basins, resulting in spatial variations in water quality. During dry periods, hypersaline conditions are common in regions of the bay that are isolated from Everglades freshwater inflow sources, tidal passes between the Florida Keys (to the south), or direct tidal exchange with the Gulf of Mexico (to the west). Given this and relatively mild tides, wind mixing is an important driver of water exchange between basins, and with the Everglades and the ocean.

Florida Bay is the epicenter of one of the largest contiguous seagrass beds in the world, running along the Florida Keys archipelago and covering the southern end of the Florida shelf. The seagrass community is primarily composed of turtle grass (*Thalassia testudinum*), manatee grass (*Syringodium filiforme*), and shoal grass (*Halodule wrightii*). As in freshwater Everglades marshes, productivity in Florida Bay is limited by P availability. We have reported a strong gradient in P availability from the east (low) to the west (higher) that also corresponds with increasing seagrass density and biomass. This is the result of P-depleted sediments and minimal P inputs from the upstream Everglades in the eastern bay, and relatively P-replete sediments and a higher P inputs from the Gulf of Mexico in the western bay.

Our long-term water quality monitoring in south Florida estuaries has been instrumental to our understanding of the connection between hydrology and nutrient dynamics. For example, between 1992 and 1999 we observed a significant decrease in DOC concentrations in the Shark River Slough mangrove estuaries (Fig. 6.8; see also Regier et al. 2016). Interestingly, we did not see this trend in DOC in the southern Everglades estuaries or in Florida Bay. Because no DOC data exist prior to 1992, we could not determine whether higher concentrations were historically normal or if this was the end of a DOC pulse of some kind. One hypothesis is that this DOC pattern was a result of Hurricane Andrew (August 1992), which devastated the Shark River Slough mangrove forests (see Chapter 7), and our data were tracking the end of a large leaching event initiated by hurricane damage. We therefore expected that the mangrove destruction caused in 2005 by Hurricanes Katrina, Rita, and Wilma would have caused a similar pulse in DOC. However, rather than a pulse of DOC, we observed a pulse of TP after 2005 (see Chapter 7).

In Florida Bay, we have used several types of studies to elucidate the sources, sinks, age, fate, transport, air–water exchange of CO_2, and phytoplankton–bacteria productivity coupling to better understand microbial C cycling (Fig. 6.9). The exchange of gases across the air–water interface affects the transport of many pollutants and biogeochemical constituents. Accurate estimates of CO_2 exchange across the air–water interface are hampered by the ability to predict the gas transfer velocity (k). However, we used meteorological techniques to map k at small spatial and temporal scales, allowing for sufficient

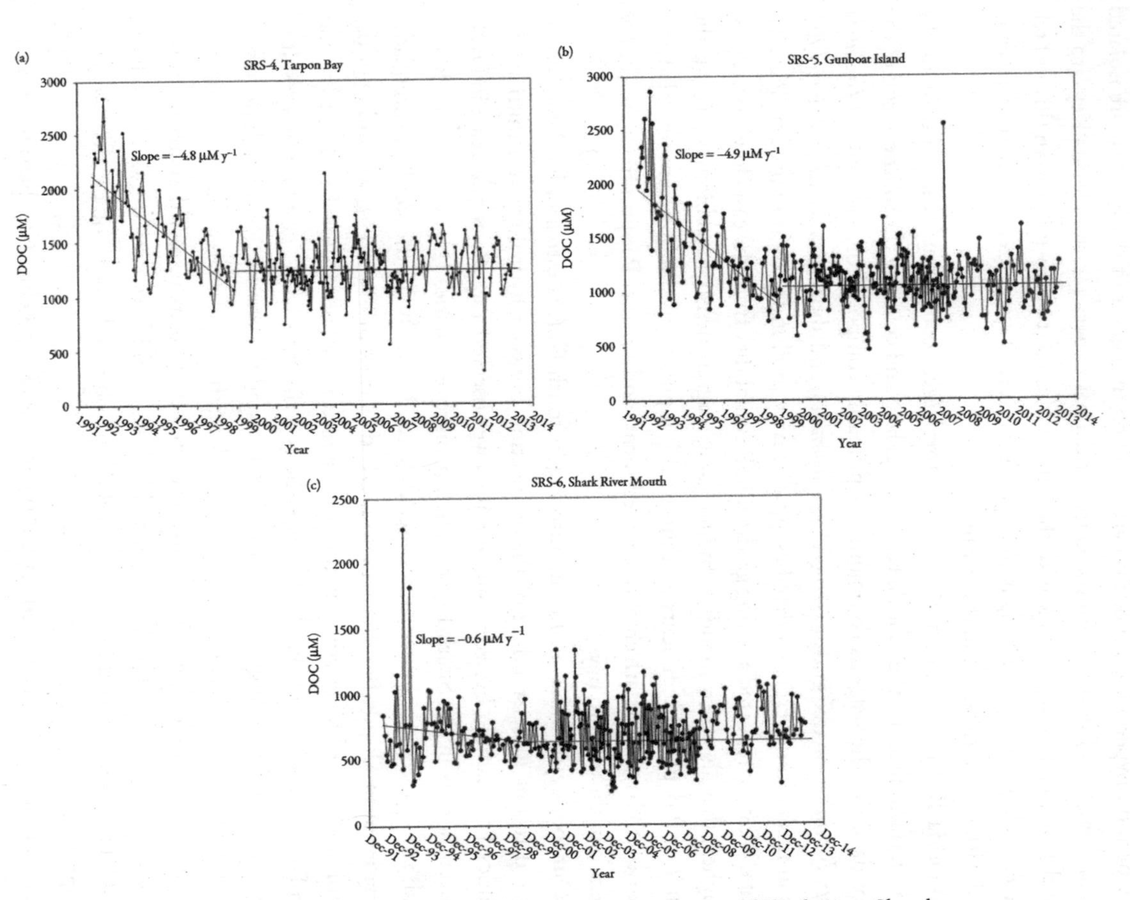

FIGURE 6.8 DOC concentrations from 1991 - 2013 in the mangrove estuaries of Shark River Slough.

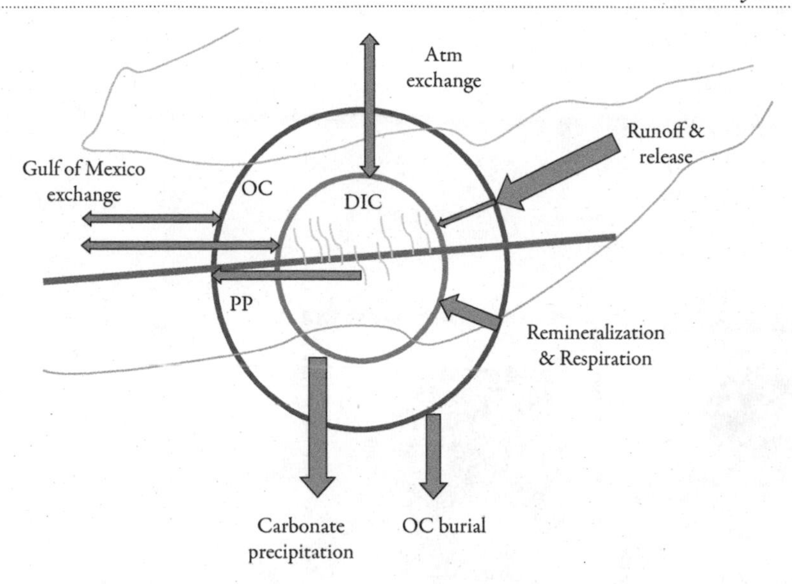

FIGURE 6.9 A conceptual diagram of the microbial C cycle in Florida Bay showing major sources, sinks, reactions, and exports through the major compartments of water, sediment, seagrass, and atmosphere.

resolution to estimate C fluxes and to develop large-scale C budgets for Florida Bay in order to explore the potential importance of this system as a moderator of terrestrial and anthropogenic constituents before their discharge to the oceans.

To characterize the distribution of pCO_2 throughout Florida Bay, we computed interpolations of the monthly pCO_2 (Fig. 6.10) to evaluate the flux of CO_2 across the surface of Florida Bay. Primary observations from these interpolations indicate a strong trend of increasing pCO_2 toward northeast Florida Bay. These values suggest that freshwater discharge from Taylor Slough and several other major freshwater sources to northeast Florida Bay is a fairly significant driving force in the C budget of the bay. This spatial variability also indicates that the flow of water into the bay is not sufficient to flush out and dilute the CO_2 concentrations in northeast Florida Bay.

We have explored bacteria–phytoplankton coupling, bacterial growth efficiency, and the influence of environmental factors on bottom-level trophodynamics in Florida Bay by measuring phytoplankton GPP, community respiration, bacterial productivity, and bacterial respiration. We found strong bacteria–phytoplankton coupling where sources of DOM other than autochthonous phytoplankton exudates were limited, similar to Cole et al. (1988), and weaker coupling where allochthonous terrestrial and coastal DOM serves as a food source for bacterioplankton. Bacterial productivity was more influenced by local TN concentrations while phytoplankton GPP was more evenly distributed. We also found that bacterial growth efficiencies were low but consistent with marine and estuarine ecosystems worldwide. These results suggest that bacterioplankton growth in Florida Bay is relatively uncoupled from phytoplankton production, which may be due

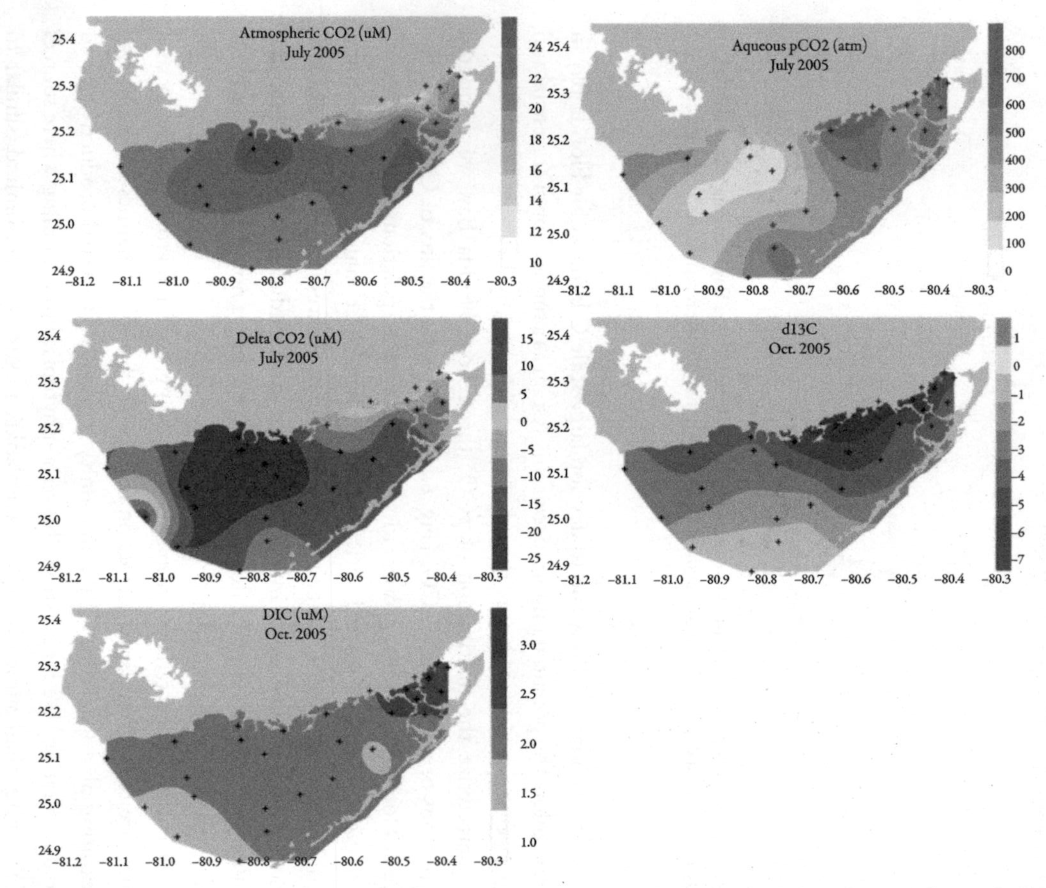

FIGURE 6.10 Atmospheric CO_2 concentrations over Florida Bay (top left). Aqueous pCO_2 in Florida Bay (top right). The difference in the CO_2 concentration across the air-water interface driving air-water carbon dioxide fluxes (center left). $\partial^{13}C$ ratios in Florida Bay (center right). The dissolved inorganic carbon concentration in Florida bay (bottom left).

in part to low phytoplankton biomass in the water column, a large amount of seagrass-derived DOM production, and the loading of TN and OM via freshwater inflow from the Everglades.

As in freshwater marshes and mangrove creeks, primary production is also an important source of plant-derived DOC in Florida Bay. Florida Bay waters contain elevated amounts of non-humic, protein-like material compared to freshwater Everglades waters. Our chemical characterization work in Florida Bay has shown that a significant portion of the DOM in Florida Bay was not associated with freshwater inputs from the Everglades; much of the bioavailable, protein-like materials in Florida Bay are derived from autochthonous sources. We believe the most likely source of this autochthonous DOM is the seagrass/benthic community. Bay-wide seasonal surveys have shown that summer DOM was more enriched in potentially labile protein-like materials exuded by seagrass primary production, whereas fall DOM showed a more refractory terrestrial signature that reflected freshwater inputs from the Everglades (Maie et al. 2014). Maie et al. (2012) also identified clear spatial variation in DOM composition across the Bay. Ya et al. (2015) used stable isotope mixing models to support these spatial and temporal patterns, showing that as much as 70% of the DOC in central Florida Bay was seagrass-derived, while for the northeast section of the bay it could be only about 30% during the wet season.

The contribution of calcifying macroalgae to the Florida Bay C cycle is complicated. Hatt and Collado-Vides et al. used a 7-year survey to show that the inorganic C contribution of the dominant species of calcifying algae in Florida Bay seagrass beds (*Halimeda* and *Penicillus* spp.) varied along a salinity gradient. In more marine sites, $CaCO_3$ production was about 80% of annual primary production, which is within the range reported for the Florida Keys and other similar environments (Davis and Fourqurean 2001; Tussenbroek and Djik 2007). In more estuarine sites of Florida Bay, $CaCO_3$ production was less than 1% of that found in marine sites and was also low compared with other reports (Hillis 1997). As algal assemblages shift toward marine-adapted species (Collado-Vides et al. 2011), we expect higher $CaCO_3$ sediment production, but responses will likely differ depending on nutrient regimes that can affect algal and seagrass competition (Davis and Fourqurean 2001) or epiphyte dynamics (Armitage et al. 2011).

Anthropogenic Effects on C Dynamics, Including SLR

In south Florida, water management includes hydrologic restoration actions to improve coastal resources, and water managers are tasked with decisions that have strong societal consequences for the region and corresponding implications for other coastal regions. Forecasting future shoreline morphology and structure has great socioeconomic significance in south Florida, where coastal wetlands provide recreation, flood control, protection of water resources, and a basis for environmental restoration projects. As the relative areas and functions of vegetation types change, the ecosystem services they provide will change (i.e., Jerath 2012). Thus, coastal stressors drive landscape change, and landscape change drives either increases or decreases in ecosystem services.

There have been considerable changes to C cycling and C storage in the Everglades, largely because of anthropogenic activities in the first half of the 20th century (Davis and Ogden 1994; McVoy et al. 2011; see Chapter 1). For example, Aich et al. (2013) and Hohner and Dreschel (2015) suggested that 760 to 920 million metric tons of C has been lost over the last roughly 120 years from peat soils in the Everglades Protection Area (a region that includes the EAA, water conservation areas, and freshwater ENP). This is a dramatic example of change in the C cycle as well as the structure of the landscape, with implications across all levels of the Everglades ecosystem.

Across the coastal Everglades, we are beginning to calculate C stocks. Jerath et al. (2016) estimated that the Everglades mangrove region holds 1.78 million metric tons C in aboveground tissues and 43.22 million metric tons C belowground (soils and roots). Across the coastal Everglades SLR is of great concern due to the repercussions of saltwater intrusion. Sea level in south Florida is conservatively predicted to rise 0.6 m by 2060 (Zhang et al. 2011), and this estimate now defines the upper range for SLR planning in south Florida (Obeysekera et al. 2015). South Florida is particularly vulnerable to SLR given the low topographic relief, shallow and highly permeable karst aquifer, and large-scale hydrologic diversions that have amplified coastal transgression (Ross et al. 2000). About 60% of ENP is at or under 0.9 m in elevation relative to mean sea level (Pearlstine et al. 2010). The inland transgression of mangroves has been suggested as a means by which subtropical and tropical coastal landscapes will "adapt" to increasing SLR. In this scenario, mangroves will replace inland marshes, stabilizing soils as they transgress (McKee et al. 2007), because historically rates of vertical soil accretion of mangrove and salt marsh wetlands have kept pace with rates of past SLR (e.g., McKee 2011). In the coastal Everglades, these salt-tolerant mangrove wetlands are expected to replace freshwater coastal peat marshes. In those peat marshes, we know that sawgrass is only weakly salt-tolerant, and we have found that its productivity is negatively affected by increasing salinity (Childers et al. 2006; Troxler et al. 2014). This may also negatively affect the soil C balance in these peat soils, leading to a positive feedback that is responsible for peat collapse. Vegetation modeling has shown that mangroves will continue to transgress inland, but freshwater marshes inland of the mangrove forests are at risk of being converted to open water systems if their peat soils collapse first (Karamperidou et al. 2013).

The SLR- and salinity-induced process of peat collapse is complex and is closely tied to processes that control soil C balance, such as redox potential, soil respiration rates, and the intensity of osmotic stress on vegetation. The term "peat collapse" has been used to describe a relatively dramatic shift in soil C balance in response to saltwater intrusion. It results from a net loss of organic C, is manifest as a rapid loss of soil elevation, and culminates in a conversion of vegetated freshwater marsh to open water (Fig. 6.11). Peat collapse has been documented to varying degrees across the United States (Cahoon

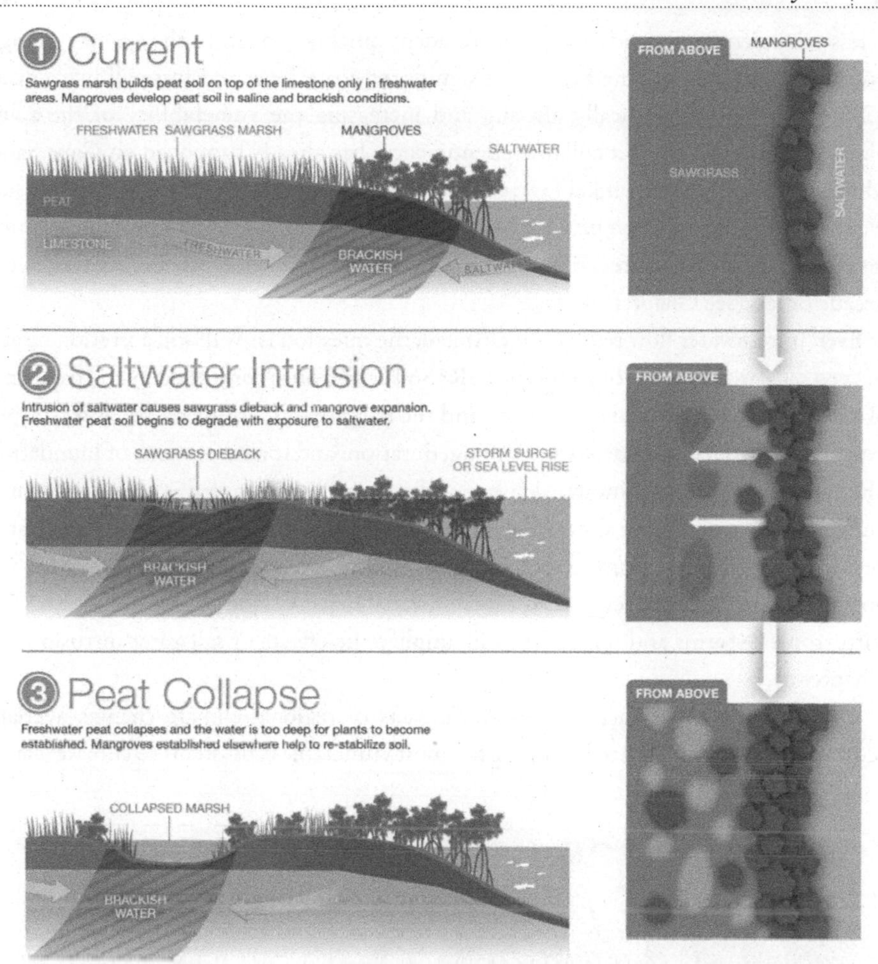

1 Current
Sawgrass marsh builds peat soil on top of the limestone only in freshwater areas. Mangroves develop peat soil in saline and brackish conditions.
FRESHWATER SAWGRASS MARSH
MANGROVES
SALTWATER
PEAT
LIMESTONE
FRESHWATER
BRACKISH WATER
FROM ABOVE
MANGROVES
SAWGRASS
SALTWATER
2 Saltwater Intrusion
Intrusion of saltwater causes sawgrass dieback and mangrove expansion. Freshwater peat soil begins to degrade with exposure to saltwater.
SAWGRASS DIEBACK
STORM SURGE OR SEA LEVEL RISE
BRACKISH WATER
FROM ABOVE
3 Peat Collapse
Freshwater peat collapses and the water is too deep for plants to become established. Mangroves established elsewhere help to re-stabilize soil.
COLLAPSED MARSH
BRACKISH WATER
FROM ABOVE

FIGURE 6.11 Conceptual schematic of the process of peat collapse (created by H2H Graphic and S. Davis). Peat development and maintenance occurs along the coastal ecotone of the Everglades. As saltwater intrusion increases with freshwater diversion and sea-level rise, the brackish mixing zone transgresses inland, moving into historically freshwater peat marsh areas. This is hypothesizied to reduce the productivity of sawgrass productivity and enhance microbial respiration, leading to a degradation of peat soil. Over time, the loss of organic matter inputs and loss of soil organic matter results in a subsidence of the interior marsh and pockets of open water. See page 2 of color insert.

et al. 2003; Nyman et al. 2006; Voss et al. 2013) and has been attributed to increased sulfate reduction, increases in other avenues of soil microbial respiration, sulfide accumulation, reduced root production, fire history, and vegetation damage from tropical storms. All these processes can synergistically contribute to the instability of freshwater marsh soils. Recent work is uncovering the mechanisms behind peat collapse (Charles 2018; Servais et al. 2019; Wilson et al. 2018a,b; Wilson et al. 2019). Additionally, past work by Chambers et al. (2014) showed that SLR—through its effect on increasing salinity and inundation—may also influence C cycling rates in our mangrove peat soils.

If less salt-tolerant wetlands are unable to adapt quickly enough to the salinity changes associated with accelerating SLR, then significant coastal wetland loss will likely occur (CISREP 2014), dramatically altering and increasing the vulnerability of the south Florida coastline. This peat collapse phenomenon has already happened on Cape Sable, where Wanless and Vlaswinkel (2005) attributed it to canals dug in the 1920s and subsequent saltwater intrusion into freshwater marshes (Fig. 6.12). The Cape Sable example suggests that significant areas of marsh may experience peat collapse over timeframes of a decade or less (see Chapter 7).

Even if freshwater flow remains unchanged, the question is: Will soil accretion be able to keep pace with accelerating rates of SLR? Soil accretion is controlled by the balance of plant productivity and soil respiration, and the balance of C influx and export. Carbon export is controlled by hydrology—timing, duration, and spatial patterns of inundation. The tension between freshwater delivery and salinity intrusion drives changes in the C budget, which potentially alters C accretion. The rate of soil C accretion will ultimately decide when salinity intrusion will win. Even before salinity intrusion wins, SLR will produce higher water levels in coastal freshwater marshes, and storm surge associated with tropical storms and hurricanes will amplify the effects of saltwater intrusion (see Chapter 7).

Complicating these uncertainties are impacts of regional climate change. Wetland ecosystems are considered to be among the most vulnerable ecosystems to climate change

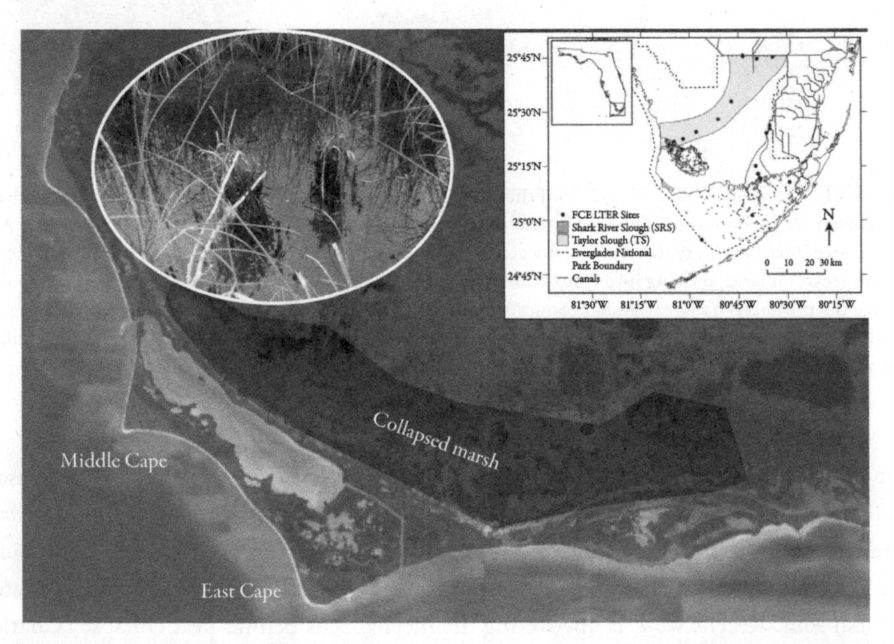

FIGURE 6.12 Map of Cape Sable illustrating area of observed peat collapse that took place between new canals being dredged in the early 1920s and 1935. Image of collapsing marsh (center). Sawgrass plant in collapsing peat marsh (upper right corner). The potential rate of vertical peat loss might be ≥ 0.3m/decade (Wanless and Vlaswinkel 2005).

(Burkett and Kusler 2000) as a result of its effects on wetland hydrology and temperature (Erwin 2009; Ferrati et al. 2005). In the Everglades, wet-season precipitation is projected to decrease by 5% to 10% (Christensen et al. 2007) while annual precipitation may be altered by −2 to +14% (IPCC 2013) and temperatures may increase by 1° to 4.2°C (IPCC 2013). These changes are projected to produce larger convective storms and stronger hurricanes (Allan and Soden 2008). While projections for temperature and precipitation are within the natural range of variation observed in the region, climate projections for south Florida suggest a future with decreased seasonality in precipitation by reducing wet-season precipitation and increasing dry-season precipitation. The wide range in climate projections for the Everglades makes estimates of future C dynamics for the region even less certain.

While the historical Everglades was a net sink for C—hence the peat soils in this relatively young wetland landscape—our research has shown that Everglades freshwater marsh ecosystems are nearly CO_2 neutral on an annual basis (Jimenez et al. 2012; Malone et al. 2014, 2015) and are likely a small source of CH_4 during periods of inundation (Malone et al. 2013). Future changes in precipitation patterns, air temperature, hydrology, wet-season length/intensity, and drought occurrence/duration may shift in patterns of CO_2 and CH_4 exchange rates to the extent that Everglades wetlands become C sources (Barr et al. 2010; Malone et al. 2013). Future changes in salinity and inundation resulting from SLR will likely also shift the balance between CO_2 uptake and CH_4 flux. Although these systems are small sources of CH_4 (Bachoon and Jones 1992), changing water levels and salinity play a substantial role in CH_4 release to the atmosphere (Barr et al. 2010, 2011; Chambers et al. 2014; Malone et al. 2013, 2015; Poffenbarger et al. 2011).

References

Abtew, W. 2004. Evapotranspiration in the Everglades; comparison of Bowen ration measurements and model estimates. South Florida Water Management District Technical Paper EMA#417.

Abtew, W., C. Pathak, R.S. Huebner, and V. Ciuca. 2009. Hydrology of the south Florida environment. In *South Florida Environmental Report*, vol. 1. The South Florida Environment. South Florida Water Management District, West Palm Beach, pp. 2.1–2.53.

Aich, S., C.W. McVoy, T.W. Dreschel. and F. Santamaria. 2013. Estimating soil subsidence and carbon loss in the Everglades Agricultural Area, Florida using geospatial techniques. *Agriculture, Ecosystems and Environment*. 171: 124–133.

Allan, R.P.R., and B.J.B. Soden. 2008. Atmospheric warming and the amplification of precipitation extremes. *Science* 321(5895): 1481–1484.

Armitage, A.R., T.A. Frankovich, and J.W. Fourqurean. 2011. Long-term effects of adding nutrients to an oligotrophic coastal environment. *Ecosystems* 14(3): 430–444.

Bachoon, D., and R.D. Jones. 1992. Potential rates of methanogenesis in sawgrass marshes with peat and marl soils in the Everglades. *Soil Biology and Biochemistry* 24(1): 21–27.

Barr, J.G., M.S. DeLonge, and J.D. Fuentes. 2014. Seasonal evapotranspiration patterns in mangrove forests. *Journal of Geophysical Research: Atmospheres* 119(7): 3886–3899.

Barr, J.G., V. Engel, J.D. Fuentes, D.O. Fuller, and H. Kwon. 2013. Modeling light use efficiency in a subtropical mangrove forest equipped with CO_2 eddy covariance. *Biogeosciences* 10(3): 2145–2158.

Barr, J.G., V. Engel, J.D. Fuentes, J.C. Zieman, T.L. O'Halloran, T.J. Smith, and G.H. Anderson. 2010. Controls on mangrove forest-atmosphere carbon dioxide exchanges in western Everglades National Park. *Journal of Geophysical Research* 115: G02020.

Barr, J.G., V. Engel, T.J. Smith, and J.D. Fuentes. 2012. Hurricane disturbance and recovery of energy balance, CO_2 fluxes and canopy structure in a mangrove forest of the Florida Everglades. *Agricultural and Forest Meteorology* 153: 54–66.

Barr, J.G., J.D. Fuentes, V. Engel, and J.C. Zieman. 2009. Physiological responses of red mangroves to the climate in the Florida Everglades. *Journal of Geophysical Research* 114: G02008.

Baustian, J. J., I.A. Mendelssohn, and M.W. Hester. 2012. Vegetation's importance in regulating surface elevation in a coastal salt marsh facing elevated rates of sea level rise. *Global Change Biology* 18(11): 3377–3382.

Belicka, L.L., E.R. Sokol, J.M. Hoch, R. Jaffe, and J.C. Trexler. 2012. A molecular and stable isotopic approach to investigate algal and detrital energy pathways in a freshwater marsh. *Wetlands* 32(3): 531–542.

Benway, H.M., and P.G. Coble. 2014. Report of The U.S. Gulf of Mexico Carbon Cycle Synthesis Workshop, March 27–28, 2013. Ocean Carbon and Biogeochemistry Program and North American Carbon Program, Woods Hole.

Bergamaschi, B.A., D.P. Krabbenhoft, G.R. Aiken, E. Patino, D.G. Rumbold, and W.H. Orem. 2012. Tidally driven export of dissolved organic carbon, total mercury, and methylmercury from a mangrove-dominated estuary. *Environmental Science & Technology* 46(3): 1371–1378.

Bouillon, S. 2011. Carbon cycle: storage beneath mangroves. *Nature Geoscience* 4(5): 282–283.

Boyer, J.N., J.W. Fourqurean, and R.D. Jones. 1997. Spatial characterization of water quality in Florida Bay and Whitewater Bay by multivariate analyses: zones of similar influence. *Estuaries and Coasts* 20(4): 743–758.

Burkett, V., and J. Kusler. 2000. Climate change: potential impacts and interactions in wetlands of the United States. *JAWRA Journal of the American Water Resources Association* 36(2): 313–320.

Cahoon, D.R., P. Hensel, J. Rybczyk, K.L. McKee, C.E. Proffitt, and B.C. Perez. 2003. Mass tree mortality leads to mangrove peat collapse at Bay Islands, Honduras after Hurricane Mitch. *Journal of Ecology* 91(6): 1093–1105.

Cai, Wei-Jun. 2011. Estuarine and coastal ocean carbon paradox: CO_2 sinks or sites of terrestrial carbon incineration? *Annual Review of Marine Science* 3: 123–145.

Cawley, K.M., D.M. McKnight, P. Miller, R. Cory, R.L. Fimmen, J. Guerard, M. Dieser, C. Jaros, Y-P. Chin, and C. Foreman. 2013. Characterization of fulvic acid fractions of dissolved organic matter during ice-out in a hyper-eutrophic, coastal pond in Antarctica. *Environmental Research Letters* 8(4): 045015.

Chambers, L.G., S.E. Davis, T. Troxler, J.N. Boyer, A. Downey-Wall, and L.J. Scinto. 2014. Biogeochemical effects of simulated sea level rise on carbon loss in an Everglades mangrove peat soil. *Hydrobiologia* 726(1): 195–211.

Chambers, L.G., K.R. Reddy, and T.Z. Osborne. 2011. Short-term response of carbon cycling to salinity pulses in a freshwater wetland. *Soil Science Society of America Journal* 75(5): 2000–2007.

Chapin, F. S III, G.M Woodwell, J.T. Randerson, G.M. Lovett, E.B. Rastetter, D.D. Baldocchi, D.A. Clark, M.E. Harmon, D.S. Schimel, R. Valentini, C. Wirth, J.D. Aber, J.J. Cole, M.L. Goulden, J.W. Harden, M. Heimann, R.W. Howarth, P.A. Matson, A.D. McGuire, J.M. Melillo, H.A. Mooney, J.C. Neff, R.A. Houghton, M.L. Pace, M.G. Ryan, S.W. Running, O.E. Sala, W.H. Schlesinger, and E.D. Schulze. 2006. Reconciling carbon-cycle concepts, terminology, and methods. *Ecosystems* 9: 1041–1050.

Charles, S. 2018. Saltwater intrusion and vegetation shifts drive changes in carbon storage in coastal wetlands. PhD dissertation. Florida International University.

Chen, M., N. Maie, K. Parish, and R. Jaffe. 2013. Spatial and temporal variability of dissolved organic matter quantity and composition in an oligotrophic subtropical coastal wetland. *Biogeochemistry* 115(1-3): 167–183.

Chen, R., and R.R. Twilley. 1999. Patterns of mangrove forest structure and soil nutrient dynamics along the Shark River estuary, Florida. *Estuaries* 22(4): 955–970.

Childers, D.L., J.N. Boyer, S.E. Davis, C.J. Madden, D.T. Rudnick, and F.H. Sklar. 2006. Relating precipitation and water management to nutrient concentration patterns in the oligotrophic "upside down" estuaries of the Florida Everglades. *Limnology and Oceanography* 51(1): 602–616.

Christensen, J.H., B. Hewitson, A. Busuioc, A. Chen, X. Gao, I. Held, R. Jones, R.K. Kolli, W.-T. Kwon, R. Laprise, V. Magaña R ueda, L. Mearns, C.G. Menéndez, J. Räisänen, A. Rinke, A. Sarr, and P. Whetton. 2007. Regional climate projections. In *Climate Change 2007: The Physical Science Basis*. Contribution of Working Group I to the Fourth Assessment Report of the Intergovernmental Panel on Climate Change, edited by S. Solomon, D. Qin, M. Manning, Z. Chen, M. Marquis, K.B. Averyt, M. Tignor, and H.L. Miller, 847–940. Cambridge and New York: Cambridge University Press.

CISREP. 2014. *Progress Toward Restoring the Everglades: The Fifth Biennial Review: 2014*. Washington, DC: National Academies Press.

Cole, J.J., S. Findlay, and M.L. Pace. 1988. Bacterial production in fresh and saltwater ecosystems: a cross-system overview. *Marine Ecology Progress Series* 43(1): 1–10.

Collado-Vides, L., V. Mazzei, T. Thyberg, and D. Lirman. 2011. Spatio-temporal patterns and nutrient status of macroalgae in a heavily managed region of Biscayne Bay, Florida, USA. *Botanica Marina* 54(4): 377–390.

Davis, B.C., and J.W. Fourqurean. 2001. Competition between the tropical alga, *Halimeda incrassata*, and the seagrass, *Thalassia testudinum*. *Aquatic Botany* 71(3): 217–232.

Davis, S.E., and D.L. Childers. 2007. Importance of water source in controlling leaf leaching losses in a dwarf red mangrove (*Rhizophora mangle L.*) wetland. *Estuarine, Coastal and Shelf Science* 71(1): 194–201.

Davis, S.E., C. Coronado-Molina, D.L. Childers, and J.W. Day Jr. 2003. Temporally dependent C, N, and P dynamics associated with the decay of *Rhizophora mangle* L. leaf litter in oligotrophic mangrove wetlands of the Southern Everglades. *Aquatic Botany* 75(3): 199–215.

Davis, S.M., and J.C. Ogden. 1994. *Everglades: The Ecosystem and Its Restoration*. Delray Beach, FL: CRC Press.

Desmond, G.B. 2003. Measuring and mapping the topography of the Florida Everglades for ecosystem restoration. U.S. Geological Survey Fact Sheet 021–03, 4.

Donar, C. M., K.W. Condon, M. Gantar, and E.E. Gaiser. 2004. A new technique for examining the physical structure of Everglades floating periphyton mat. *Nova Hedwigia* 78(1-2): 107–119.

Erwin, K.L. (2009). Wetlands and global climate change: the role of wetland restoration in a changing world. *Wetlands Ecology and Management* 17(1): 71–84.

Ewe, S.M.L., E.E. Gaiser, D.L. Childers, D. Iwaniec, V.H. Rivera-Monroy, and R.R. Twilley. 2006. Spatial and temporal patterns of aboveground net primary productivity (ANPP) along two freshwater-estuarine transects in the Florida Coastal Everglades. *Hydrobiologia* 569(1): 459–474.

Ferrati, R., G. Ana Canziani, and D. Ruiz Moreno. 2005. Esteros del Ibera: hydrometeorological and hydrological characterization. *Ecological Modelling* 186(1): 3–15.

Florida Oceans and Coastal Council (FOCC). 2010. Climate change and sea-level rise in Florida: an update of the "The effects of climate change on Florida's ocean and coastal resources." [2009 report] Tallahassee, Florida, 26p.

Gaiser, E.E., D.L. Childers, R.D. Jones, J.H. Richards, L.J. Scinto, and J.C. Trexler. 2006. Periphyton responses to eutrophication in the Florida Everglades: cross-system patterns of structural and compositional change. *Limnology and Oceanography* 51(1): 617–630.

Gaiser, E.E., J.C. Trexler, and P. Wetzel. 2012. The Everglades. In *Wetland Habitats of North America: Ecology and Conservation Concerns*, edited by D. Batzer and A. Baldwin, 231–252. Berkeley: University of California Press.

Gleason, P.J., and P. Stone. 1994. Age, origin, and landscape evolution of the Everglades peatland. In *Everglades: The Ecosystem and Its Restoration*, edited by S.M. Davis and J.C. Ogden, 149–197. Boca Raton, FL: CRC Press.

Gorham, E. 1991. Northern peatlands: role in the carbon cycle and probable responses to climatic warming. *Ecological Applications* 1(2): 182–195.

Hatt, D., and L. Collado-Vides. A comparative analysis of the organic and inorganic carbon content of *Halimeda* and *Penicillus* (Chlorophyta, Bryopsidales) in a coastal subtropical lagoon. *Botanica Marina*. In Press.

He, D., R.N. Mead, L. Belicka, O. Pisani, and R. Jaffé. 2014. Assessing biomass contributions to particulate organic matter in a subtropical estuary: a biomarker approach. *Organic Geochemistry* 75:129–139.

Hillis, L. 1997. Coralgal reefs from a calcareous green alga perspective, and a first carbonate budget. *Proceedings of the 8th International Coral Reef symposium* 1: 761–766.

Ho, D.T., S. Ferrón, V.C. Engel, L.G. Larsen, and J.G. Barr. 2014. Air-water gas exchange and CO_2 flux in a mangrove-dominated estuary. *Geophysical Research Letters* 41(1): 108–113.

Hohner, S.M., and T.W. Dreschel. 2015. Everglades peats: using historical and recent data to estimate predrainage and current volumes, masses and carbon contents. *Mires and Peat* 16: 1–15.

Hopkinson, C.S., W. Cai, and A. Hu. 2012. Carbon sequestration in wetland dominated coastal systems—a global sink of rapidly diminishing magnitude. *Current Opinion in Environmental Sustainability* 4(2): 186–194.

IPCC. 2013. *Climate Change 2013: The Physical Science Basis*. Contribution of Working Group I to the Fifth Assessment Report of the Intergovernmental Panel on Climate Change, edited by Stocker, T.F., D. Qin, G.-K. Plattner, M. Tignor, S.K. Allen, J. Boschung, A. Nauels, Y. Xia, V. Bex, and P.M. Midgley. Cambridge and New York: Cambridge University Press.

Jaffe, R., J.N. Boyer, and J.W. Fourqurean. 2007. *Technical Annual Report for NOAA Project NA04NOS4780187*. Miami: Florida International University.

Jaffé, R., J.N. Boyer, X. Lu, N. Maie, C. Yang, N.M. Scully, and S. Mock. 2004. Source characterization of dissolved organic matter in a subtropical mangrove-dominated estuary by fluorescence analysis. *Marine Chemistry* 84(3): 195–210.

Jerath, M. 2012. An economic analysis of carbon sequestration and storage service by mangrove forests in Everglades National Park, Florida. FIU Electronic Theses and Dissertations. Paper 702.

Jerath, M., M. Bhat, V.H. Rivera-Monroy, E. Castañeda-Moya, M. Simard, and R.R. Twilley. 2016. The role of economic, policy, and ecological factors in estimating the value of carbon stocks in Everglades mangrove forests, South Florida, USA. *Environmental Science & Policy* 66: 160–169.

Jimenez, K.L., G. Starr, C.L. Staudhammer, J.L. Schedlbauer, H.W. Loescher, S.L. Malone, and S.F. Oberbauer. 2012. Carbon dioxide exchange rates from short- and long-hydroperiod Everglades freshwater marsh. *Journal of Geophysical Research* 117(G4): G04009.

Karamperidou, C., V. Engel, U. Lall, E. Stabenau, and T.J. Smith III. 2013. Implications of multiscale sea level and climate variability for coastal resources. *Regional Environmental Change* 13(1): 91–100.

Koch, G.R., D.L. Childers, P.A. Staehr, R.M. Price, S.E. Davis, and E.E. Gaiser. 2012. Hydrological conditions control P loading and aquatic metabolism in an oligotrophic, subtropical estuary. *Estuaries and Coasts* 35(1): 292–307.

Kramer, P.J., and J.S. Boyer. 1995. *Water Relations of Plants and Soils*. San Diego, CA: Academic Press.

Larsen, L.G., J.W. Harvey, G.B. Noe, and J.P. Crimaldi. 2009. Predicting organic floc transport dynamics in shallow aquatic ecosystems: Insights from the field, the laboratory, and numerical modeling. *Water Resources Research* 45(1): W01411.

Larsen, S.C., M. Foulkes, C.J. Sorenson, and A. Thompson. 2011. Environmental learning and the social construction of an exurban landscape in Fremont County, Colorado. *Geoforum* 42(1): 83–93.

Light, S.S., and J.W. Dineen. 1994. Water control in the Everglades: a historical perspective. In *Everglades: The Ecosystem and its Restoration*, edited by S.M. Davis and J.C. Ogden, 47–84. Delray Beach, FL: CRC Press.

Lovett, G.M., J.J. Cole, and M.L. Pace. 2006. Is net ecosystem production equal to ecosystem carbon accumulation? *Ecosystems* 9(1): 152–155.

Lu, X.Q., N. Maie, J.V. Hanna, D.L. Childers, and R. Jaffe. 2003. Molecular characterization of dissolved organic matter in freshwater wetlands of the Florida Everglades. *Water Research* 37(11): 2599–2606.

Maie, N., R. Jaffe, M. Toshikazu, and D.L. Childers. 2006. Quantitative and qualitative aspects of dissolved organic carbon leached from senescent plants in an oligotrophic wetland. *Biogeochemistry* 78(3): 285–314.

Maie, N., S. Sekiguchi, A. Watanabe, K. Tsutsuki, Y. Yamashita, L. Melling, K.M. Cawley, E. Shima, and R. Jaffe. 2014. Dissolved organic matter dynamics in the oligo/meso-haline zone of wetland-influenced coastal rivers. *Journal of Sea Research* 91: 58–69.

Maie, N. Y. Yamashita, R.M. Cory, J.N. Boyer, and R. Jaffé. 2012. Application of excitation emission matrix fluorescence monitoring in the assessment of spatial and seasonal drivers of dissolved organic matter composition: sources and physical disturbance controls. *Applied Geochemistry* 27(4): 917–929.

Maie, N., C. Yang, T. Miyoshi, K. Parish, and R. Jaffe. 2005. Chemical characteristics of dissolved organic matter in an oligotrophic subtropical wetland/estuarine ecosystem. *Limnology and Oceanography* 50(1): 23–35.

Malone, S.L., G. Starr, C.L. Staudhammer, and M.G. Ryan. 2013. Effects of simulated drought on the carbon balance of Everglades short-hydroperiod marsh. *Global Change Biology* 19: 2511–2523.

Malone, S.L., C.L. Staudhammer, H. Loescher, P.C. Olivas, S. Oberbauer, M.G. Ryan, J. Schedlbauer, and G. Starr. 2015. Ecosystem resistance in the face of climate change: a case study from the freshwater marshes of the Florida Everglades. *Ecosphere* 6(4): Article 57.

Malone, S.L., C.L. Staudhammer, S. Oberbauer, P.C. Olivas, M.G. Ryan, J. Schedlbauer, H. Loescher, and G. Starr. 2014. El Niño Southern Oscillation (ENSO) enhances CO_2 exchange rates in freshwater marsh ecosystems in the Florida Everglades. *PLoS ONE* 9(12): e115058.

McCormick, P.V., and J.A. Laing. 2003. Effects of increased phosphorus loading on dissolved oxygen in a subtropical wetland, the Florida Everglades. *Wetlands Ecology and Management* 11(3): 199–216.

McKee, K.L. 2011. Biophysical controls on accretion and elevation change in Caribbean mangrove ecosystems. *Estuarine, Coastal and Shelf Science* 91(4): 475–483.

McKee, K.L., D.R. Cahoon, and I.C. Feller. 2007. Caribbean mangroves adjust to rising sea level through biotic controls on change in soil elevation. *Global Ecology and Biogeography* 16(5): 545–556.

Mcleod, E., G.L. Chmura, S. Bouillon, R. Salm, M. Björk, C.M. Duarte, C.E. Lovelock, W.H. Schlesinger, and B.R. Silliman. 2011. A blueprint for blue carbon: toward an improved understanding of the role of vegetated coastal habitats in sequestering CO_2. *Frontiers in Ecology and the Environment* 9(10): 552–560.

McVoy, C.W., W. Park Said, J. Obeysekera, J. VanArman, and T.W. Dreschel. 2011. *Landscapes and Hydrology of the Predrainage Everglades*. Gainesville: University Press of Florida.

Myers, R.L., and J.J. Ewel (Editors). 1990. *Ecosystems of Florida*. Orlando: University of Central Florida Press.

Najjar, R.G., M. Friedrichs, and W.-J. Cai. 2012. Report of the U.S. East coast carbon cycle synthesis workshop. January 19–20, 2012, Ocean Carbon and Biogeochemistry Program and North American Carbon Program, 34 pp.

Najjar, R.G., C.R. Pyke, M.B. Adams, D. Breitburg, C. Hershner, M. Kemp, R. Howarth, M.R. Mulholland, M. Paolisso, D. Secor, K. Sellner, D. Wardrop, and R. Wood. 2010. Potential climate-change impacts on the Chesapeake Bay. *Estuarine, Coastal and Shelf Science* 86(1): 1–20.

Neto, R.R., R.N. Mead, J.W. Louda, and R. Jaffe. 2006. Organic biogeochemistry of detrital flocculent material (floc) in a subtropical, coastal wetland. *Biogeochemistry* 77(3): 283–304.

Nyman, J.A., R.J. Walters, R.D. Delaune, and W.H. Patrick, Jr. 2006. Marsh vertical accretion via vegetative growth. *Estuarine, Coastal and Shelf Science* 69(3-4): 370–380.

Obeysekera, J., J. Barnes, and M. Nungesser. 2015. Climate sensitivity runs and regional hydrologic modeling for predicting the response of the Greater Florida Everglades Ecosystem to climate change. *Environmental Management* 55(4): 749–762.

Pearlstine, L.G., E.V. Pearlstine, and N.G Aumen. 2010. A review of the ecological consequences and management implications of climate change for the Everglades. *Journal of the North American Benthological Society* 29(4): 1510–1526.

Pisani, O., J.W. Louda, and R. Jaffe. 2013. Biomarker assessment of spatial and temporal changes in the composition of flocculent material (floc) in the subtropical wetland of the Florida Coastal Everglades. *Environmental Chemistry* 10(5): 424–436.

Pisani, O., Y. Yamashita, and R. Jaffe. 2011. Photo-dissolution of flocculent, detrital material in aquatic environments: Contributions to the dissolved organic matter pool. *Water Research* 45(13): 3836–3844.

Poffenbarger, H.J., B.A. Needelman, and J.P. Megonigal. 2011. Salinity influence on methane emissions from tidal marshes. *Wetlands* 31: 831–842.

Price, R.M., P.K. Swart, and J.W. Fourqurean. 2006. Coastal groundwater discharge–an additional source of phosphorus for the oligotrophic wetlands of the Everglades. *Hydrobiologia* 569(1): 23–36.

Regier, P., H. Briceño, and R. Jaffé. 2016. Long-term environmental drivers of DOC fluxes: linkages between management, hydrology and climate in a subtropical coastal wetland. *Estuarine, Coastal & Shelf Science* 182. doi:10.1016/j.ecss.2016.09.017.

Romigh, M.M., Davis, S.E., V. H. Rivera-Monroy, and R.R. Twilley. 2006. Flux of organic carbon in a riverine mangrove wetland in the Florida Coastal Everglades. *Hydrobiologia* 569(1): 505–516.

Ross, M.S., J.F. Meeder, J.P. Sah, P.L. Ruiz, and G.J. Telesnicki. 2000. The Southeast Saline Everglades revisited: 50 years of coastal vegetation change. *Journal of Vegetation Science* 11(1): 101–112.

Rubio, G., and D.L. Childers. 2006. Controls on herbaceous litter decomposition in the estuarine ecotones of the Florida Everglades. *Estuaries and Coasts* 29(2): 257–268.

Schedlbauer, J.L., J.W. Munyon, S.F. Oberbauer, E.E. Gaiser, and G. Starr. 2012. Controls on ecosystem carbon dioxide exchange in short- and long-hydroperiod Florida Everglades freshwater marshes. *Wetlands* 32(5): 801–812.

Schedlbauer, J.L., S.F. Oberbauer, G. Starr, and K.L. Jimenez. 2010. Seasonal differences in the CO_2 exchange of a short-hydroperiod Florida Everglades marsh. *Agricultural and Forest Meteorology* 150(7-8): 994–1006.

Schlesinger, W.H. 1997. *Biogeochemistry: An Analysis of Global Change* (2nd ed.). San Diego, CA: Elsevier Science.

Servais, S.M., J.S. Kominoski, S.P. Charles, E.E. Gaiser, V. Mazzei, T.G. Troxler, and B.J. Wilson. 2019. Saltwater intrusion and soil carbon loss: Testing effects of salinity and phosphorus loading on microbial functions in experimental freshwater wetlands. *Geoderma* 337(1): 1291–1300.

Simard, M., K. Zhang, V.H. Rivera-Monroy, M.S. Ross, P.L. Ruiz, E. Castañeda-Moya, R.R. Twilley, and E. Rodriguez. 2006. Mapping height and biomass of mangrove forests in Everglades National Park with SRTM elevation data. *Photogrammetric Engineering and Remote Sensing* 72(3): 299–311.

Smith, K.A., T. Ball, F. Conen, K.E. Dobbie, J. Massheder, and A. Rey. 2003. Exchange of greenhouse gases between soil and atmosphere: interactions of soil physical factors and biological processes. *European Journal of Soil Science* 54(4): 779–791.

Smoak, J.M., J.L. Breithaupt, T.J. Smith, and C.J. Sanders. 2013. Sediment accretion and organic carbon burial relative to sea-level rise and storm events in two mangrove forests in Everglades National Park. *Catena* 104: 58–66.

Snyder, G.H., and J.M. Davidson. 1994. Everglades agriculture: past, present, and future. In *Everglades: The Ecosystem and its Restoration*, edited by S.M. Davis and J.C. Ogden, 85–115. Delray Beach, FL: CRC Press.

Timko, S.A., C. Romera-Castillo, R. Jaffe, and W.J. Cooper. 2014. Photo-reactivity of natural dissolved organic matter from fresh to marine waters in the Florida Everglades, USA. *Environmental Science: Processes & Impacts* 16(4): 866–878.

Troxler, T.G., E. Castaneda, E. Standen, and M. Martinez. 2017. Ecological Monitoring of Southern Everglades Wetlands, Mangrove Transition Zone and "White Zone" Interactions with Florida Bay. Annual report submitted to the South Florida Water Management District, West Palm Beach, FL. 71p.

Troxler, T.G., D.L. Childers, and C. J. Madden. 2014. Drivers of decadal-scale change in southern Everglades wetland macrophyte communities of the coastal ecotone. *Wetlands* 34(1): 81–90.

Troxler, T.G., C. Coronado-Molina, D. Rondeau, S. Krupa, S. Newman, M. Manna, R.M. Price, and F.H. Sklar. 2014. Interactions of local climatic, biotic and hydrogeochemical processes facilitate phosphorus dynamics along an Everglades forest-marsh gradient. *Biogeosciences* 11: 899–914.

Troxler, T.G., E. Gaiser, J. Barr, J.D. Fuentes, R. Jaffe, D.L. Childers, L. Collado-Vides, V.H. Rivera-Monroy, E. Castañeda-Moya, W. Anderson, R. Chambers, M. Chen, C. Coronado-Molina, S.E. Davis, V. Engel, C. Fitz, J. Fourqurean, T. Frankovich, J. Kominoski, C. Madden, S.L. Malone, S. F. Oberbauer, P. Olivas, J. Richards, C. Saunders, J. Schedlbauer, F.H. Sklar, T. Smith, J.M. Smoak, G. Starr, R.R. Twilley, and K. Whelan. 2013. Integrated carbon budget models for the Everglades terrestrial-coastal-oceanic gradient: current status and needs for inter-site comparisons. *Oceanography* 26(3): 98–107.

Tussenbroek, B.I., and J.K. Van Dijk. 2007. Spatial and temporal variability in biomass and production of psammophytic *Halimeda Incrassata* (Bryopsidales, Chlorophyta) in a Caribbean reef lagoon. *Journal of Phycology* 43(1): 69–77.

Voss, C.M., R.R. Christian, and J.T. Morris. 2013. Marsh macrophyte responses to inundation anticipate impacts of sea-level rise and indicate ongoing drowning of North Carolina marshes. *Marine Biology* 160(1): 181–194.

Wanless, H.R., and B.M. Vlaswinkel. 2005. Coastal landscape and channel evolution affecting critical habitats at Cape Sable, Everglades National Park, Florida. Everglades National Park Service, U.S. Department of Interior.

Webster, K.L., J.W. McLaughlin, Y. Kim, M.S. Packalen, and C.S. Li. 2013. Modelling carbon dynamics and response to environmental change along a boreal fen nutrient gradient. *Ecological Modelling* 248: 148–164.

Weston, N.B., R.E. Dixon, and S.B. Joye. 2006. Ramifications of increased salinity in tidal freshwater sediments: geochemistry and microbial pathways of organic matter mineralization. *Journal of Geophysical Research* 111: G01009.

Whelan, K. 2005. The successional dynamics of lightning-initiated canopy gaps in the mangrove forests of Shark River, Everglades National Park, USA. Florida International University, Miami.

Whiting, G.J., and J.P. Chanton. 1993. Primary production control of methane emission from wetlands. *Nature* 364(6440): 794–795.

Wilson B.J., S. Servais, V. Mazzei, S. Davis, E. Gaiser, J.S. Kominoski, J. Richards, F. Sklar, and T. Troxler. 2018a. Changes in ecosystem carbon cycling with increased salinity exposure in the coastal Florida Everglades. *Ecological Applications* 28(8): 2092–2108.

Wilson, B.J., S. Servais, S.P. Charles, S.E. Davis, E. Gaiser, J.S. Kominoski, J. Richards, and T. G. Troxler. 2018b. Declines in Plant Productivity Drive Carbon Loss from Brackish Coastal Wetland Mesocosms Exposed to Saltwater Intrusion. *Estuaries and Coasts* 41: 2147–2158.

Wilson, B.J., Servais, S., Charles, S. P., Mazzei, V., Kominoski, J.S., Gaiser, E., Richards, J., and Troxler, T. 2019. Phosphorus alleviation of salinity stress: effects of saltwater intrusion on an Everglades freshwater peat marsh. *Ecology* 00(00): e02672.

Ya, C., W.T. Anderson, and R. Jaffe. 2015. Assessing dissolved organic matter dynamics and source strengths in a subtropical estuary: Application of stable carbon isotopes and optical properties. *Continental Shelf Research* 92: 98–107.

Yamashita, Y., L.J. Scinto, N. Maie, and R. Jaffe. 2010. Dissolved organic matter characteristics across a subtropical wetland's landscape: Application of optical properties in the assessment of environmental dynamics. *Ecosystems* 13(7): 1006–1019.

Zhang, K., J. Dittmar, M. Ross, and C. Bergh. 2011. Assessment of sea level rise impacts on human population and real property in the Florida Keys. *Climatic Change* 107(1-2): 129–146.

Zieman, J.C., and R.T. Zieman. 1989. The ecology of the seagrass meadows of the west coast of Florida: a community profile. *U.S. Fish & Wildlife Service Biological Report* 85(7.25).

7

Exogenous Drivers

WHAT HAS DISTURBANCE TAUGHT US?

Stephen E. Davis III, Edward Castañeda-Moya, and Ross Boucek with Randolph Chambers, Ligia Collado-Vides, Carl Fitz, Jose D. Fuentes, Evelyn Gaiser, Michael Heithaus, Jennifer Rehage, Victor Rivera-Monroy, Jay Sah, Fred Sklar, and Tiffany Troxler

In a Nutshell

- The Everglades ecosystem has been shaped over millennia by press and pulse disturbance events and, more recently, by human activities that have altered hydrologic conditions and habitats system-wide.
- Exotic plant and animal species also represent a disturbance to the Everglades, altering community structure and food webs.
- Current research on the impacts of sea level rise and saltwater intrusion in the oligohaline ecotone will help predict the stability of peat soils as coastal wetlands migrate landward, or transgress.
- Coastal mangrove wetlands are a significant barrier to storm surge in the Everglades, are a trap for storm surge–delivered sediments, and are important for carbon sequestration and soil building as sea level rise continues to accelerate.
- Hurricanes have both immediate and lasting impacts on ecosystem function via storm surge, high wind, and rainfall, leading to vegetation loss, the inland deposition of marine sediment, and changes in water quality.

- Cold-spell events impact tropical species that are at the northern extent of their range in south Florida, leading to mass mortality, changes in behavior, and long-term changes in community structure and function.

Introduction

A common discovery from long-term research is that disturbance is a fundamental exogenous driver shaping and maintaining ecosystems. For this reason, disturbance has been one of five common core research areas for the U.S. Long Term Ecological Research (LTER) Network since its inception in 1980. Pickett and White (1985) defined disturbance as "any relatively discrete event in time that disrupts ecosystem, community, or population structure and changes resources, substrate availability, or the physical environment." This definition points to the initial "disruption" in population or community structure as much as it alludes to subsequent changes in the feedbacks between abiotic and biotic components of an ecosystem. It also recognizes scales of space and time in considering the influence of disturbance.

While we can rather easily identify discrete disturbance events in space and time, it is not always easy to characterize and classify these events in terms of their impacts on biotic and abiotic components of an ecosystem. Furthermore, it is often a challenge to understand how disturbance impacts play out over space and time (i.e., initial response vs. recovery), how discrete disturbance events interact to produce a unique outcome (e.g., two hurricane events in succession), and the complex interactions among environmental policies and management that modulate or amplify disturbance and ecosystem resilience. Holling's (2001) *adaptive cycle* provides a mechanism for understanding ecosystem reconfiguration following disturbance. However, understanding the interactions and shifts between different stages of this cycle requires consistent long-term data collection over a large area before, during, and following disturbance events. This is something to which LTER sites are ideally suited.

As described in the preface and Chapter 1, the Florida Coastal Everglades (FCE) is a social-ecological system subjected to a range of press and pulse dynamics (*sensu* Collins et al. 2011). The Everglades landscape has been shaped over millennia by changes in sea level, hurricanes, fires, floods, droughts, frosts, and human manipulations. At one extreme, the Everglades itself was born about 6,000 years ago as rising sea levels stabilized after the Wisconsin Glaciation. Sea level rise (SLR) has always been, and is, a foundational global-scale press. On a much shorter temporal scale the Everglades experiences a semi-annual pulse driven by its subtropical climate. Each year includes a relatively cool dry season and a wet season, the latter of which is characterized by episodic rainfall events that drive intra-annual fluctuations in water levels and hydroperiod, primary productivity, and predator–prey dynamics across the landscape.

Disturbances at different spatial scales are also critical in shaping landscapes, influencing recruitment, and driving materials exchange. Lightning strikes in the coastal Everglades create small, circular gaps in the mangrove canopy that allow for localized light penetration and mangrove seedling recruitment; these gaps increase landscape heterogeneity (Whelan 2005). At larger scales, the over-grazing of seagrass by sea urchins in Florida Bay leads to destabilization of sediments, more turbid water, and reduced light penetration, and ultimately to changes in resource availability and species composition (Peterson et al. 2002). Because Everglades wetland habitats are connected by the flow of water, both hydroperiod and water quality are fundamental drivers of ecological processes such as faunal distribution across the entire landscape (Sokol et al. 2014; Trexler et al. 2006). Large-scale water quality trends, as described in Chapters 4 and 5, follow global-scale forcings that are superimposed on seasonal oscillations (Briceño and Boyer 2010). In the mangrove ecotone, such seasonal patterns are periodically interspersed with storm events expressed as pulsed changes in source-dependent water quality, followed by a more gradual return to pre-event conditions (Davis et al. 2003, 2004). Across spatial scales, water quality in the coastal Everglades is largely a function of proximity to water control structures and to end-member sources (e.g., marine vs. freshwater; Childers et al. 2006).

While natural disturbance has always shaped the communities and habitats of the Everglades, we are now witnessing the more indelible mark of human-driven press–pulse changes across the landscape. Over the past century, human behaviors and actions—played out as changes in policy, infrastructure, water management, and widespread land use change—have altered the volume, timing, distribution, and quality of water in the Everglades (see Chapter 1). This has greatly affected the pulsed-system dynamics and pattern of natural disturbance (after Odum et al. 1995). Consequently, the Everglades must be viewed as a social-ecological system and needs to be considered in light of the more recently developed press–pulse model of Collins et al. (2011). A suite of actions—water diversion, impoundment, over-drainage, nutrient enrichment, and exotic species introductions— has led to shifts in macrophyte communities, fish abundances, tree island health, wading bird populations, fertility gradients, and landscape pattern (e.g., Armentano et al. 2006; Rehage and Trexler 2006; Sklar et al. 2002) and has contributed to changes in the overall health and resilience of the Everglades landscape. On a more global scale, human-induced climate change and accelerating rates of SLR are chronic press disturbances that are coming to the forefront of discussions on management and restoration of this ecosystem as well as its connection with south Florida's built environment.

In this chapter, we discuss the interaction of external drivers and local ecosystem processes through time in the coastal Everglades. We describe ecological responses to an array of press–pulse dynamics, including hurricanes, fires, and extreme cold spells. These events typify the variety of natural disturbances that shaped the Everglades from its earliest days. However, ecosystem responses to these press–pulse dynamics and recovery must be viewed through the social-ecological lens of human-driven changes in hydrology, water quality, and trophic dynamics. In effect, the latter forms the context for the former, resulting in potentially unknown outcomes. Given this, we speculate on changes in ecosystem responses and resilience to the increasing role that human decisions play in shaping the Everglades today, compared to the past, in order to address the impacts of large-scale Everglades restoration.

A Model for Understanding the Role of Disturbance

Conceptualizing this complex set of natural exogenous drivers and human-influenced interactions across scales of space and time is challenging. However, over the past 15 years we have refined our research questions and experimental designs to understand the spatial and temporal dynamics of the coastal Everglades, and how it interfaces with the 7 million Floridians who depend on it and live adjacent to it. We have positioned our research to capture the effects of a number of natural disturbance events and how the impacts of those events change through time. History tells us that the Everglades has experienced a number of different large-scale disturbance events over the past century, with some years experiencing a combination of drought, hurricanes, or cold events (Fig. 7.1). Since the establishment of the FCE LTER, we have seen only a single year with hurricane impacts (2005), a few years of drought, and the most extreme cold event on record (early 2010; see Fig. 7.1).

Through our observation of conditions before, during, and after these events, our collective experience has led to the development of a conceptual ecosystem-based model that depicts the interactions between external drivers (presses and pulses, both natural and human) and local ecological processes across the estuarine gradient from marine water to freshwater supply (after Collins et al. 2011). This Everglades panarchical model captures the human component in terms of market forces, demographics, and policies that affect the health, management, and restoration of the ecosystem. The model also places these sets of conditions within the greater context of time, allowing for an understanding of how things have changed as well as an ability to make projections under different scenarios of future states. However, to predict future states of the coastal Everglades, we must first look to its past in order to understand how human actions and policies have shaped the present conditions and trajectories of change.

Human-Induced Press Disturbances

RESHAPING THE EVERGLADES

In the late 1800s and the early 1900s, long stretches of canals were dug in attempts to drain the relatively pristine Everglades for agriculture and early development. Hurricanes and devastating floods led to construction of the Herbert Hoover Dike around Lake

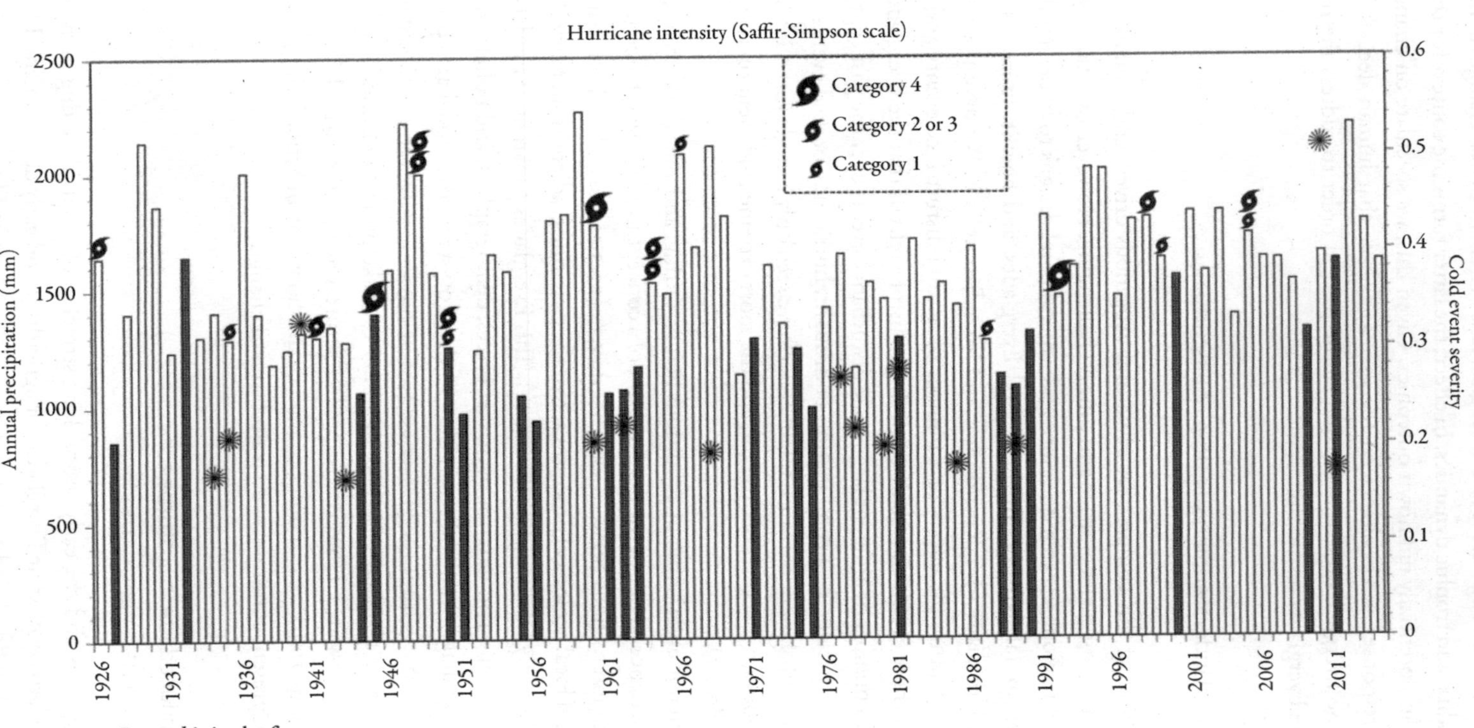

FIGURE 7.1 Legend is in the figure.

Okeechobee and federal authorization (1948) of the Central and South Florida (C&SF) Project, creating an elaborate network of canals, levees, and water control structures to improve regional flood control and water supply (Light and Dineen 1994; see Chapter 1 for details). The C&SF Project combined with more recent urban and agricultural development has led to an approximately 50% reduction in spatial extent of the Everglades and has fragmented once-continuous Everglades wetlands into a series of large impoundments with dramatically altered flows and hydropatterns (see Chapter 4). With agricultural and urban runoff, much of the inflow to the Everglades carried higher loads of nutrients into historically oligotrophic wetlands, leading to large-scale ecosystem degradation (McCormick et al. 2002). Collectively, these impingements to the Everglades have created an altered regime of flooding, hydroperiod, fire, and eutrophication that has affected nearly all trophic and landscape components of the ecosystem (Sklar et al. 2005; see Chapter 5).

As described in Chapter 8, a variety of projects are under way to restore the Everglades by optimizing management of hydrology and water quality, two fundamental drivers of Everglades ecology. Central to these efforts is the Comprehensive Everglades Restoration Plan (CERP), a multibillion-dollar, multi-decadal project that is attempting to restore the remnant Everglades. Under the CERP, there will be significant decompartmentalization (i.e., removing levees that impound parts of the Everglades) and increased water storage and flows in the Greater Everglades system (USACE and SFWMD 1999). The most significant opportunity to jumpstart restoration will be the Central Everglades Planning Project (CEPP), an integration of multiple CERP projects that will focus on the core Everglades ecosystem, including Everglades National Park (ENP; Davis et al. 2014; USACE and SFWMD 2014). As CEPP and other CERP and non-CERP restoration projects (e.g., Tamiami Trail bridging to increase freshwater flow to ENP) are built, we expect to see incremental improvements in hydrologic conditions in the coastal Everglades. Our research will continue to elucidate ecosystem responses to disturbance and restoration and, ultimately, will allow us to understand how improved conditions will enhance ecosystem resilience.

FRESHWATER DIVERSION AND DROUGHT

Although restoration is progressing, a significant amount of fresh water is still being diverted away from the Everglades (CISRERP 2014). This diversion leads to reduced flows, shortened hydroperiods, and increased incidence and frequency of drought-like conditions across much of the landscape. Drought conditions can be expressed in a variety of ways. The Palmer Drought Severity Index is one tool often used to characterize drought from a water supply perspective. Going back to 1926, there have been at least 22 years with multiple months of severe or extreme drought between Lake Okeechobee and the coastal Everglades, including 2015 (see Fig. 7.1). These events are a function of regional rainfall patterns and factors influencing evapotranspiration over several months and sometimes spanning multiple years.

In the coastal Everglades, water diversion and extended periods of unusually low rainfall, low water stages, and related low-flow conditions are characterized by fires, a decreased capacity for marsh carbon storage, and hypersalinity in Florida Bay (Malone et al. 2013; Rudnick et al. 2005; Smith et al. 2015). These drought expressions are tightly coupled, but antecedent storage conditions and water management operations can either exacerbate or moderate the ecological effects of a drought year (Abtew et al. 2012). Similarly, we have identified two functional types of drought: (1) several consecutive years of low rainfall or (2) a single, abnormally dry year with a prolonged dry season. These often result in different ecosystem responses.

Drought effects on water levels and flows are often exacerbated in ENP by water management infrastructure upstream, and a primary objective of restoration is to alleviate this. It is important to note that ecological responses to droughts differ from regular dry season patterns. Everglades wetlands are adapted to the natural seasonal variation in rainfall such that ecosystem function, and population and community dynamics, do not vary considerably across normal hydrologic years (Boucek and Rehage 2014a; David 1996; D'Odorico et al. 2011; Trexler et al. 2005). While both seasonal drying and droughts create harsh conditions for some organisms, drought conditions often exceed the resistance and resilience capacity of some of these species, causing large mortality events and leading to the complete restructuring of many communities. As a result, following droughts, many Everglades ecosystem characteristics differ from pre-drought conditions, sometimes for several years (Boucek and Rehage 2014a; David 1996; Trexler et al. 2005). We posit that droughts are distinguishable from normal dry seasons because droughts are rainfall deficit events that result in long-term changes to Everglades ecosystems.

Droughts and extended dry conditions alter Everglades ecosystems in a number of ways. At the downstream end in Florida Bay, extended periods of reduced freshwater inflow have had dramatic impacts on vertebrate fauna ranging from small prey base fish to roseate spoonbills and the American crocodile (Lorenz 2014). In the freshwater marshes, drought can result in severe disturbance to long-hydroperiod, peat-forming marshes as droughts of even short duration can lead to accelerated rates of soil oxidation in the exposed peats, increasing the potential for peat fires (Bruland et al. 2007; Malone et al. 2013). In contrast, dry soil conditions are a regular, seasonal phenomenon in short-hydroperiod, marl-forming marshes, and these marshes seem to maintain carbon uptake capacity under seasonally dry conditions (Schedlbauer et al. 2010). Thus, short-hydroperiod, marl-forming marshes in the Everglades may be more resilient to drought disturbance. This resilience is evidenced by the rapid response of periphyton mats in these marshes to desiccation and recovery following rehydration (Gottlieb et al. 2005; Thomas et al. 2006).

Water quality in the coastal Everglades is affected by drought, with impacts ranging from increases in salinity to changes in the availability of ecologically important nutrients such as phosphorus (P). These changes in water quality can also influence patterns of aquatic metabolism (Koch et al. 2012). During dry years or at the height of the dry season, water column nutrient concentrations, particularly total P, are highest (Briceño et al. 2014; Koch et al. 2012). This increase is likely linked to upstream oxidation of peat soils, the efflux and mobilization of soluble or leachable components following rewetting, P-rich groundwater discharge, or a shift toward labile carbon limitation in the mangrove ecotone (Bruland et al. 2007; Davis and Childers 2007; Gottlieb et al. 2005; Price et al. 2006). Likewise, high rates of evapotranspiration during droughts also accelerate shallow groundwater flow from marsh soils to tree island soils, contributing to nutrient and salt accumulation in island soils (Ross and Sah 2011; Troxler et al. 2014; Wetzel et al. 2011). Nutrients derived from groundwater and soils can then be rapidly sequestered at the onset of the wet season by the benthic community (e.g., periphyton mats), returning the water column to very low concentrations (Thomas et al. 2006).

Drought also affects aquatic faunal communities, most notably through altered dry season prey concentration required by economically important species such as wading birds and popular sportfish, specifically snook (Lorenz 2014). Drought-induced drying out of the marsh is responsible for large mortality of small fishes in the marsh (Boucek and Rehage 2014a; Trexler et al. 2005). In the year following a drought, prey fish numbers (e.g., sunfishes) are reduced by 60% to 70%. These reductions in prey biomass are likely due to high mortality of marsh fishes coupled with reduced growth in the wet season following drought, both of which reduce sunfish population sizes (Boucek and Rehage 2014a). We have also found longer-term legacy effects of drought on dry season prey concentrations. For instance, 2 years following drought, the biomass of sunfish found in refugia returned to predisturbance conditions, but sunfish size structure was skewed toward larger-bodied (>10 cm), spring-spawned individuals. This size structure shift was likely related to different predation pressures on sunfish by recovering populations of marsh piscivorous fishes (Boucek and Rehage 2014b) and perhaps wading birds.

We have observed similar effects of drought on the density of forage fishes consumed by wading birds in other regions of the Everglades. For instance, in marshes where the return frequency of drought conditions was less than 5 years, we postulated that densities of both forage fish and piscivorous fish were relatively low, as both predators and prey were in a continuous phase of recovery from disturbance. By comparison, in marshes where droughts and marsh drying occurred once every 10 years, communities of large-bodied piscivores were in a continuous phase of recovery while prey were not (Trexler et al. 2005). This relaxes predation pressures on smaller forage fishes, thus increasing forage fish densities and prey availability to wading birds. In Everglades marshes where the return frequencies of drought were greater than 10 years (WCA 3A), piscivore populations were in higher abundance, and subsequent densities of small fishes that supplement wading bird reproduction were lower relative to marshes that dried more frequently. To this end, droughts likely play an important role in regulating the dynamics of prey species that subsidize important consumers, including both wading birds and important estuarine gamefish such as snook. With forecasted increases in the intensity and frequency of drought in the Everglades, this disturbance will likely become increasingly important in affecting predator–prey dynamics and the provisioning of iconic Everglades species.

CLIMATE CHANGE AND SLR

The impacts of climate variability on the coastal Everglades has been studied through long-term analyses and retrospective studies, and these findings are being coupled with climate models to forecast the future of the landscape using a scenarios analysis approach (Obeysekera et al. 2015). The National Climate Assessment Report (Melillo et al. 2014) predicts increased warming, leading to decreased water availability. A recent modeling effort by Obeysekera et al. (2011) shows that projected climate change and SLR will have significant implications on the water supply and flood control capacity for the 7 million-plus residents of south Florida. Translating this to the Everglades ecosystem, climate change scenarios indicate a warmer, drier ecosystem—the ecological implications of which we understand through long-term studies on the impacts of water management and drought (e.g., Sklar et al. 2002). However, more research is needed to understand the coupled influence of changing thermal and moisture regimes with accelerating SLR (Nungesser et al. 2015).

With the rate of SLR increasing from warming oceans and melting ice caps, the impact of this human-induced press disturbance on the future Everglades is unknown. The landscape is too young for there to be a past analog. We know that the present-day Everglades ecosystem formed and is continually being shaped by SLR and other disturbance events such as hurricanes and frontal passages (Wanless et al. 1994). We have developed a better understanding of how SLR and water management have shifted coastal ecotone habitats landward, and we are currently studying the effect of increasing sea levels and saltwater intrusion on the stability of coastal ecosystems and the fate of carbon sequestered in their soils.

Over the past 100 years, sea levels in south Florida have risen approximately 22 cm (CISRERP 2014). This rate is based on more than 100 years of data from the Key West tide gauge—one of the longest-running records of sea level in the world. Since the early 1990s, the global rate of mean SLR has increased to about 3.2 mm every year, with projections of up to 52 to 98 cm additional rise by 2100 (IPCC 2013). And this is a conservative projection; for example, it does not account for growing instability in several major continental ice sheets. This accelerating rate of SLR is a significant resource management issue given that nearly two-thirds of ENP is less than 1 m above sea level. Recent work by Smoak et al. (2013) indicated that mangroves in ENP are barely keeping pace with the current rate of SLR, as mangrove soil accretion rates range from 2.5 to 3.6 mm per year.

Conventional wisdom holds that Everglades coastal habitats (e.g., mangroves) will gradually transgress upslope into freshwater sawgrass marshes with increases in sea level. We have documented this landward migration of mangroves and other coastal habitats over the past 50 to 100 years (Fuller and Wang 2014; Gaiser et al. 2006; Ross et al. 2000; Smith et al. 2013), and there is also strong evidence that water management (through reduction in freshwater flow) has accelerated this process in some areas (Davis et al. 2005; Smith et al. 2013). The construction of canals in areas such as Cape Sable (ENP) in the 1920s accelerated the penetration of salt water into freshwater marshes there, providing a glimpse into the future of the fate of freshwater marshes in the southern Everglades and lower Shark River Slough (Wanless and Vlaswinkel 2005). Without restoration of freshwater flow to the Everglades, saltwater intrusion-induced peat collapse will be enhanced, and this catastrophic loss of soil will likely preclude the landward migration of mangroves in areas with an abundance of freshwater peat soils (see Fig. 6.11; Chambers et al. 2014; Weston et al. 2011).

Wetlands across the Everglades have been accreting soils for the last several thousand years. Mangrove peat in the Everglades can be as deep as 5 m in some areas along lower Shark River, suggesting that mangroves and other soil-accreting ecosystems have been in those same locations for roughly 5 m of SLR. Indeed, pollen analysis and paleoreconstruction of a deep soil core from the mouth of the Shark River estuary indicated that mangrove forests started to appear at about 3,800 calendar years before the present (cal yr BP), and a fully established dense *Rhizophora mangle* forest had grown by about 1150 cal yr BP (Yao et al. 2015). Similar to mangroves along the coast, freshwater sawgrass marshes also build peat soil—albeit much more slowly (Gleason and Stone 1994; Smoak et al. 2013). In fact, peat soils in some Everglades freshwater sawgrass marshes are more than 1 m deep in some areas of Shark River Slough, and freshwater marsh soils are much deeper further north in the landscape.

Wanless and Vlaswinkel (2005) reported that freshwater marsh peats in Cape Sable and lower Shark River Slough that were exposed to salt water rapidly destabilized, then catastrophically collapsed. This may be due in part to salinity stress on the freshwater marsh plants that diminishes plant growth and thus the production of soil organic matter. On the other side of the balance, evidence from other coastal wetlands has shown a shift in soil redox potential as salt water intrudes and dramatic increases in soil decomposition rates via enhanced sulfate reduction (e.g., Weston et al. 2011). Recent evidence from Chambers et al. (2014) supported the occurrence of this shift, and we are currently conducting field and mesocosm experiments to explore the rates of peat collapse, the mechanisms behind it, and potential areas of vulnerability across the coastal Everglades.

The end result of a tipped soil balance is a collapse of the freshwater marsh (in some cases by 0.5 m or more over a period of a few decades or less) and conversion to an open-water, mangrove-free environment (Wanless and Vlaswinkel 2005). On the surface, these collapsed areas begin as a small pool surrounded by sawgrass with some intermittent mangrove trees (Fig. 7.2). Collectively, they represent a larger area of destabilized peat soil that can coalesce over time, growing into a larger aggregation of collapsed marsh. In Cape Sable, the collapsed area has exceeded 4,000 ha in size and continues to expand. Aside from the massive nutrient release associated with this scale of collapse, the collapsed areas

Close-up view showing
depth of collapse
View from the edge
of a collapsed pool

FIGURE 7.2 Scales of peat collapse in freshwater marshes of lower Shark River Slough experiencing saltwater intrusion. Images show about 20 cm of soil collapse around sawgrass plants that are likely < 10 years old. Collapsed pools rarely support vegetation and contain soft, unconsolidated organic sediments/soils. See page 14 of color insert.

are often too deep for mangroves to become established. As a consequence, these areas will remain as open-water habitats as sea level continues to rise and will ultimately shape the future geomorphology of south Florida.

Wanless and Vlaswinkel (2005) have hypothesized that the formation of Whitewater Bay in ENP is a past example of disturbance-driven, large-scale Everglades marsh collapse. What is now Whitewater Bay was originally near the southern terminus of Shark River Slough and the area was characterized by freshwater marsh dominated by water lily sloughs, sawgrass marshes, and thick peat soils (Gleason and Stone 1994). A rapid rise in sea levels 2,500 years ago led to the formation of Cape Sable around the southwest tip of the peninsula, causing a natural rerouting of Everglades flow from Shark River

Slough through the present-day Shark and Harney Rivers. They hypothesized that the abandoned network of marsh drainage channels above Cape Sable became vectors for saltwater inundation from storm surges and gradual SLR. Ultimately, this plus other disturbances such as freeze events and fire led to the breakdown and collapse of freshwater peat soils, with the end result being the formation of Whitewater Bay—a large estuarine embayment supporting vast seagrass beds and scattered mangrove islands (Wanless and Vlaswinkel 2005).

ESTABLISHMENT OF INVASIVE EXOTIC SPECIES

The introduction of invasive exotic species is a common threat to ecosystem structure and function in terrestrial and aquatic ecosystems (Pysek et al. 2008; Vellend et al. 2007). There are approximately 250 non-native plant species and 192 non-native animal species documented in the Greater Everglades Ecosystem (CISRERP 2014). Terrestrial plant invasions in and around the Everglades have been well documented and are associated with intentional introductions (e.g., *Melaleuca quinquenervia*) or invasions from the ornamental plant industry (e.g., *Schinus terebinthifolius*). For aquatic fauna in the Everglades, many species invasions are associated with intentional or accidental releases (e.g., aquarium fish, amphibians, snails, pet snakes). Regardless of the mode of introduction, plant species such as *M. quinquenervia* have greatly altered habitat quality across much of the Everglades ecosystem, and predatory species such as Burmese pythons continue to increase in population size and are altering food web structure across the Everglades—particularly in the small mammalian community (CISRERP 2014; Dorcas et al. 2012).

The scale of effort needed to control some species and difficulty of finding others (e.g., it is estimated that we are able to find and capture fewer than 1 out of every 100 pythons) in the Everglades has made invasive species management even more challenging (CISRERP 2014). The National Academy of Sciences, in their most recent 2014 biennial review of Everglades restoration for Congress, highlighted the need for more communication and information on invasive and exotic species. They recommended improved coordination among resource management agencies, enhanced control efforts, and expanded research to understand the impacts of species invasions—particularly as they pertain to the success of the restoration effort (CISRERP 2014).

In the aquatic realm, south Florida is one of the most invaded areas in the United States, with more than 500 species reportedly introduced (Florida Fish and Wildlife Conservation Commission 2014). About 34 non-native freshwater fish species are now established in Florida (Kline et al. 2014; Shafland et al. 2008). In ENP, 17 non-native fish species have established populations, and several of these species are abundant and dominant members of wetland communities (Kline et al. 2014; Trexler et al. 2000). In particular, aquatic habitats in the Rocky Glades, the coastal Everglades, and canals are highly invaded (Kobza et al. 2004; Rehage et al. 2014; Trexler et al. 2000). In canals, for instance, the degree of invasion is highly variable spatially, with non-native species representing less than 1% to 70% of all individuals, with an average of 18% being non-native (Rehage and Gandy 2014).

Since the Everglades freshwater fish fauna is relatively species depauperate (35 native species; Loftus 2000), these invasions represent close to a 50% increase in fish diversity, and thus account for major changes in the structure of these fish communities. However, despite this high invasion rate, our understanding of the effects of these invasions on the function of these communities remains limited. To date, few studies have documented any significant ecological effects of invasive and exotic fish in the Everglades, which has led to conflicting perspectives on the overall impact of non-native aquatic taxa across the landscape (Shafland 1996; Trexler et al. 2000). Empirical work that mechanistically examines interactions between native and non-native aquatic taxa and the implications for ecosystem functioning is also limited (but see Porter-Whitaker et al. 2012).

Marine species introductions are also a problem in the coastal Everglades, as both Florida Bay and Biscayne Bay are adjacent to large urban areas and in proximity to large ports (Miami and Fort Lauderdale). The most common vectors of invasive and exotic marine species are the aquarium trade, ballast water releases, and biofouling on ships (Booth et al. 2007; Cormaci et al. 2004; Geller et al. 2010; Gravili et al. 2010; Wallentinus 2002; Williams and Smith 2007). In the coastal Everglades, there are two primary invasive exotic species: the lionfish (*Pterois volitans* Linnaeus) and the green seaweed (*Caulerpa brachypus* Harvey). Both species live in Atlantic coral reefs (Florida Fish and Wildlife Conservation Commission 2014; Lapointe and Bedford 2010). Major detrimental effects of the lionfish are associated with its high fecundity, a lack of competitors or predators, and its carnivorous diet.

Two other invasive exotic seaweed species, the red algae *Laurencia caduciramulosa*, which grows primarily in seagrass beds (Collado-Vides et al. 2014), and the green tide forming *Ulva ohnoi* M. Hiraoka and S. Shimada were recently found in Biscayne Bay. These invasive exotic algal species are having detrimental and persistent impacts on the native macroalgae *Anadyomene stellate* (Wulfen) C. Agardh; another introduced species, *Anadyomene* sp., has displaced seagrasses in Biscayne Bay (Collado-Vides et al. 2013). We must continue to monitor *U. ohnoi* as well as the possible expansion of *Halophila stipulacea* (Forsskål) Ascherson, a non-native seagrass that is rapidly expanding across the Caribbean (Vera et al. 2014; Willette et al. 2014; Williams 2007). Both species are aggressive invaders and pose a threat to Florida Bay (Leliaert et al. 2009; Willette et al. 2014).

Hurricanes: A Frequent, Significant Natural Disturbance in South Florida

Hurricanes epitomize large-scale, pulsed disturbances that regularly affect south Florida, including the mangrove coastline and subtropical marshes of the coastal Everglades. Hurricanes passing within 50 miles of the south Florida coast have a recurrence frequency of once in every 6 to 8 years, which is one of the highest rates in the U.S. Atlantic Basin (National Hurricane Center 2014). From 1900 to 2012, the south Florida counties of Broward, Miami-Dade, and Monroe were struck by more major hurricanes than the rest of the state and any other county in the Atlantic and Gulf Coasts of the United States (Table 7.1). In addition to intense rainfall, hurricanes impact the Everglades through strong winds and storm surges that damage vegetation, scour seagrass beds, and transport and deposit salt water and sediments. These phenomena have lasting effects on water quality, landscape pattern, and vegetation structure, affecting the flow of water and habitat availability across the Everglades (Castañeda-Moya et al. 2010; Davis et al. 2004; Deng et al. 2010; Smith et al. 2009).

More noticeable and directly quantifiable impacts of hurricanes are associated with human and economic harm. The economic impacts of hurricanes in Florida are far-reaching, exceeding billions of dollars for repairs and reconstruction of private property and public infrastructure (Hebert and Taylor 1992). In fact, the economic losses in Florida due to hurricane impacts are estimated at $460 billion (normalized to the 2005 dollar) since the early 1900s. The Great Miami Hurricane (1926) ranks first in the top-10 list (1900–2007) with an estimated total damage to Florida of $129 billion. Hurricane Andrew comes in second with a damage tag of $52.3 billion (Pielke et al. 2008). Notably, three of the top-10 costliest Florida hurricanes occurred in 2004 (Hurricanes Ivan and Charley) and 2005 (Hurricane Wilma; Malmstadt et al. 2009).

Hurricane Wilma

The hurricane season of 2005 was one of the most active on record in the Atlantic Basin, with 27 named storms (Farris et al. 2007). Four of these storms made landfall in the northern Gulf of Mexico (Dennis, July 10; Katrina, August 25; Rita, September 24; Wilma, October 24). Hurricane Wilma made landfall as a Category 3 storm on the southwest Florida coast between Everglades City and Cape Romano on October 24, 2005, and tracked inland directly along our Shark River Slough (SRS) sampling transect (Fig. 7.3; Zhang et al. 2008). Maximum sustained winds were estimated to be about 190 km/h (120 mph; Pasch et al. 2006). The eye of the hurricane had a diameter of 89 to 105 km and a wind speed of 155 to 166 km/h (see Fig. 7.3; Zhang et al. 2008). At one point, Wilma was the most intense tropical cyclone ever recorded in the Atlantic, with a minimum central pressure of 882 mb (Pasch et al. 2006). Damage from Wilma's winds, storm surge, and flooding was extensive over south Florida, and with more than $21 billion in damages it was the fifth-costliest hurricane in the United States since 1851 (Blake et al. 2011).

TABLE 7.1.

Total Number of Hurricane Strikes and the Number of Major Hurricane Strikes by County in Florida, 1900–2010

County	Total Strikes	Major Strikes
Bay	14	6
Brevard	17	6
Broward	22	12
Charlotte	11	6
Citrus	3	1
Collier	16	6
Dixie	3	1
Duval	4	0
Escambia	14	8
Flagler	6	0
Franklin	11	3
Glades	14	4
Gulf	12	4
Hendry	16	4
Hernando	5	2
Hillsborough	8	4
Indian River	14	6
Jefferson	3	1
Lee	12	7
Levy	5	2
Manatee	6	3
Martin	18	7
Miami-Dade	25	14
Monroe	32	15
Nassau	3	0
Okaloosa	11	7
Okeechobee	12	4
Palm Beach	18	8
Pasco	4	2
Pinellas	8	4
Santa Rosa	13	7
Sarasota	7	4
St. Johns	4	0
St. Lucie	16	6
Taylor	2	0
Volusia	7	1

TABLE 7.1. CONTINUED

County	Total Strikes	Major Strikes
Wakulla	7	1
Walton	13	6

South Florida counties containing much of the Everglades ecosystem are shaded to indicate that most hurricane strikes occur in this part of the state.

Source: http://www.nhc.noaa.gov/climo/#uss

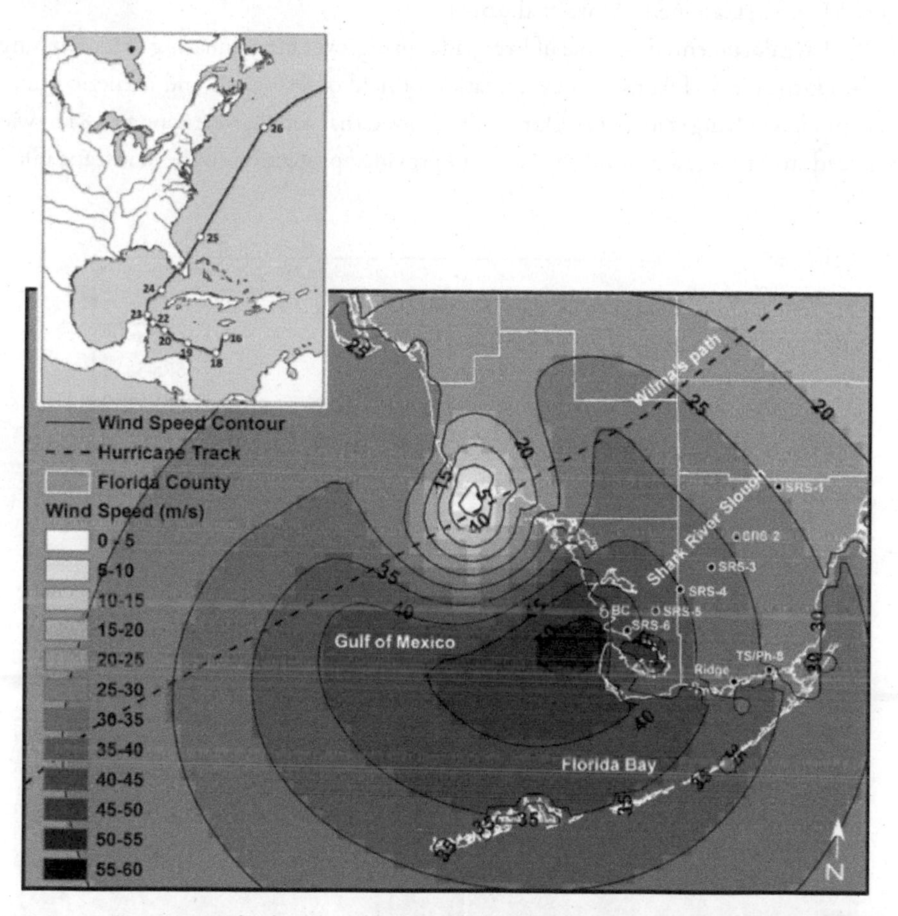

FIGURE 7.3 Hurricane Wilma's path and wind fields during its passage across south Florida (from Castañeda-Moya et al. 2010) and proximity to FCE sites in Shark River Slough (SRS-1-6) and Taylor Slough (TS/Ph-8). The insert shows the complete track of Wilma since it formed as a tropical storm in the Caribbean Sea on October 16, 2005.

STORM SURGE AND HYDROLOGY

Hurricane Wilma had numerous effects on the coastal Everglades, particularly on the mangrove forests. Generally, hurricane impacts on mangrove forests depend on the position of the hurricane track relative to the forest, the stature of the forest itself, the physical characteristics of the storm (i.e., intensity, radius to maximum wind speed, storm movement velocity), and the degree of protection offered by topographic features (Davis et al. 2004; Krauss et al. 2005; Piou et al. 2006; Zhang et al. 2008). Wilma produced extensive inundation along the southwestern Florida coast, with a maximum storm surge of 3 to 5 m in some mangrove stands (Castañeda-Moya et al. 2010; Krauss et al. 2009; Smith et al. 2009). Wilma's storm surge was attenuated significantly inland, from about 4 m near the mouth of Shark River (SRS-6) to less than 0.5 m 14 km further upstream (SRS-4), illustrating the valuable ecosystem service that mangroves provide to coastal areas (Fig. 7.4; Castañeda-Moya et al. 2010).

We have documented the role of Everglades mangroves in attenuating the storm surge from Hurricane Wilma using a combination of field observations and numeric simulation models (Zhang et al. 2012). Our results showed that a mangrove zone 7 to 8 km wide reduced storm surge amplitude by 80% and provided protection to the landscape inland

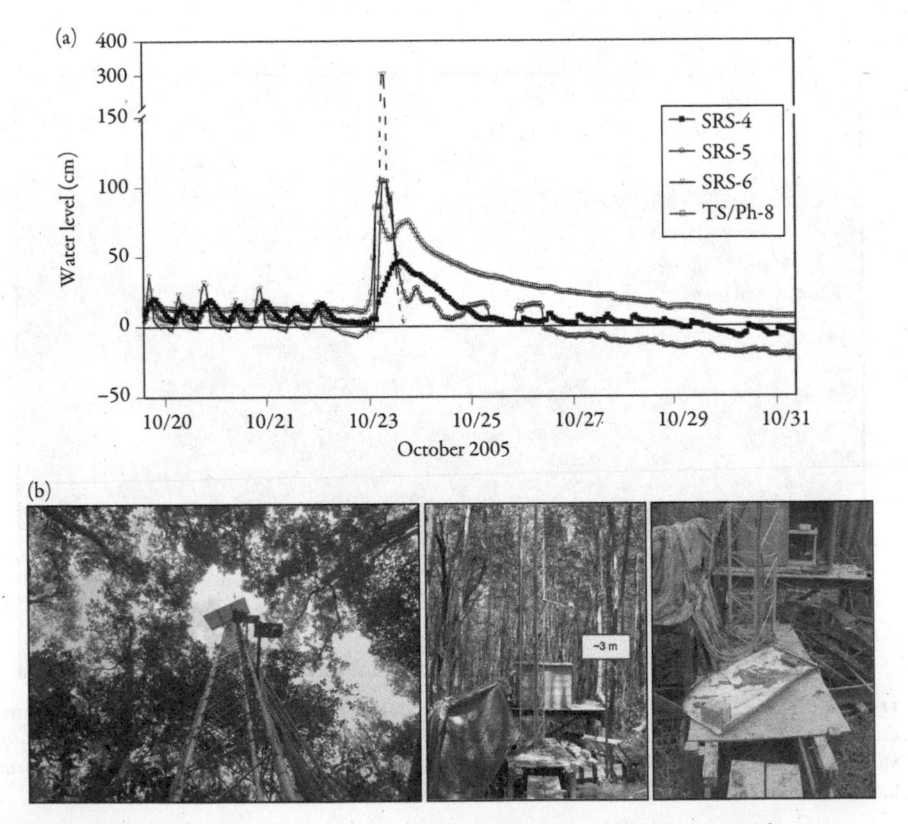

FIGURE 7.4 Water levels at FCE mangrove sites during the passage of Hurricane Wilma on October 24, 2005 (0 = the soil surface at each site). Hurricane water levels ranged from 3-4 m at downstream locations (SRS-6, 4.1 km from the mouth) to 0.46 m 18.2 km from the mouth (SRS-4). Water levels at SRS-6 were estimated from field observations after the storm, including water marks on tree trunks and sediment deposition observed on equipment at the eddy flux tower located 100 m from the shore; photos). The water level instrument installed at this site was 1.5 m above the soil surface, preventing any measurements above that level. The storm surge lasted approximately 7-8 h (Castañeda-Moya et al. 2010). See page 12 in color insert.

of the mangrove zone. These model results also showed a surge attenuation rate of 40 to 50 cm/km when mangroves were present, but only 20 cm/km with a mixture of mangrove islands and open water. In comparison, without the presence of a mangrove zone, the surge attenuated at a rate of only 6 to 10 cm/km (Zhang et al. 2012). These results highlight the buffering capacity of mangroves against storm surges and their significant role in shoreline protection. This is particularly significant given the projected increases in sea level in the future.

Further upstream, in freshwater marshes of Shark River Slough, Deng et al. (2010) noted direct wind-induced shifts in patterns of sheet flow as Hurricane Wilma passed. They also found that strong storms such as Wilma influence the mobilization and distribution of organic sediment ("floc") across the marsh, thus affecting vegetation patterning in the landscape (Deng et al. 2010; see Chapter 6). Harvey et al. (2009) suggested that strong storms such as Wilma could have long-term effects on flow direction and velocity in some areas of the Everglades, ultimately affecting the interaction between wetland hydrology and a number of important ecological processes (Deng et al. 2010).

SEDIMENT DEPOSITION

Hurricane Wilma's storm surge transported large amounts of mineral sediment from the coastal shelf into the mangrove forests of southwestern ENP, along a 70-km stretch of coastline from Lostman's River to Flamingo (Castañeda-Moya et al. 2010; Smith et al. 2009; Fig. 7.5). The thickness of Wilma-derived sediment deposits in mangrove forests varied depending on location but in general was less than 10 cm thick (see photos in Fig. 7.5). Maximal deposition occurred near the mouths of estuaries (e.g., Big Sable Creek, Shark River, Lostman's River), and there was no evidence of deposition further than 15.5 to 18.2 km from the Gulf of Mexico (Castañeda-Moya et al. 2010; Smith et al. 2009). A distinct pattern of deposition was also evident from edge to interior mangrove forests (Castañeda-Moya et al. 2010). We found the highest sediment deposition within 250 m of the shoreline; it gradually decreased in the interior forest. A similar pattern of deposition was associated with Hurricane Andrew in 1992 (Risi et al. 1995; Smith et al. 1994). However, we estimate that the total area impacted by Wilma's storm deposits (400 km^2) was about 3.5 times larger than the area affected by Andrew. Notably, Hurricane Andrew's storm track was east to west while Wilma's was southwest to northeast (Smith et al. 2009).

Sediment deposition was also evident in mangrove forests along the Buttonwood Ridge of northeastern Florida Bay, but we found no observable storm deposits in the scrub mangrove wetlands inland of the ridge. The Buttonwood Ridge (1 km wide; ~0.5 m in height) is a depositional berm that stretches about 60 km across the southern tip of Florida. It acts as a barrier and is a more favorable depositional environment during a storm surge, and thereby isolates these scrub mangrove areas from the direct influence of Florida Bay during storm events (Castañeda-Moya et al. 2010; Davis et al. 2004). We

FIGURE 7.5 Storm-derived inorganic sediment (gray) deposited on mangrove peats (dark brown) across mangrove forests in the Florida Coastal Everglades after Hurricane Wilma. See page 13 of color insert.

observed similar depositional patterns at these locations in 1999 following the passage of Hurricane Irene—a weaker storm than Wilma that also approached the Florida peninsula from the southwest (Davis et al. 2004).

MANGROVE FOREST STRUCTURE

Mangrove damage from Hurricane Wilma, including defoliation, tree snapping, and uprooting, was related to storm surge and proximity to the eyewall of the storm (Zhang et al. 2008). The spatial extent of damage spanned mangrove areas from Big Sable Creek, south of Shark River, to approximately 10 km north of the Ranger Station at Lostman's River. Scattered areas of damage were observed outside this region (Smith et al. 2009). Along lower Shark River (e.g., SRS-6 and mouth of the estuary), Wilma caused severe damage to the mangrove forest, with 99% of the canopy defoliated and several large trees (i.e., tree height >15 m) downed, broken, or uprooted (Fig. 7.6). Hurricane effects were less severe toward the freshwater Everglades (SRS-4), where wind fields caused approximately 20% to 30% canopy defoliation but no downed or broken trees (E. Castañeda-Moya, unpublished data). Our Light Detection and Ranging (LIDAR) analyses of the Shark River and Broad River areas before and after the 2005 hurricane season showed that the total area of gaps in the study area increased from 1% to 2% to 12% after the

FIGURE 7.6 Mangrove forests at SRS-6 in Shark River estuary before (a) and after (b and c) the passage of Hurricane Wilma. Panel d shows hurricane damage to the mangrove forest at the mouth of this estuary. See page 13 of color insert.

storms (Zhang et al. 2008). These results highlight the importance of hurricane-induced disturbance in regulating mangrove forest structure in the coastal Everglades.

Analysis of tree density and mortality rates from 2001 to 2009 revealed distinct trends along Shark River before and after Hurricane Wilma (see example in Fig. 7.7). Before the storm (i.e., the period from 2001 to 2004), total tree densities ranged from 111 to 117 individuals/ha; after the storm, tree density declined to 95 or 96 individuals ha^{-1} in 2006

FIGURE 7.7 Panoramic view of the SRS-6 mangrove site in Shark River estuary before (left: December 2004) and after (right: December 2005) Hurricane Wilma. Note the location of the sampling platform.

and then to 75 to 77 individuals ha^{-1} in 2009, indicating delayed mortality. Annual average mortality increased from 2.3 ± 0.9% (2001–2004) to 11.2 ± 1.9% (2006–2009) as a result of Wilma impacts (Rivera-Monroy, unpublished data). Based on analysis of aerial photographs (0.33-m resolution), our first-order estimate was that 1,250 ha of mangrove forest was impacted by Hurricane Wilma (Smith et al. 2009).

CARBON AND NUTRIENT INPUTS

Woody debris represents a significant pool of carbon and nutrient storage in mangrove forests (Krauss et al. 2005; Robertson and Daniel 1989; Romero et al. 2005), particularly in areas with high hurricane recurrence such as south Florida. Nevertheless, estimates of downed wood in coastal Everglades mangroves after hurricane disturbances are limited to few studies (Krauss et al. 2005; Castañeda-Moya, unpublished data). After Wilma, our December 2005 estimates of total woody debris volume across mangrove areas at our SRS sites ranged from 170 to 259 m^3 ha^{-1}. We found approximately twice the volume of woody debris in mangrove areas closest to Wilma's eyewall (Castañeda-Moya, unpublished data). This is consistent with results reported by Krauss et al. (2005) after the passage of Hurricane Andrew in 1992. In contrast, woody debris estimates from Micronesian mangrove forests are considerably lower (35–104 m^3 ha^{-1}) due to differences in age and structure (i.e., lower canopies; Allen et al. 2000). Furthermore, our wood decay experiments have demonstrated the significant role of woody debris on nutrient dynamics (Romero et al. 2005). These results showed that between 17% and 68% of the total phosphorus (TP) in woody debris leached out during the first two months. Newly deposited wood from living trees was a short-term source of N into the soil, but the wood became a net sink of N after 2 years of decomposition. These findings underscore the significant input of nutrients from decomposing wood into mangrove soils and the potential role of large pools of woody debris following hurricane disturbances. Thus, we expect that hurricane-induced input of woody debris into mangrove soils will continue to have a lasting impact on carbon and nutrient storage and budgets in the coastal Everglades.

Nutrient subsidies from hurricane events have been recognized as a critical component in the nutrient biogeochemistry of mangroves and near-shore areas of the coastal Everglades (Castañeda-Moya et al. 2010; Davis et al. 2004). Allochthonous mineral inputs (i.e., calcium-bound P) from the Gulf of Mexico, transported and deposited during storm events, control soil formation and the observed patterns in community structure and productivity in this landscape (Castañeda-Moya et al. 2013; Chen and Twilley 1999a, 1999b; Davis et al. 2004). Storm-derived deposits from Hurricane Wilma made significant contributions to the TP pool of mangrove soils at our mangrove sites (Castañeda-Moya et al. 2010). Specifically, we found that these sediments increased soil TP content by 20% to 54%, and the mean TP content of the deposited sediment was 1.6 times higher (0.36 ± 0.02 mg cm^{-3}) than the underlying mangrove soils (top 10 cm: 0.22 mg cm^{-3}). Calcium-bound P was the largest bioavailable fraction of P in storm deposits across all mangrove sites, accounting for up to 25% to 29% of TP (Castañeda-Moya et al. 2010).

P deposition from Hurricane Wilma supplemented existing gradients in productivity that we have documented across the coastal Everglades landscape (per Childers et al. 2006). In the southwestern Everglades, mangrove areas near the mouths of estuarine rivers have higher forest development and productivity as a result of being closer to the marine source of P (Castañeda-Moya et al. 2013; Chen and Twilley 1999a; Simard et al. 2006). In contrast, inland mangrove areas in this region and those to the north and east of Florida Bay have lower P content in their soils, thus resulting in P-limited environments and restricted forest growth and development (Castañeda-Moya et al. 2013). This trend of lower P in mangroves along the eastern region of Florida Bay parallels Fourqurean et al. (1992), who reported a similar P gradient in seagrass communities across the bay. These patterns are explained in large part by the "upside-down" nature of Everglades estuaries and the marine end-member source of P that controls primary production in these oligotrophic, P-limited ecosystems (Childers et al. 2006; Gaiser et al. 2015).

Our long-term monitoring of TP in Shark River revealed a multi-year legacy effect of Hurricane Wilma's P-rich sediment deposition on water quality in Everglades mangrove waterways (Fig. 7.8). Before Wilma, mean TP in Shark River ranged from 0.29 to 0.46 μM, with increasing concentrations towards the Gulf of Mexico. Following passage of

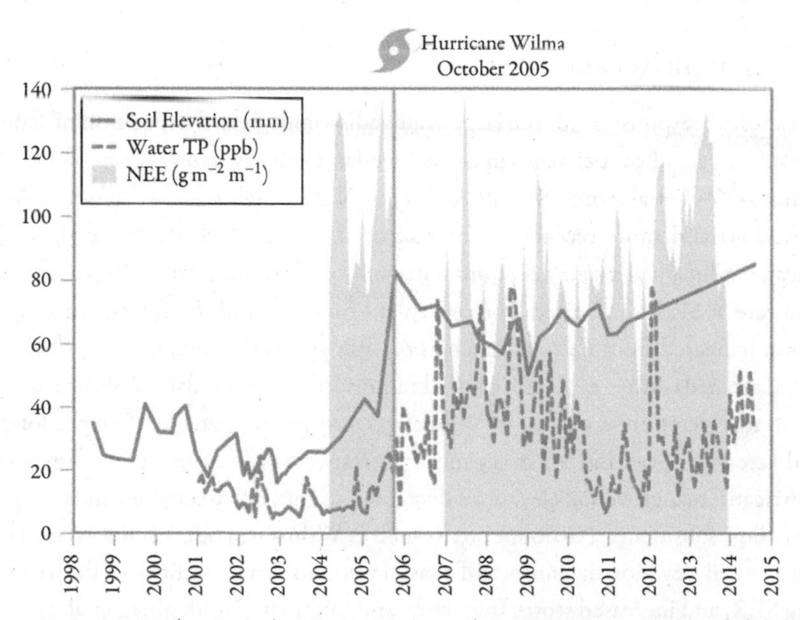

FIGURE 7.8 Long-term (1998-2015) variation in water column TP concentrations, soil elevation, and mangrove forest net ecosystem exchange (NEE; see Chapter 6) at the SRS-6 mangrove site before and after Hurricane Wilma.

Wilma, mean TP in the surface water increased sharply over the next few years until 2008, when concentrations began to decline and eventually reached pre-storm levels in 2011 (see Fig. 7.8). Similarly, we found significant long-term increases in soil pore water soluble reactive P (SRP) concentrations at the same sites in response to Wilma—particularly at the SRS-6 site, where pore water SRP increased twofold to threefold during the dry and wet seasons of 2006 relative to pre-storm years (Castañeda-Moya et al. 2013). After 2007, pore water SRP returned to pre-Wilma values (Rivera-Monroy, unpublished data).

Our pre- and post-Wilma observations illustrate the immediate and lasting changes in nutrient pools across the coastal Everglades as a result of hurricane disturbances, and highlight the significance of these events in regulating patterns of productivity across this oligotrophic coastal landscape. Multiple storms in the same season may have an even greater, perhaps synergistic, effect on nutrient storage and exchange in the Everglades. Davis et al. (2004) showed that, independent of sediment deposition, a combination of storm events can have a significant impact on the net flux of nutrients (N and P) and fresh water between relatively non-tidal estuarine creeks of the Southern Everglades and northeastern Florida Bay. In 1999, Tropical Storm Harvey and Hurricane Irene (both wet storms) struck south Florida within a month of one another. Because 1999 was an otherwise a dry year for south Florida, these two storms accounted for 60% to 65% of that year's total loads of N and P to Florida Bay, and precipitation from them sustained high discharge and low salinity levels throughout the remainder of the 1999 wet season (Davis et al. 2004).

LONG-TERM SOIL ACCRETION

Hurricanes are responsible for the large-scale redistribution and deposition of sediments that results in significant changes in coastal wetland soil elevations (Cahoon et al. 1995; Castañeda-Moya et al. 2010; Nyman et al. 1995; Whelan et al. 2009). Work by Whelan et al. (2009) and more recently by Smoak et al. (2013) in Shark River showed that Hurricane Wilma had a significant influence on soil elevation. After Wilma, the soil accretion rate at SRS-6 increased from 6.5 to 11.5 mm yr^{-1}, and the long-term rate of soil elevation increased from 1.4 to as much as 6.5 mm yr^{-1} (Smoak et al. 2013; Whelan et al. 2009). Castañeda-Moya et al. (2010) used radioisotope (^{137}Cs) data to show that the increase in soil accretion as a result of Wilma was 8 to 17 times greater than the long-term annual rate (3.0 ± 0.3 and 2.7 ± 0.3 mm yr^{-1}, respectively). However, within a year of the hurricane, mangrove soil elevations decreased 10 mm due to erosion and compaction from shallow subsidence. No doubt events such as Wilma will be critical drivers of coastal Everglades soil elevation dynamics and coastal wetland sustainability in a future of accelerating SLR, and increased storm frequency and intensity (Goldenberg et al. 2001; Saha et al. 2011).

The Post-Hurricane Resilience of Coastal Everglades Mangrove Forests

We have been measuring long-term patterns of mangrove ecosystem productivity using both direct biometric approaches (e.g., litterfall and aboveground wood production) and eddy covariance gas exchange methods since 2000. These data have greatly contributed to our understanding of ecosystem responses to pulse–press disturbances (Rivera-Monroy et al. 2011, 2013). Following damage to the SRS-6 tower from Wilma (see Fig. 7.4), eddy covariance gas exchange estimates of carbon (C) fluxes before and after the storm have allowed us to understand functional responses of this mature mangrove forest to a disturbance of this magnitude. High winds from Wilma caused about 99% forest defoliation in the upper canopy and widespread tree mortality. The defoliated canopy allowed greater amounts of sunlight to reach the soil surface. As a result, there was a greater coupling of the forest surface to the atmosphere. The leafless trees allowed more efficient transport of carbon dioxide and water vapor from the soil surface to the upper canopy layer. We also found significant increases in nighttime net ecosystem exchange after the storm as soil respiration increased and dead tree biomass decomposed. The dramatic reductions in foliage meant lower gross primary production (GPP) compared with pre-storm conditions (Fig. 7.9).

One year after Wilma, changes in the vertical distribution and the degree of clumping in newly emerged leaves corresponded with progressive increases in GPP. Wilma's

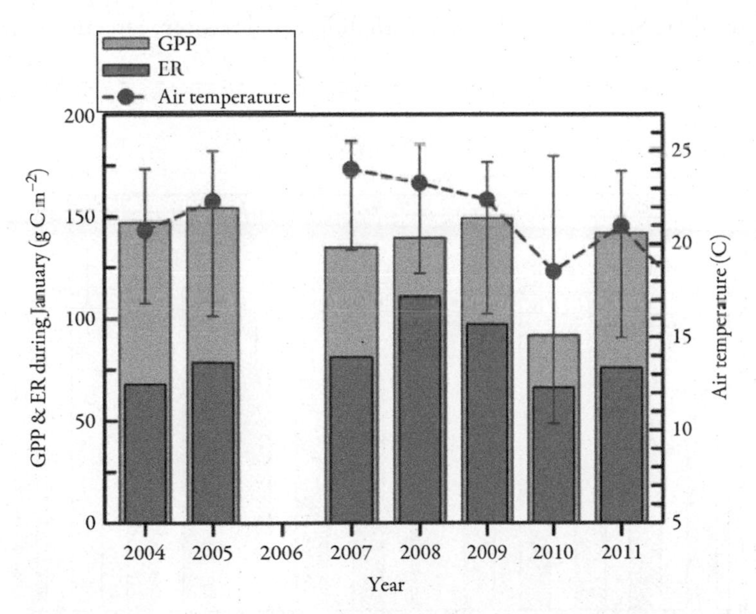

FIGURE 7.9 Gross primary productivity (GPP), ecosystem respiration rates (ER), and air temperature from the SRS-6 mangrove flux tower site from of 2004 to 2011 illustrate the impacts of Hurricane Wilma, which moved over the Everglades on 24 October 2005. Net ecosystem exchange (NEE) data are required to estimate GPP and ER (GPP = - NEE + ER). Missing data in 2006 are because the tower was destroyed and took time to replace.

long-term impacts became apparent over subsequent years, as C assimilation rates were still 250 g C m^{-2} yr^{-1} lower in 2009 compared to 2004–2005. We also found that dry season C assimilation rates were more influenced by the disturbance than those of the wet season (Barr et al. 2012). Complex leaf regeneration dynamics (resprouting and epicormic growth) on damaged trees during ecosystem recovery likely resulted in the variable dry versus wet season impacts on daytime GPP. In contrast, nighttime ecosystem respiration steadily increased after 2005, driven largely by decomposition of litter and coarse woody debris generated by the storm (see Fig. 7.9). The largest pre- to post-storm differences in GPP coincided with the delayed tree mortality that we observed in subsequent years following the storm.

Litterfall is one of the most common methods to measure aboveground net primary productivity (NPP_L) in mangrove forests. We have been using this approach at our Shark River sites since 2001. Litterfall production in these forests is composed mostly of leaves (>70% total annual) followed by twigs and reproductive parts (Castañeda-Moya et al. 2013). We saw an immediate impact of Hurricane Wilma on mangrove litterfall, with the largest decrease in production at SRS-6 (2006), followed by an extended recovery (Fig. 7.10). Wilma's impact on this forest was reflected in the significant 65% reduction in the canopy C contribution to the forest floor in the year after the storm. After 2006, annual NPP_L gradually increased to pre-storm levels (4.8 Mg C ha^{-1} yr^{-1}) by 2009 (see Fig. 7.10). This pattern of litterfall production recovery after the storm was evident at other mangrove sites along Shark River, although with differential recovery rates due to less severe

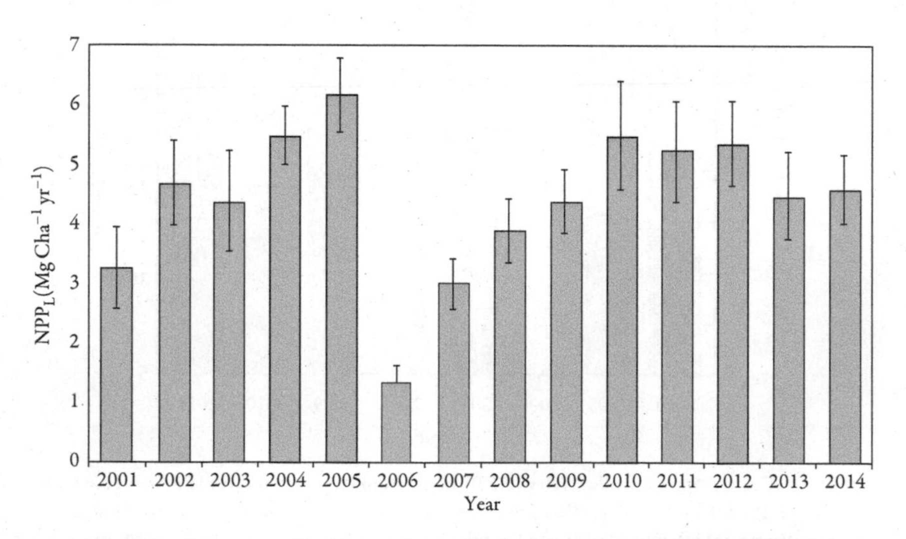

FIGURE 7.10 Annual mangrove forest net primary productivity measured as Litterfall (NPP_L; Mg C ha^{-1} ± SE) at SRS-6 from 2001-2013 (Rivera-Monroy, unpublished data). Data from 2001-2005 are from Castañeda-Moya et al. (2013).

hurricane impacts at the intermediate (SRS-5) and upstream (SRS-4) regions along this estuary (Danielson et al. 2017).

From a species-specific perspective, the dominant mangrove species at our SRS-6 site, *Laguncularia racemosa* (43%), showed the fastest leaf recovery following Hurricane Wilma. The litterfall contribution of this species decreased significantly (-1.5 Mg C ha^{-1} yr^{-1}) immediately after the storm, but this trend reversed within a year to 1.74 ± 0.1 Mg C ha^{-1} yr^{-1} (2007–2013) that eventually offset C losses caused by the hurricane (Danielson et al. 2017). Further, *L. racemosa* maintained a high leaf production rate after the disturbance and reached a steady state after 2010, with rates comparable to pre-Wilma values (1.94 ± 0.1 Mg C ha^{-1} yr^{-1}). The combination of high light regimes due to canopy defoliation and high fertility conditions in lower Shark River favored *L. racemosa*'s competitive dominance over *Rhizophora mangle* and *Avicennia germinans*, particularly when average annual interstitial soil salinities were less than 35 ppt (Cardona et al. 2006; Castañeda-Moya et al. 2013). The forest structure dominance by *L. racemosa* suggests that a history of regular hurricane disturbances is an important driver of community structure in this forest (Castañeda-Moya et al. 2013; Chen and Twilley 1999b).

Since NPP is associated with the ecosystem service of C sequestration and may be easily used as an ecosystem performance measure (Twilley and Rivera-Monroy 2005), these data will also help us to assess the long-term effects of Everglades restoration on the coastal Everglades. Indeed, we expect that upstream restoration efforts will influence seasonal hydrology and nutrient dynamics across the coastal Everglades, although the long-term impact on mangrove ecosystem structure and function will also be greatly influenced by accelerating rates of SLR. Continued long-term ecological research on mangrove ecosystem structure and function in response to disturbances of this type will be essential to that end.

Fire, Flood, and Ice

Aside from hurricanes, other extreme and episodic weather disturbances also play major roles in shaping coastal Everglades ecosystems. We define disturbances, both extreme (more severe and less frequent) and episodic (less severe and more frequent), as discrete weather events that exhibit climatic parameters (e.g., temperature, daily rainfall) that are statistical outliers in the period of record and create conditions that often exceed the acclimation capacity of some species, thus causing large mortality events (Gutschick and Bassir Rad 2003). Besides hurricanes, the extreme and episodic disturbances that have strong footprints on the Everglades are droughts, fires, flooding, and cold spells. Drought effects were described earlier in the chapter; in this section, we describe fires, flooding, and cold spells and summarize their ecological effects on the Everglades landscape.

FIRE AND FLOODING

Everglades plant communities are adapted to both flooding and fire. As we have already described, seasonal flooding exerts a consistent, pervasive control over ecosystem process and pattern. Fire is also quite common and widespread throughout the Everglades. Smith et al. (2015) synthesized fire data from ENP between 1948 and 2010. They described 2,588 fires that burned more than 1.8 million acres in the park and found a significant negative relationship between area burned each year from natural and suppressed fires and peak dry season water level. The rate and extent of vegetation recovery from fire depends on vegetation type, soil characteristics, fire intensity, and both pre- and post-fire hydrologic conditions. For instance, in seasonally flooded marl prairies, vegetation recovery to pre-burn structure and composition after a single burn event takes about 3 to 4 years (La Puma et al. 2007). But sequential disturbances, such as fire followed by immediate flooding, either delay the vegetation recovery process or cause a change in community structure by removing dominant species and facilitating the growth of opportunistic species (Sah et al. 2011, 2012). We have used trajectory analysis of pre-burn and post-burn vegetation composition data to show that recovery takes a more circuitous route back to pre-fire conditions when fire is immediately followed by flooding as opposed to gradual reflooding of a site. Sites with extreme post-fire flooding remain different from pre-burn conditions even 8 years after the fire (Sah et al. 2012).

Tree islands are also sensitive to fire and flooding. Following a large marl prairie fire around a tree island, post-fire hydrology and tree island size were the primary drivers that determined tree island recovery (Ruiz et al. 2013). Even a month of extended post-fire hydroperiod slowed down recovery of woody vegetation, ultimately affecting the contraction or even continued existence of tree islands.

Interacting multiple disturbances, such as fire and flooding, are not always deleterious. Rather, the synergistic effects of these two disturbances may lead to enhanced spatial and temporal heterogeneity across the landscape, ultimately increasing plant diversity. In fact, fire and flooding may be used as a tool to restore short-hydroperiod marsh and tree island habitats in the Everglades. For instance, in marl prairies or short-hydroperiod marshes that have been overgrown by mesic woody vegetation, prescribed fire followed by flooding may reset the plant communities to earlier stages of succession where herbaceous species again dominate. However, maintenance of such communities would require regular use of fire and flooding to prevent the recurrence of woody vegetation.

COLD SPELLS

Extreme cold spells are an episodic result of equatorial shifts in the jet stream that deflect arctic air into the subtropics, causing short durations of frigid temperatures and often resulting in large mortality events in the Everglades (Boucek and Rehage 2014b; Downton and Miller 1992). The frequency and intensity of these events are correlated with Pacific North American teleconnection patterns, and 80% of Florida's ecologically impactful cold spells occur during the positive phase of the Pacific North American teleconnection (Sheridan 2003; see Chapter 4). These teleconnections with the northern Pacific Ocean result in extreme cold events (i.e., the top 1% most impactful cold spells) with return frequencies of approximately 30 to 40 years. These extreme events have occurred in January 1940, 1981, and 2010 (Boucek and Rehage 2014b; see Fig. 7.1). At the onset of these climate extremes, winter minimum air temperatures decrease almost overnight from average minimum temperatures of 18° to 20°C to temperatures at or below 0°C, producing severe thermal stress for many species.

Because of the direct and univariate response to cold front stressors, the immediate effects of cold spells on ecological communities are easier to predict compared to other disturbances (Boucek and Rehage 2014b). In the Everglades, these disturbances may have catastrophic, yet predictable, impacts on species abundances and community structure. Everglades communities, both floral and faunal, include cold-intolerant tropical species at the northern fringe of their range, and a less diverse group of temperate, cold-resistant species at their equatorial range limit (Lodge 2005). Tropical species affected by cold spells range from primary producers, such as mangroves, to long-lived large-bodied top predators such as Burmese pythons and bull sharks (Matich and Heithaus 2012; Mazzotti et al. 2011; Ross et al. 2009). Following cold-spell disturbances, there is often a temporary increase in temperate species dominance as the presence of tropical species decreases.

While tropical species do recover from cold spells, ecosystem processes temporarily change. Mangrove responses to cold snaps include significant defoliation, major reductions in water and vapor fluxes, and resultant impacts on water quality (see earlier in this chapter). For instance, as a result of the 2010 cold spell we measured mangrove litterfall rates of 40 to 58 g C m^{-2} mo^{-1} at our Shark River sites, which was two to three times the long-term average leaf litter production rate (Danielson et al. 2017). Cold spells also have major effects on trophic structure and food web dynamics. For instance, the cold spell of 2010 had significant effects on the behavior and apparent survival of some large predators, while it appears to have had a limited impact on others. Juvenile bull sharks (*Carchahinus leucas*), for example, were hard-hit by the cold spell (Matich and Heithaus 2012). During a period of several weeks, all sharks being tracked inside the Shark River estuary either emigrated from the system into warmer coastal waters or died in the estuary. Only one tagged shark returned to the system after the disturbance, and abundances of sharks were depressed for many months. In addition, we only captured young-of-the-year sharks (born after the cold snap) in the first year after the cold spell (Matich and Heithaus 2012). After several years, the age structure of sharks in the estuary was still different from that of multiple years preceding the cold snap. Although fewer data are available, it appears that American alligators (*Alligator mississippiensis*) were less impacted by this event, and it is even possible that they benefited from the huge pulse of food in the system as thousands of tropical fish died (Boucek and Rehage 2014b).

Interestingly, one of the important benefits of cold spells in the Everglades is their ability to regulate populations of some invasive exotic species. Many non-native species in south Florida are tropical, including Brazilian red pepper, some pollinating bees, cichlid fishes, Burmese pythons, and iguanas. We have found that extreme cold events, such as the 2010 event, cause extensive population declines in many tropical invasive species. For non-native fishes, cold-induced mortality is considerably higher in shallow marsh habitats. Man-made canals, on the other hand, provide thermal refuges during extreme cold events and likely play a key role in the long-term persistence and recovery of non-native populations throughout the landscape. One difference between cold-spell effects on non-native versus tropical native fishes is that non-natives recover at a slower pace than natives (Boucek and Rehage 2014b).

Whether ecologically harmful cold spells will increase in frequency and intensity in the future will largely depend on climate change–related alterations to mid-tropospheric circulation dynamics over the Northern Pacific (Sheridan 2003; see Chapter 4). Kodra et al. (2011) predicted that, at the very least, the intensity and frequency of extreme cold spells impacting Florida will remain constant and may actually increase. In combination with climate warming, future ecological effects of extreme cold spells may become more severe. For instance, as average air temperatures increase, we may expect the range of temperate species in South Florida to retract north and away from South Florida. This range shift will increase the dominance of tropical species, increasing their influence on ecosystem function. Thus, following future cold spells, we may expect to see severe changes to ecosystems as novel tropical species communities essentially reset.

Disturbance Interactions

Positive, negative, or synergistic effects of multiple disturbances can be predicted using disturbance-resilience theory (Gunderson 2000). Resilience theory predicts that shifts between ecosystem states depend on the internal resilience of a given state and the magnitude of external forces (disturbances). For example, disturbances—small and frequent or large and infrequent—may or may not cause an ecosystem to cross a state change threshold and shift to a new, possibly irreversible, state. However, if a system has diminished resilience as it recovers from a disturbance, it may be more sensitive to subsequent disturbances that will either reset the recovery process or cause a regime shift to a new system state. Thus, the interactions of disturbances that occur simultaneously or in close sequence often have multiplicative, rather than additive, effects on biologic communities and ecosystems. Interacting multiple disturbances often result in changes in community characteristics that are different from, and less predictable than, the effects of individual disturbances in isolation. Moreover, the effects of compounding multiple perturbations are controlled by their sequential order, intensity, and spatiotemporal variability, including the intervals between disturbances (Collins and Smith 2006; Fukami 2001; Paine et al. 1998). Rapidly compounded disturbances with short recurrence intervals can have severe effects on community structure, occasionally resulting in ecosystem state change.

The duration and depth of inundation, phosphorus availability, and salinity are key long-term drivers of ecosystem structure and function in the coastal Everglades (Childers et al. 2003; Noe et al. 2001). As we have explained, episodic events such as hurricanes, drought, fire, and cold spells are acute disturbances that can result in dramatic, rapid shifts with long-term effects. The interacting influences of these drivers vary across spatial and temporal scales and may result in positive, negative, or synergistic outcomes. Furthermore, the effects of long-term drivers and acute disturbances are exacerbated by human influences on water management and climate. In fact, the interaction of human and natural disturbance defines and continues to shape the present-day Everglades landscape.

Summary

The coastal Everglades is a disturbance-prone landscape that provides an ideal context for understanding how disturbances can interact to influence ecosystem resilience. It is characterized by long-term press disturbances (e.g., SLR, climate change, drainage and water diversion) interacting with short-term pulses (e.g., tropical storms, fire, floods, cold snaps) that provide an interesting mosaic for exploring disturbance ecology. Waide and Willig (2012) suggested that simultaneous disturbances of different spatiotemporal scales may change the distribution of ecological characteristics in space in ways that create emergent, or surprising, landscape properties. Our work on press–pulse interactions in the Everglades is aligning with other studies suggesting that presses reduce resilience to pulses (*sensu* Lugo and Scatena 1995). For example, the long-term press of SLR is changing the susceptibility of mangrove forests to storm surges, but resilience is enhanced where soil accretion keeps up with SLR. Tropical storms are heterogeneous in their characteristics and their impacts (Foster et al. 1998), but ecosystem response is also dependent on antecedent conditions as well as system resilience.

There is a complicated mosaic of shorter-term disturbances in the Everglades that have also provided insight into how disturbances influence resilience in space and time. Repeated fires increase resilience to the next, somewhat like the frequently burned prairies of the shortgrass steppe (Turner et al. 2003), and when combined with floods or alterations in hydrology these fires may actually increases landscape-level diversity, similar to the interactions of fire and grazing in tall-grass prairies (Knapp et al. 1999).

Long-term reductions in freshwater flows have accelerated the effects of SLR and saltwater intrusion, reducing the stability of coastal sawgrass marshes and their peat soils. The collapse of peat soils in response to these interacting presses would not have been captured, or its causes understood, without long-term attention to this system. It is an example of an "ecological surprise" that is appearing more often in the ecological literature as datasets are extended and drivers of change become more complex with human modification (Dodds et al. 2012). The interaction of climate change and exotic species is another arena where surprises, or novel outcomes, may prevail, especially when observed over long timeframes. Our long-term, persistent study of the interactions among urban and wildland systems in south Florida, through a press–pulse lens, will become increasingly important as the combination of disturbances facing the planet and our species becomes increasingly complex and unpredictable.

References

Abtew, W., C. Pathak, and V. Ciuca. 2012. Chapter 2: Regional Hydrology. 2012 South Florida Environmental Report, edited by G. Redfield. Pp. 2-1–2-38. West Palm Beach: South Florida Water Management District.

Allen, J.A., K.C. Ewel, B.D. Keeland, T. Tara, and T.J. Smith. 2000. Downed wood in Micronesian mangrove forests. *Wetlands* 20:169–176.

Armentano, T.V., J.P. Sah, M.S. Ross, D.T. Jones, H.C. Cooley, and C.S. Smith. 2006. Rapid response of vegetation to hydrological changes in Taylor Slough, Everglades National Park, Florida, USA. *Hydrobiologia* 569:293–309.

Barr, J.G., V. Engel, T.J. Smith, and J.D. Fuentes. 2012. Hurricane disturbance and recovery of energy balance, CO_2 fluxes and canopy structure in a mangrove forest of the Florida Everglades. *Agricultural and Forest Meteorology* 153:54–66.

Blake, E.S., C.W. Landsea, and E.J. Gibney. 2011. The deadliest, costliest, and most intense United States tropical cyclones from 1851 to 2010 (and other frequently requested hurricane effects). NOAA Technical Memorandum NWS NHC-6. National Hurricane Center, Miami.

Booth, D., J. Provan, and C.A. Maggs. 2007. Molecular approaches to the study of invasive seaweeds. *Botanica Marina* 50: 385–396.

Boucek, R.E., and J.S. Rehage. 2014a. Climate extremes drive changes in functional community structure. *Global Change Biology* 20:1821–1831.

Boucek, R.E., and J.S. Rehage. 2014b. Examining the effectiveness of consumer diet sampling as a nonnative detection tool in a subtropical estuary. *Transactions of the American Fisheries Society* 143(2): 489–494.

Briceno, H.O., and J.N. Boyer. 2010. Climatic controls on phytoplankton biomass in a subtropical estuary, Florida Bay, USA. *Estuaries and Coasts* 33(2): 541–553.

Briceño, H.O., G. Miller, and S.E. Davis. 2014. Relating freshwater flow with estuarine water quality in the southern Everglades mangrove ecotone. *Wetlands* 34(1): 101–111.

Bruland, G.L., T.Z. Osborne, K.R. Reddy, S. Grunwald, S. Newman, and W.F. DeBusk. 2007. Recent changes in soil total phosphorus in the Everglades: Water Conservation Area 3. *Environmental Monitoring and Assessment* 129: 379–395.

Cahoon, D.R., D.J. Reed, J.W.J. Day, G.D. Steyer, R.M. Boumans, J.C. Lynch, D. McNally, and N. Latif. 1995. The influence of Hurricane Andrew on sediment distribution in Louisiana coastal marshes. *Journal of Coastal Research* 21: 280–294.

Cardona-Olarte, P., R.R. Twilley, K.W. Krauss, and V.H. Rivera-Monroy. 2006. Responses of neotropical mangrove seedlings grown in monoculture and mixed culture under treatments of hydroperiod and salinity. *Hydrobiologia* 569: 325–341.

Castañeda-Moya, E., R.R. Twilley, and V.H. Rivera-Monroy. 2013. Allocation of biomass and net primary productivity of mangrove forests along environmental gradients in the Florida Coastal Everglades, USA. *Forest Ecology and Management* 307: 226–241.

Castañeda-Moya, E., R.R. Twilley, V.H. Rivera-Monroy, K. Zhang, S.E. Davis, III, and M. Ross. 2010. Sediment and nutrient deposition associated with Hurricane Wilma in mangroves of the Florida Coastal Everglades. Estuaries and Coasts 33: 45–58.

Chambers, L.C., S.E. Davis, T.G. Troxler, J.N. Boyer, A. Downey-Wall, and L.J. Scinto. 2014. Biogeochemical effects of simulated sea level rise on carbon loss in an Everglades mangrove peat soil. *Hydrobiologia* 726(1): 195–211.

Chen, R., and R.R. Twilley. 1999a. A simulation model of organic matter and nutrient accumulation in mangrove wetland soils. *Biogeochemistry* 44: 93–118.

Chen, R., and R.R. Twilley. 1999b. Patterns of mangrove forest structure and soil nutrient dynamics along the Shark River Estuary, Florida. *Estuaries* 22: 955–970.

Childers, D.L., J.N. Boyer, S.E. Davis, C.J. Madden, D.T. Rudnick, and F.H. Sklar. 2006. Relating precipitation and water management to nutrient concentration patterns in the oligotrophic "upside down" estuaries of the Florida Everglades. *Limnology and Oceanography* 51(1): 602–616.

Childers, D.L., R.F. Doren., G.B. Noe, M. Rugge, and L.J. Scinto. 2003. Decadal change in vegetation and soil phosphorus patterns across the Everglades landscape. *Journal of Environmental Quality* 32: 344–362.

CISRERP. 2014. Progress Toward Restoring the Everglades: The Fifth Biennial Review, 2014. Committee on Independent Scientific Review of Everglades Restoration Progress, National Research Council of the National Academies.

Collado-Vides, L., C. Avila, S. Blair, F. Leliaert, D. Rodriguez, T. Thyberg, S. Sheneider, J. Rojas, P. Sweeney, C. Drury, and D. Lirman. 2013. A persistent bloom of *Anadyomene* J. V. Lamouroux (Anadyomenaceae, Chlorophyta) in Biscayne Bay, Florida. *Aquatic Botany* 111: 95–103.

Collado-Vides, L., V. Cassano, J. Díaz-Larrea, A. Duran, A. Da Silva Medeiros, A. Senties, and M. Toyota Fujii. 2014. Spread of the introduced species *Laurencia caduciramulosa* (Rhodomelaceae, Rhodophyta) to the northwest Atlantic: a morphological and molecular analysis. *Phytotaxa* 183(2): 93–107.

Collins, S.L., S.R Carpenter, S.M. Swinton, D.E. Orenstein, D.L. Childers, T.L. Gragson, N.B. Grimm, J.M. Grove, S.L. Harlan, J.P. Kaye, A.K. Knapp, G.P. Kofinas, J.J. Magnuson, W.H. McDowell, J.M. Melack, L.A. Ogden, G.P. Robertson, M.D. Smith, and A.C. Whitmer. 2011. An integrated conceptual framework for long-term social-ecological research. *Frontiers in Ecology and the Environment* 9(6): 351–357.

Collins, S.L., and M.D. Smith. 2006. Scale-dependent interaction of fire and grazing on community heterogeneity in tallgrass prairie. *Ecology* 87 (8): 2058–2067.

Cormaci, M., G. Furnari, G. Giaccone, and D. Serio. 2004. Alien macrophytes in the Mediterranean Sea: a review. *Recent Research in Developmental and Environmental Biology* 1: 153–202.

Danielson, T., V.H. Rivera-Monroy, E. Castañeda-Moya, H.O. Briceno, R. Travieso, B.D. Marx, E.E. Gaiser, and L. Farfan. 2017. Assessment of Everglades mangrove forest resilience: Implications for above-ground net primary productivity and carbon dynamics. *Forest Ecology and Management* 404: 115–125.

David, P.G. 1996. Changes in plant communities relative to hydrologic conditions in the Florida Everglades. *Wetlands* 16: 15–23.

Davis, S.E., J.E. Cable, D.L. Childers, C. Coronado-Molina, J.W.J. Day, C.D. Huttle, C.J. Madden, E. Reyes, D. Rudnick, and F. Sklar. 2004. Importance of storm events in controlling ecosystem structure and function in a Florida Gulf coast estuary. *Journal of Coastal Research* 20: 1198–1208.

Davis, S.E., and D.L. Childers. 2007. Importance of water source in controlling leaf leaching losses in a dwarf red mangrove (*Rhizophora mangle* L.) wetland. *Estuarine, Coastal and Shelf Science* 71(1-2): 194–201.

Davis, S.E., D.L. Childers, J.W. Day, D.T. Rudnick, and F.H. Sklar. 2003. Factors affecting the concentration and flux of materials in two southern Everglades mangrove wetlands. *Marine Ecology Progress Series* 253: 85–96.

Davis, S.E., G.M. Naja, and A. Arik. 2014. Restoring the heart of the Everglades: the challenges and benefits. *National Wetlands Newsletter* 36: 5–9.

Davis, S.M., D.L. Childers, J.J. Lorenz, H.R. Wanless, and T.E. Hopkins. 2005. A conceptual model of ecological interactions in the mangrove estuaries of the Florida Everglades. *Wetlands* 25(4): 832–842.

Deng, Y., H. Solo-Gabriele, M. Laas, L. Leonard, D.L. Childers, G. He, and V. Engel. 2010. Impacts of hurricanes on surface flow of water within a wetland. *Journal of Hydrology* 392: 164–173.

Dodds, W., C. Robinson, E. Gaiser, G. Hansen, H. Powell, J. Smith, N. Morse, S. Gregory, T. Bell, T. Kratz, and W. McDowell. 2012. Surprises and insights from long-term aquatic data sets. *BioScience* 62: 709–721.

D'Odorico P., V. Engel, J.A. Carr, S.F. Oberbauer, M.S. Ross, and J.P. Sah. 2011. Tree-grass coexistence in the Everglades freshwater system. *Ecosystems* 14: 298–310.

Dorcas, M.E., J.D. Willson, R.N. Reed, R.W. Snow, M.R. Rochford, M.A. Miller, W.E. Meshaka, Jr., P.T. Andreadis, F.J. Mazzotti, C.M. Romagosa, and K.M. Hart. 2012. Severe mammal declines coincide with proliferation of invasive Burmese pythons in Everglades National Park. *Proceedings of the National Academy of Sciences USA* 109(7): 2418–2422.

Downton, M.W., and K.A. Miller. 1992. The freeze risk of Florida citrus. Part II: Temperature variability and circulation patterns. *Journal of Climate* 6: 364–372.

Farris, G.S., G.J. Smith, M.P. Crane, C.R. Demas, L.L. Robbins, and D.L. Lavoie (eds.). 2007. Science and the storms—the USGS response to the hurricanes of 2005. U.S. Geological Survey Circular 1306.

Florida Fish and Wildlife Conservation Commission. 2014. Nonnative Marine Species. http://myfwc.com/wildlifehabitats/non-natives/marine-species/.

Foster, D., D. Knight, and J. Franklin. 1998. Landscape patterns and legacies resulting from large infrequent forest disturbances. *Ecosystems* 1: 497–510.

Fourqurean, J.W., J.C. Zieman, and G.V.N. Powell. 1992. Phosphorus limitation of primary production in Florida Bay: evidence from C:N:P ratios of the dominant seagrass *Thalassia testudinum*. *Limnology and Oceanography* 37: 162–171.

Fukami, T. 2001. Sequence effects of disturbance on community structure. *Oikos* 92(2): 215–224.

Fuller, D.O., and Y. Wang. 2014. Recent trends in satellite vegetation index observations indicate decreasing vegetation biomass in the southeastern saline Everglades wetlands. *Wetlands* 34: 67–77.

Gaiser, E.E., E.P. Anderson, E. Castañeda-Moya, L. Collado-Vides, J.W. Fourqurean, M.R. Heithaus, R. Jaffe, D. Lagomasino, N. Oehm, R.M. Price, V.H. Rivera-Monroy, R. Roy Chowdhury, and T. Troxler. 2015. New perspectives on an iconic landscape from comparative international long-term ecological research. *Ecosphere* 6(10): 181.

Gaiser, E., A. Zafiris, P. Ruiz, F. Tobias, and M. Ross. 2006. Tracking rates of ecotone migration due to salt-water encroachment using fossil mollusks in coastal south Florida. *Hydrobiologia* 569: 237–257.

Geller, J.B., J.A. Darling, and J.T. Carlton. 2010. Genetic perspectives on marine biological invasions. *Annual Review of Marine Sciences* 2: 367–393.

Gleason P.J., and P. Stone. 1994. Age, origin, and landscape evolution of the Everglades peatland. In *Everglades: The Ecosystem and Its Restoration*, edited by S.M. Davis and J.C. Ogden, 149–197. Delray Beach, FL: St. Lucie Press.

Goldenberg, S.B., C.W. Landsea, A.M. Mesta-Nunez, and W.M. Gray. 2001. The recent increase in Atlantic hurricane activity: causes and implications. *Science* 293: 474–479.

Gottlieb, A.G., J. Richards, and E. Gaiser. 2005. Effects of dessication duration on the community structure and nutrient retention of short and long hydroperiod Everglades periphyton mats. *Aquatic Botany* 82: 99–112.

Gravili, C., G. Belmontea, E. Cecereb, F. Denittoa, A. Giangrandea, P. Guidettia, C. Longoc, F. Mastrototaroc, S. Moscatelloa, A. Petrocellib, S. Pirainoa, A. Terlizzia, and F. Boeroa. 2010. Nonindigenous species along the Apulian coast, Italy. *Chemistry and Ecology* 26: 121–142.

Gunderson, L.H. 2000. Ecological resilience—in theory and application. *Annual Reviews of Ecology, Evolution and Systematics* 31: 425–499

Gutschick, V.P., and H. Bassir-Rad. 2003. Extreme events as shaping physiology, ecology, and evolution of plants: toward a unified definition and evaluation of their consequences. *New Phytologist* 160: 21–42.

Harvey, J.W., R.W. Schaffranek, G.B. Noe, L.G. Larsen, D.J. Nowacki, and B.L. O'Connor. 2009. Hydroecological factors governing surface water flow on a low-gradient floodplain. *Water Resources Research* 45: W03421.

Hebert, P.J., and G. Taylor. 1992. The deadliest, costliest, and most intense United States hurricanes of this century (and other frequently requested hurricane facts). NOAA Technical Memorandum NWS NHC-31, updated Feb. 1992. National Hurricane Center, Coral Gables, FL.

Holling, C.S. 2001. Understanding the complexity of economic, ecological, and social systems. *Ecosystems* 4: 390–405.

IPCC. 2013. Climate Change 2013: The Physical Science Basis. Contribution of Working Group I to the Fifth Assessment Report of the Intergovernmental Panel on Climate Change, edited by T.F. Stocker, D. Qin, G.-K. Plattner, M. Tignor, S.K. Allen, J. Boschung, A. Nauels, Y. Xia, V. Bex, and P.M. Midgley. Cambridge and New York: Cambridge University Press.

Kline, J.L., W.F. Loftus, K. Kotun, J.C. Trexler, J.S. Rehage, J.J. Lorenz, and M. Robinson. 2014. Recent fish introductions into Everglades National Park: an unforeseen consequence of water management. *Wetlands* 34: S175–S187.

Knapp, A.K., J.M. Blair, J.M. Briggs, S.L. Collins, D.C. Hartnett, L.C. Johnson, and E.G. Towne. 1999. The keystone role of bison in North American tallgrass prairie: bison increase habitat heterogeneity and alter a broad array of plant, community, and ecosystem processes. *Bioscience* 49(1): 39–50.

Kobza, R.M., J.C. Trexler, W.F. Loftus, and S.A. Perry. 2004. Community structure of fishes inhabiting aquatic refuges in a threatened karst wetland and its implications for ecosystem management. *Biological Conservation* 116: 153–165.

Koch, G.R., D.L. Childers, P.A. Staehr, R.M. Price, S.E. Davis, and E.E. Gaiser. 2012. Hydrological conditions control P loading and aquatic metabolism in an oligotrophic, subtropical estuary. *Estuaries and Coasts* 35: 292–307.

Kodra, E., K. Steinhaeuser, and A.R. Ganguly. 2011. Persisting cold extremes under 21st-century warming scenarios. *Geophysical Research Letters* 38: 1–5.

Krauss, K.W., T.W. Doyle, T.J. Doyle, C.M. Swaarzenski, A.S. From, R.H. Day, and W.H. Conner. 2009. Water level observations in mangrove swamps during two hurricanes in Florida. *Wetlands* 29: 142–149.

Krauss, K.W., T.W. Doyle, R.R. Twilley, and T.J. Smith. 2005. Woody debris in the mangrove forests of south Florida. *Biotropica* 37: 9–15.

Lapointe, B.E., and J.B. Bedford. 2010. Ecology and nutrition of invasive *Caulerpa brachypus* f. *parvifolia* blooms on coral reefs off southeast Florida, USA. *Harmful Algae* 9: 1–12.

La Puma, D.A., J.L. Lockwood, and M.J. Davis. 2007. Endangered species management requires a new look at the benefit of fire. The Cape Sable seaside sparrow in the Everglades ecosystem. *Biological Conservation* 136: 398–407.

Leliaert, F., X. Zhang, N. Ye, E.-J. Malta, A.H. Engelen, F. Mineur, H. Verbruggen, and O. De Clerck. 2009. Identity of the Qindao algal bloom. *Phycological Research* 57: 147–151.

Light, S.S., and J.W. Dineen. 1994. Water control in the Everglades: a historical perspective. In *Everglades: The Ecosystem and its Restoration*, edited by S.M. Davis and J.C. Ogden, 47–84. Delray Beach, FL: St. Lucie Press.

Lodge, T.E. 2005. *Everglades Handbook: Understanding the Ecosystem*, 2nd ed. Boca Raton, FL: CRC Press.

Loftus, W.F. 2000. Inventory of fishes of Everglades National Park. *Florida Scientist* 63: 27–47.

Lorenz, J.J. 2014. A review of the effects of altered hydrology and salinity on vertebrate fauna and their habitats in northeastern Florida Bay. *Wetlands* 34(1): 189–200.

Lugo, A., and F. Scatena. 1995. Ecosystem-level properties of the Luquillo Experimental Forest with emphasis on the tabonuco forest. In *Tropical Forests: Management and Ecology*, edited by A. Lugo and C. Lowe, 59–108. New York: Springer-Verlag.

Malmstadt, J., K. Scheitlin, and J. Elsner. 2009. Florida hurricanes and damage costs. *Southeastern Geographer* 49: 108–131.

Malone, S.L., G. Starr, C. Staudhammer, and M.G. Ryan. 2013. Effects of simulated drought on the carbon balance of Everglades short-hydroperiod marsh. *Global Change Biology* 19: 2511–2523.

Matich, P., and M.R. Heithaus. 2012. Effects of an extreme temperature event on the behavior and age structure of an estuarine top predator, *Carcharhinus leucas. Marine Ecology Progress Series* 447: 165–178.

Mazzotti, F.J., M.S. Cherkiss, K.M. Hart, R.W. Snow, M.R. Rochford, M.E. Dorcas, and R.N. Reed. 2011. Cold-induced mortality of invasive Burmese pythons in south Florida. *Biological Invasions* 13: 143–151.

McCormick, P.V., S. Newman, S. L. Miao, D. E. Gawlik, D. Marley, K. R. Reddy, and T. D. Fontaine. 2002. Effects of anthropogenic phosphorus inputs on the Everglades. In *The Everglades, Florida Bay, and Coral Reefs of the Florida Keys: An Ecosystem Sourcebook*, edited by J.W. Porter and K.G. Porter, 83–126. Boca Raton, FL: CRC Press.

Melillo, J.M., T.C. Richmond, and G.W. Yohe (eds.). 2014. Climate Change Impacts in the United States: The Third National Climate Assessment. U.S. Global Change Research Program. doi:10.7930/J0Z31WJ2.

Melton, T., L. Collado-Vides, and J.M. Lopez-Bautista (in preparation). Molecular data reveal a broader range of the tropical/subtropical green tide species *Ulva ohnoi* (Ulvophyceae, Chlorophyta): a new record for the Western Atlantic. *Phycologia*.

National Hurricane Center. 2014. Tropical Cyclone Climatology—Hurricane Return Periods. http://www.nhc.noaa.gov/climo/#returns.

Noe, G., D.L. Childers, and R.D. Jones. 2001. Phosphorus biogeochemistry and the impacts of phosphorus enrichment: why are the Everglades so unique? *Ecosystems* 4: 603–624.

Nungesser, M., C. Saunders, C. Coronado-Molina, J. Obeysekera, J. Johnson, C. McVoy, and B. Benscoter. 2015. Potential effects of climate change on Florida's Everglades. *Environmental Management* 55: 824–835.

Nyman, J.A., C.R. Crozier, and R.D. DeLaune. 1995. Roles and patterns of hurricane sedimentation in an estuarine marsh landscape. *Estuarine, Coastal and Shelf Science* 40: 665–679.

Obeysekera, J., J. Barnes, and M. Nungesser. 2015. Climate sensitivity runs and regional hydrologic modeling for predicting the response of the greater Florida Everglades ecosystem to climate change. *Environmental Management* 55: 749–762.

Obeysekera, J., M. Irizarry, J. Park, J. Barnes, and T. Dessalegne. 2011. Climate change and its implications for water resources management in south Florida. *Stochastic Environmental Research and Risk Assessment* 25: 495–516.

Odum, W.P., E.P. Odum, and H.T. Odum. 1995. Nature's pulsing paradigm. *Estuaries* 18(4): 547–555.

Paine, R.T., M.J. Tegner, and E.A. Johnson. 1998. Compounded perturbations yield ecological surprises. *Ecosystems* 1: 535–545.

Pasch, R.J., E.S. Blake, H.D. Cobb, and D.P. Roberts. 2006. Tropical cyclone report Hurricane Wilma. National Weather Service. National Hurricane Center, Miami, FL.

Peterson, B.J., C.D. Rose, L.M. Rutten, and J.W. Fourqurean. 2002. Disturbance and recovery following catastrophic grazing: studies of a successional chronosequence in a seagrass bed. *Oikos* 97(3): 361–370.

Pickett, S.T., and P.S. White. 1985. *The Ecology of Natural Disturbance and Patch Dynamics*. New York: Academic Press.

Pielke, R.A., J. Gratz, C.W. Landsea, D. Collins, M.A. Saunders, and R. Musulin. 2008. Normalized hurricane damage in the United States: 1900–2005. *Natural Hazards Review* 9: 29–42.

Piou, C., I.C. Feller, U. Berger, and F. Chi. 2006. Zonation patterns of Belizean offshore mangrove forests 41 years after a catastrophic hurricane. *Biotropica* 38: 365–374.

Porter-Whitaker, A., J.S. Rehage, W.F. Loftus, and S.E. Liston. 2012. Multiple predator effects and native prey responses to two non-native Everglades cichlids. *Ecology of Freshwater Fishes* 21: 375–385.

Price, R.M., P.K. Swart, and J.W. Fourqurean. 2006. Coastal groundwater discharge-an additional source of phosphorus for the oligotrophic wetlands of the Everglades. *Hydrobiologia* 569: 23–36.

Pysek, P., D.M. Richardson, J. Pergl, V. Jarosik, Z. Sixtova, and E. Weber. 2008. Geographical and taxonomic biases in invasion ecology. *Trends in Ecology and Evolution* 23: 237–244.

Rehage, J.S., and D.A. Gandy. 2014. Assessing the impact and potential for containment of non-native fishes across Everglades habitats. Annual reports to the National Park Service under cooperative agreement H5000065040.

Rehage, J.S., S.E. Liston, K.J. Dunker, and W.F. Loftus. 2014. Fish community responses to the combined effects of decreased hydroperiod and nonindigenous fish invasions in a karst wetland: are Everglades solution holes sinks for native fishes? *Wetlands* 34: S159–S173.

Rehage, J.S., and J.C. Trexler. 2006. Assessing the net effect of anthropogenic disturbance on aquatic communities in wetlands: community structure relative to distance form canals. *Hydrobiologia* 569: 359–373.

Risi, J.A., H.R. Wanless, L.P. Tedesco, and S. Gelsanliter. 1995. Catastrophic sedimentation from Hurricane Andrew along the southwest Florida coast. *Journal of Coastal Research* 21: 83–102.

Rivera-Monroy, V.H., E. Castaneda-Moya, J.G. Barr, V. Engel, J.D. Fuentes, T.G. Troxler, R.R. Twilley, S. Bouillon, T.J. Smith, III, and T.L. O'Halloran. 2013. Current methods to evaluate net primary production and carbon budgets in mangrove forests. In *Methods in Biogeochemistry of Wetlands*, edited by R.D. DeLaune, R.D. Reddy, K.R. Megonigal, and C. Richardson, 243–288. Madison, WI: Soil Science Society of America Book Series.

Rivera-Monroy, V.H., R.R. Twilley, S.E. Davis, III., D.L. Childers, M. Simard, R.M. Chambers, R. Jaffe, J.N. Boyer, D. Rudnick, K. Zhang, E. Castaneda-Moya, S.M.L. Ewe, R.M. Price, C. Coronado-Molina, M. Ross, T.J. Smith, III., B. Michot, E.A. Meselhe, W. Nuttle, T.G. Troxler, and G.B. Noe. 2011. The role of the Everglades mangrove ecotone region (EMER) in regulating nutrient cycling and wetland productivity in south Florida. *Critical Reviews in Environmental Science and Technology* 41: 633–699.

Robertson, A.I., and P.A. Daniel. 1989. Decomposition and the annual flux of detritus from fallen timber in tropical mangrove forests. *Limnology and Oceanography* 34: 640–646.

Romero, L.M., T.J. Smith, and J.W. Fourqurean. 2005. Changes in mass and nutrient content of wood during decomposition in a south Florida mangrove forest. *Journal of Ecology* 93: 618–631.

Ross, M., J. Meeder, J. Sah, P. Ruiz, and G. Telesnicki. 2000. The southeast saline Everglades revisited: 50 years of coastal vegetation change. *Journal of Vegetation Science* 11:101–112.

Ross, M.S., P.L. Ruiz, S.P. Sah, E.J. Hanan. 2009. Chilling damaging in a changing climate in coastal landscapes of the subtropical zone: a case study from south Florida. *Global Change Biology* 15: 1817–1832.

Ross, M.S., and J.P. Sah. 2011. Forest resource islands in a sub-tropical marsh: soil-site relationships in Everglades hardwood hammocks. *Ecosystems* 14(4): 632–645.

Rudnick, D.T., P.B. Ortner, J.A. Browder, and S.M. Davis. 2005. A conceptual ecological model for Florida Bay. *Wetlands* 25(4): 870–883.

Ruiz, P.L., J.P. Sah, M.S. Ross, and A.A. Spitzig. 2013. Tree island response to fire and flooding in the short-hydroperiod marl prairie grasslands of the Florida Everglades. *Fire Ecology* 9(1): 38–54.

Sah, J.P., M.S. Ross, P.L. Ruiz, and J.R. Snyder. 2012. Fire and flooding interactions: vegetation trajectories in the southern Everglades Marl Prairies, FL, USA. The 9th INTECOL International Wetlands Conference Abstracts, p. 509.

Sah, J.P., M.S. Ross, P.L. Ruiz, J.R. Snyder, D. Rodriguez, and W.T. Hilton. 2011. Cape Sable seaside sparrow habitat—Monitoring and Assessment—2010. Final Report submitted to U. S. Army Corps of Engineers, Jacksonville, FL. (Cooperative Agreement # W912HZ-10-2-0025).

Saha, A.K., S. Saha, J. Sadle, J. Jiang, M.S. Ross, R.M. Price, L.O. Stenberg, and K.S. Wendelberger. 2011. Sea level rise and south Florida coastal forests. *Climate Change* 107(1-2): 81–108.

Schedlbauer, J.L., S.F. Oberbauer, G. Starr, and K.L. Jimenez. 2010. Seasonal differences in the CO2 exchange of a short-hydroperiod Florida Everglades marsh. *Agricultural and Forest Meteorology* 150(7-8): 994–1006.

Shafland, P.L. 1996. Exotic fishes of Florida, 1994. *Reviews in Fisheries Science* 4: 101–122.

Shafland, P.L., K.B. Gestring, and M.S. Stanford. 2008. Florida's exotic freshwater fishes—2007. Florida Scientist 71(3): 220–245.

Sheridan, S.C. 2003. North American weather-type frequency and teleconnection indices. *International Journal of Climatology* 23: 27–45.

Simard, M., K. Zhang, V.H. Rivera-Monroy, M.S. Ross, P.L. Ruiz, E. Castaneda-Moya, R.R. Twilley, and E. Rodriguez. 2006. Mapping height and biomass of mangrove forests in Everglades National Park with SRTM elevation data. *Photogrammetric Engineering and Remote Sensing* 72: 299–311.

Sklar, F.H., M.J. Chimney, S. Newman, P.V. McCormick, D. Gawlik, S. Miao, C. McVoy, W. Said, J. Newman, C. Coronado, G. Crozier, M. Korvela, and K. Rutchey. 2005. The ecological-societal underpinnings of Everglades restoration. *Frontiers in Ecology and the Environment* 3:161–169.

Sklar, F.H., C. McVoy, R. Vanzee, D.E. Gawlik, K. Tarboton, D. Rudnick, and S. Miao. 2002. The effects of altered hydrology on the ecology of the Everglades. In *The Everglades, Florida Bay, and Coral Reefs of the Florida Keys: An Ecosystem Sourcebook*, edited by Porter, J.W. and K. G. Porter, 39–82. Boca Raton, FL: CRC Press.

Smith, T. J., G.H. Anderson, K. Balentine, G. Tiling, and G.A. Ward. 2009. Cumulative impacts of hurricanes on Florida mangrove ecosystems: sediment deposition, storm surge and vegetation. *Wetlands* 29: 24–34.

Smith, T.J., A.M. Foster, and J.W. Jones. 2015. Fire history of Everglades National Park and Big Cypress National Preserve, southern Florida. U.S. Geological Survey Open-File Report 2015–1034. http://dx.doi.org/10.3133/ofr20151034.

Smith, T.J., A.M. Foster, G. Tiling-Range, and J.W. Jones. 2013. Dynamics of mangrove-marsh ecotones in subtropical coastal wetlands: fire, sea-level rise, and water levels. *Fire Ecology* 9(1): 66–77.

Smith, T.J., M.B. Robblee, H.R. Wanless, and T.W. Doyle. 1994. Mangroves, hurricanes, and lightning strikes. *BioScience* 44: 256–262.

Smoak, J.M., J.L. Breithaupt, T.J. Smith, and C.J. Sanders. 2013. Sediment accretion and organic carbon burial relative to sea-level rise and storm events in two mangrove forests in Everglades National Park. *Catena* 104: 58–66.

Sokol, E.R., J.M. Hoch, E.E. Gasier, and J.C. Trexler. 2014. Metacommunity structure along resource and disturbance gradients in Everglades wetlands. *Wetlands* 34(1): 135–146.

Thomas, S., E. Gaiser, M. Gantar, and L. Scinto. 2006. Quantifying the response of calcareous periphyton crusts to rehydration: a microcosm study (Florida Everglades). *Aquatic Botany* 84: 317–323.

Troxler, T.G., C. Coronado-Molina, D.N. Rondeau, S. Krupa, S. Newman, M. Manna, R.M. Price, and F.H. Sklar. 2014. Interactions of local climatic, biotic and hydrogeochemical processes facilitate phosphorus dynamics along an Everglades forest-marsh gradient. *Biogeosciences* 11: 899–914.

Trexler, J.C., E.E. Gaiser, and D.L. Childers. 2006. Interaction of hydrology and nutrients in controlling ecosystem function in oligotrophic coastal environments of south Florida. *Hydrobiologia* 569(1): 1–2.

Trexler, J.C., W.F. Loftus, F.C. Jordan, J.J. Lorenz, J.H. Chick, and R.M. Kobza. 2000. Empirical assessment of fish introductions in a subtropical wetland ecosystem: an evaluation of contrasting views. *Biological Invasions* 2: 265–277.

Trexler, J.C., W.F. Loftus, and S.A. Perry. 2005. Disturbance frequency and community structure: a twenty-five-year intervention study. *Oecologia* 145: 140–152.

M.G. Turner, S.L. Collins, A.L. Lugo, J.J. Magnuson, T.S. Rupp, and F.J. Swanson, 2003. Disturbance dynamics and ecological response: the contribution of long-term ecological research. *BioScience* 53(1): 46–56.

Twilley, R.R., and V.H. Rivera-Monroy. 2005. Developing performance measures of mangrove wetlands using simulation models of hydrology, nutrient biogeochemistry, and community dynamics. *Journal of Coastal Research* 40: 79–93.

USACE and SFWMD. 1999. Central and Southern Florida Project, Comprehensive Review Study, Final Integrated Feasibility Report and Programmatic Environmental Impact Statement. U.S. Army Corps of Engineers and South Florida Water Management District.

USACE and SFWMD. 2014. Central Everglades Planning Project: Final Integrated Project Implementation Report and Environmental Impact Statement. U.S. Army Corps of Engineers and South Florida Water Management District.

U.S. Drought Monitor. 2015. U.S. Drought Monitor. http://droughtmonitor.unl.edu.

Vellend, M., L.J. Harmon, J.L. Lockwood, M.M. Mayfield, A.R. Hughes, J.P. Wares, and D.F. Sax. 2007. Effects of exotic species on evolutionary diversification. *Trends in Ecology and Evolution* 22: 481–488.

Vera, B., L. Collado-Vides, C. Moreno, and I.B. van Tussenbroek. 2014. *Halophila stipulacea* (Hydrocharitaceae): a recent introduction to the continental waters of Venezuela. *Caribbean Journal of Science* 48: 66–70.

Waide, R., and M. Willig. 2012. Conceptual overview: disturbance, gradients, and ecological response. In *A Caribbean Forest Tapestry: The Multidimensional Nature of Disturbance and Response*, edited by N. Brokaw, T. Crowl, A. Lugo, W. McDowell, F. Scatena, R. Waide, and M. Willig, 42–71. New York: Oxford University Press.

Wallentinus, I. 2002. Introduced marine algae and vascular plants in European aquatic environments. In *Invasive Aquatic Species of Europe. Distribution, Impacts and Management*, edited by E. Leppäkoski, E., S. Gollasch, S., and S. Olenin, 27–52. Dordrecht: Kluwer Academic Publishers.

Wanless, H.R., R.W. Parkinson, and L.P. Tedesco. 1994. Sea level control on stability of Everglades wetlands. In *Everglades: The Ecosystem and Its Restoration*, edited by S.M. Davis and J.C. Ogden, 199–223. Delray Beach, FL: St. Lucie Press.

Wanless, H.R., and B.M. Vlaswinkel. 2005. Coastal landscape and channel evolution affecting critical habitats at Cape Sable, Everglades National Park, Florida. Final Report submitted to Everglades National Park.

Weston, N.B., M.A. Vile, S.C. Neubauer, and D.J. Velinksy. 2011. Accelerated microbial organic matter mineralization following salt-water intrusion into tidal freshwater marsh soils. *Biogeochemistry* 102: 135–151.

Wetzel, P.R., F.H. Sklar, C.A. Coronado, T.G. Troxler, S.L. Krupa, P.L. Sullivan, S. Ewe, R.M. Price, S. Newman, and W.H. Orem. 2011. Biogeochemical processes on tree islands in the Greater Everglades: initiating a new paradigm. *Critical Reviews in Environmental Science and Technology* 41(1): 670–701.

Whelan, K. 2005. The successional dynamics of lightning-initiated canopy gaps in the mangrove forests of Shark River, Everglades National Park, USA. Ph.D. dissertation, Florida International University.

Whelan, K.R.T., T.J. Smith, G.H. Anderson, and M.L. Ouellette. 2009. Hurricane Wilma's impact on overall soil elevation and zones within the soil profile in a mangrove forest. *Wetlands* 29: 16–23.

Willette, D.A., J. Chalifour, D. Debrot, M.S. Engel, J. Miller, H.A. Oxenford, F.T. Short, S.C.C. Steiner, and F. Védie. 2014. Continued expansion of the trans-Atlantic invasive marine angiosperm *Halophila stipulacea* in the Eastern Caribbean. *Aquatic Botany* 112: 98–102.

Williams, C.J., J.N. Boyer, and F.J. Jochem. 2008. Indirect hurricane effects on resource availability and microbial communities in a subtropical wetland-estuary transition zone. *Estuaries and Coasts* 31: 204–214.

Williams, S.L. 2007. Introduced species in seagrass ecosystems: status and concerns. *Journal of Experimental Marine Biology and Ecology* 350: 89–110.

Williams, S.L., and J.E. Smith. 2007. Distribution, taxonomy, and impacts of introduced seaweeds. *Annual Review of Ecology, Evolution, and Systematics* 38: 327–359.

Yao, Q., K-b Liu, W.J. Platt, and V.H. Rivera-Monroy. 2015. Palynological reconstruction of environmental changes in coastal wetlands of the Florida Everglades since the mid-Holocene. *Quaternary Research* 83(3): 449–458.

Zhang, K., H. Liu, Y. Li, H. Xu, J. Shen, J. Rhome, and T.J. Smith. 2012. The role of mangroves in attenuating storm surges. *Estuarine, Coastal and Shelf Science* 102–103: 11–23.

Zhang, K., M. Simard, M.S. Ross, V.H. Rivera-Monroy, P. Houle, P.L. Ruiz, R.R. Twilley, and K.R.T. Whelan. 2008. Airborne laser scanning quantification of disturbances from hurricanes and lightning strikes to mangrove forests in Everglades National Park, USA. *Sensors* 8: 2262–2292.

8

Back to the Future

REBUILDING THE EVERGLADES

Fred Sklar with James Beerens, Laura Brandt, Carlos Coronado, Stephen E. Davis III, Tom Frankovich, Christopher Madden, Agnes McLean, Joel Trexler, and Walter Wilcox

In a Nutshell

- Bringing back the past is constrained by legacy effects, irrevocable damages, and anthropogenic trends not seen in the past. Restoration in the purest sense of the word is not possible; a more accurate term is *rehabilitation*.
- Rehabilitation requires an understanding of what is ecologically, economically, and legally possible. It requires models to extrapolate and synthesize long-term datasets, adaptive management to test alternative hypotheses of ecosystem processes, and a system of governance that allows for debate on the socioeconomic, legal, and ecological constraints of future alternatives.
- For simulations of the future to be credible, they require long-term datasets that capture cyclical patterns; an understanding of feedbacks, drivers, and forcing functions; calibration; validation; and clear peer-reviewed documentation.
- The ecological uncertainties of restoration or rehabilitation are reduced by focusing on the impacts of Anthropocene trends in ecotones, where change is most dynamic.

Welcome to the Anthropocene

Few places, if any, exist on the planet that have not been directly or indirectly altered, influenced, degraded, or destroyed by humans (Klein 2014; Kolbert 2014). Since societies value the landscapes, seascapes, plants, and animals that were prevalent 100 to 200 years ago, there has been a concerted effort by scientists, decision makers, and nongovernmental organizations to look at the (seemingly) pristine past and develop plans to move the past into the future (Dengler 2007; Doyle and Drew 2008; Miller and Hobbs 2007). The Society of Ecosystem Restoration defines restoration on its webpage as "the process of assisting the recovery of an ecosystem that has been degraded, damaged, or destroyed. An ecosystem has recovered—and is restored—when it contains sufficient biotic and abiotic resources to continue its development without further assistance or subsidy." Because of its sheer size and its importance for water supply and flood protection, the Everglades landscape will never fit this rigorous definition at all times and in all places. Regardless, justification for pursuing "restoration" of the Everglades can be found in the Final Environmental Impact Statement for the Central Everglades Planning Project (CEPP) that was released by the U.S. Army Corps of Engineers (USACE) in August 2014. We will reference this 3,000-page document often (USACE 2014) because, as in the movie series *Back to the Future*, CEPP is the "flux-capacitor" that generates the ability to analyze the past and travel to a potential future.

Native Americans in South Florida have always had a deep appreciation and respect for the Everglades (Carr 2002; see Chapter 1). However, from a utilitarian perspective (Greer 2011), the past was no bed of roses in south Florida. Everything was wet, alligators were dangerous, mosquitos were everywhere, cultivation during the rainy season was impossible, and industry saw no reason to leave the coal-burning, energy-rich states in the north. Early in its free-market economic growth, around 1910, south Florida was a dichotomy of two economic sectors—one built around the commerce of tourism along the coasts, and the other built around the biologic resources of a rich, unique inland ecosystem capable of providing habitat, fresh water, fashion items such as feathers and alligator skins, and protein (Carr 2002; Simmons and Ogden 2010; Ogden 2011). Not many decades into the 20th century, agricultural development began to transform the interior and northern Everglades, reflecting the nation's priorities for greater development (McCally 2000). Yet even in this time period, the ability of the Everglades to provide critical ecosystem services, such as water supply, carbon sequestration, recreation, storm protection, flood control, climate buffering, nutrient management, biodiversity, and food, was not yet strained or irreversibly compromised. This was probably because relatively few people were living in south Florida. However, it took but a millisecond in geologic time to create the Anthropocene in the Everglades (see Ogden et al. 2013 for a broad description of the Anthropocene).

Today, south Florida remains dependent on tourism and environmental resources, and these two sectors have become inseparable, particularly when looking at the recreational aspects of the economy (e.g., hotels and coral reefs). The differences between today and 100 years ago are largely a function of the number of people both visiting and living in south Florida (see Chapter 1 for more detail). In 1880, when drainage of the Everglades began in earnest (McVoy et al. 2011), the population in all of Florida was only about 500,000. Today, Florida is home to 19.8 million people, about half of whom reside in south Florida. As 20th-century entrepreneurs built levees and canals to constrain the Everglades on a grand scale, rich visitors took steamboat rides in the coastal ecotones and sent postcards of the beautiful and soon invasive water hyacinths (Fig. 8.1). The other difference between now and then is the importance of agriculture as a biologic resource and an economic engine. Florida is now the third-largest producer of cattle in the nation, is the largest producer of sugarcane in the nation, and is known worldwide for its citrus, nursery products, and winter vegetable crops (Cattelino 2010; Ewing 2013).

The future of the coastal Everglades will continue to be shaped by water management decisions in light of human demands for fresh water, accelerating sea level rise (SLR), and environmental policy. In this chapter we do not attempt to explore all the possible outcomes of this future dynamic; such a task would be hindered by too many uncertainties. Rather, our modest goal is to focus on two predictions of the future for the Everglades and south Florida. One is a future *with* Everglades restoration, as envisioned in the CEPP. The other is a future *without* restoration, where current conditions prevail; we will refer to this future as FWO ("future without restoration"). Our "back-to-the-future" vehicle was fueled by knowledge generated by the Florida Coastal Everglades Long Term Ecological Research Program (FCE LTER), collaborative state–federal research, the hydrologic models developed by the South Florida Water Management District (SFWMD), and the socioeconomic restoration planning process developed by

FIGURE 8.1 This old Florida postcard from the mid-20th century depicts a culture not yet aware of the fragility of the Everglades.

USACE (2014). With this vehicle, we will focus on the coastal Everglades to the extent possible and explore the scientific and political complexities of creating a rehabilitated future.

Managing the Science of Restoration—A Conceptual Model

There are two critical information flow paths for Everglades restoration in the Comprehensive Everglades Restoration Plan (CERP) conceptualization (Ogden et al. 2005) of how research and society interact to produce a vision of the future (Fig. 8.2). One is called the design flow path (the left side of Fig. 8.2), where the visions that society creates are incorporated into models that include legal and economic constraints. The other is called the assessment flow path (the right side of Fig. 8.2), where the visions that science creates are incorporated into the pilot studies, monitoring, and experimental approaches needed to reduce the uncertainties associated with restoration. These two flow paths are the "yin and yang" of restoration: both are integral to a successful adaptive management approach (Holling 1978; RECOVER 2011).

The design flow path is where everyone—government, stakeholders, scientists, nongovernmental organizations, and the general public—gets to develop, discuss, and evaluate their visions of the future. This flow path builds the physical and ecological models needed to test the effectiveness of these visions. These models use the scientific information generated by the assessment flow path, along with the legal and economic constraints

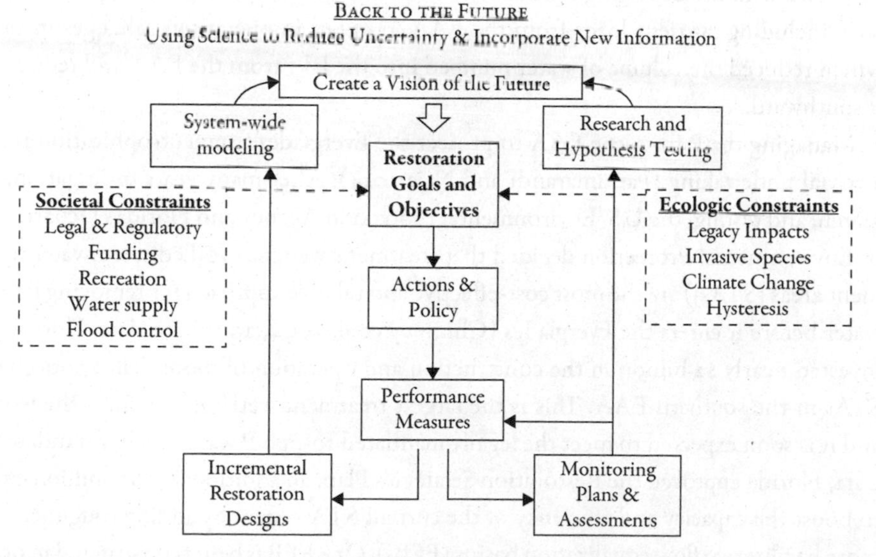

FIGURE 8.2 A conceptual model used in CERP to describe how society and science come together to create an adaptive vision of the future for the Everglades. The left side is the Design Flow Path; the right side is the Assessment Flow Path.

imposed by society (Table 8.1). As with any large restoration effort, for the Everglades these constraints attempt to minimize the tax burden, maximize economic sustainability, preserve intrinsic natural resources, and protect societal interests. Constraints on Everglades restoration are written into state statutes or in the USACE Programmatic Regulations. The most significant of these are the total phosphorus (TP) rule, threatened and endangered species, recreation, and water management (water supply and flood control).

THE TP RULE

As described in previous chapters of this book, the Everglades landscape has endured more than a century of impacts from drainage and excess loading of phosphorus (P). The Everglades has been and continues to be center stage for litigation related to P pollution and its likely irreversible, hysteretic impacts on Everglades ecosystems. Sound science and federal court rulings have established that water entering the Everglades must be clean (i.e., TP concentrations of <10 ppb) in order to protect sensitive Everglades wetlands. Some of this pollution problem is associated with Lake Okeechobee, where TP concentrations are often 12 to 15 times higher than 10 ppb. This lake pollution was derived from the upstream Kissimmee River basin and from decades of backpumping from the Everglades Agricultural Area (EAA) to the lake (Davis 1994). The regulation schedule for Lake Okeechobee has undergone several modifications since the construction of the Herbert Hoover Dike. Initially, the lake was managed in the low range of 13 to 15 feet. In the 1970s, the dike allowed for higher stages in response to concerns about water supply. Problems with the drowning of nearshore vegetation and declining water quality in the lake, including nutrient loads from the EAA, resulted in operational changes in 1979, which reduced the volume of water pumped into the lake from the EAA and redirected it southward.

Managing the P from the EAA to protect the Everglades from eutrophication is not a trivial undertaking (Faridmarandi and Naja 2014). After many years of litigation, research, and testing, the U.S. Environmental Protection Agency and Florida's Department of Environmental Protection decided that treatment wetlands (called stormwater treatment areas [STAs]) are the most cost-effective and reliable approach for removing P from water before it enters the Everglades (Chimney et al. 2004, 2007). To date, Florida has invested nearly $2 billion in the construction and operation of more than 27,000 ha of STAs in the southern EAA. This is the largest treatment wetland system in the world, and it is soon expected to meet the legally mandated 10-ppb P water quality standard. In 2012, Florida approved the Restoration Strategies Plan, an additional $880 million effort to boost the capacity and efficiency of the current STA system by adding roughly 2,000 more ha plus two flow equalization basins (FEBs). One FEB is being constructed in order to meet the water quality standard of existing inflows to the Everglades. The other FEB is being constructed to meet the additional water treatment needs of the CEPP. Visions

TABLE 8.1.

Approval of a Restoration Plan for the Everglades Requires Compliance with These Laws, Policies, and Regulatory Constraints
Anadromous Fish Conservation Act
Archaeological Resources Protection Act of 1979
American Indian Religious Freedom Act
Bald and Golden Eagle Protection Act
Clean Air Act of 1972
Clean Water Act of 1972
Coastal Barrier Resources Act and Coastal Barrier Improvement Act of 1990
Coastal Zone Management Act of 1972
Endangered Species Act of 1973
Estuary Protection Act of 1968
Federal Water Project Recreation Act/Land and Water Conservation Fund Act
Fish and Wildlife Coordination Act of 1958, as amended.
Farmland Protection Policy Act of 1981
Magnuson-Stevens Fishery Conservation and Management Act
Marine Mammal Protection Act of 1972
Marine Protection, Research and Sanctuaries Act
National Environmental Policy Act of 1969
National Historic Preservation Act of 1966
Native American Graves Protection and Repatriation Act
Resource Conservation and Recovery Act, as Amended by the Hazardous and Soils, Waste Amendments of 1984. CERCLA as Amended by the 5.26.21 Superfund Amendments and Reauthorization Act of 1996, Toxic Substances Control Act of 1976.
Rivers and Harbors Act of 1899
Submerged Lands of 1953
Wild and Scenic River Act of 1968
E.O. 11514, Protection of the Environment.
E.O. 11593 Protection and Enhancement of the Cultural Environment
E.O. 11988 Flood Plain Management
E.O. 11990 Protection of Wetlands
E.O. 12962, Recreational Fisheries
E.O. 12898 Environmental Justice
E.O 13007 Indian Sacred Sites
E.O. 13045 Protection of Children
E.O. 13089 Coral Reef Protection
E.O. 13122 Invasive Species
E.O. 13175 Consultation and Coordination with Indian Tribal Governments
E.O. 13186, Responsibilities of Federal Agencies to Protect Migratory Birds

(*continued*)

TABLE 8.1. CONTINUED

Memorandum on Government to Government Regulations with Native American Tribal Governments
Seminole Indian Claims Settlement Act of 1987

of more flow to the ecosystem, beyond CEPP, will need to consider further EAA and Kissimmee River modifications.

THREATENED AND ENDANGERED SPECIES

The Endangered Species Act of 1973 requires concurrence of the CEPP plan by the U.S. Fish and Wildlife Service (USFWS) and the National Marine Fisheries Service on endangered species and critical habitat that may be present in the project area. After a series of consultations, biological assessments, and requests for additional information, the USFWS provided a "Biological Opinion" that provides preliminary terms and conditions to support species management and recovery in anticipation of CEPP project implementation and consultations under the Endangered Species Act.

RECREATION

Authorization and guidance for the development of ancillary recreation resources, as required by a 1998 USACE memorandum called "Recreation Development at Ecosystem Restoration Projects," is designed to contribute to community health and well-being (CEPP 2014). The recreation resources that are being proposed as part of the CEPP were designed to comply with this philosophy. However, resources for recreation required a relaxation of restoration targets as originally envisioned in the CERP to accommodate bass fishing and deer hunting (USACE 1999). For example, moving forward with a CEPP vision for northern Water Conservation Area (WCA)-3A required a relaxation of the restoration targets for ridge and slough vegetation in areas of the Everglades that are now dry enough for recreational hunting because of historical drainage, peat fires, and peat soil losses.

WATER MANAGEMENT FOR FLOOD PROTECTION AND WATER SUPPLY

Any vision for the future must ensure that Everglades restoration does not reduce flood protection or water supply. CEPP planners took a vision of restoration to "Model Land" (see the section "Caution: Entering Model Land"), where computer simulation tools are used to evaluate system-wide operational changes, hydrologic conditions, and discharges. They discovered that many homeowners and developers were highly concerned that the water needed for this restoration vision involved changes to currently approved water management rules (e.g., the Lake Okeechobee Regulation Schedule [2008] and the Everglades Restoration Transition Plan [2012]) and that the quantity of water available for irrigation and water supply had been reduced by the vision. Furthermore, the models indicated that some of the water utilized by Lake Okeechobee water users would be transferred south to the WCAs and Everglades National Park (ENP). According to a savings clause written into the original approval of CERP, this is not possible. As a result, CEPP identified an additional source of water of comparable quantity and quality available to replace the water that would be transferred south. (A sidebar note on complexity: as with any limited resource, this created a cascade of additional challenges, such as the need to replace the replacement water with another CERP project [e.g., the Indian River Lagoon-South—C-44 Reservoir/STA], which in turn was confounded by the need to enhance the Herbert Hoover Dike that surrounds Lake Okeechobee for flood control.)

The assessment flow path of Figure 8.2 is the CERP science program known as RECOVER (REstoration, COordination and VERification), a multi-agency planning process where targets for full restoration have been created, assessed, and reevaluated in light of the physical, biogeochemical, hydrologic, and ecological constraints that are intrinsic to the Everglades. To create realistic expectations for the future, these constraints must separate natural trends from anthropogenic trends; evaluate the degree of degradation; and appraise the biologic, chemical, and physical requirements of restoration targets. The two most significant ecological constraints in the Everglades are those associated with the loss of the ridge and slough landscape and those associated with historical degradation patterns (i.e., legacy impacts).

Ridge and Slough Landscapes and Tree Islands

The ridge/slough/tree island "corrugated" patterning in the Everglades is a special kind of paradox and constraint because ridges and tree islands are now 0.3 to 1 m lower than they were 100 years ago (McVoy et al. 2011; see Chapter 4), but the low-nutrient condition of the extant Everglades is not conducive to a "rapid" restoration and accretion of peat soil on the ridges and tree islands. This is a special case of hysteresis, where restoring clean water, historical water depths, and hydroperiods may, in certain locations, make conditions worse, especially for tree islands that are now only 0.15 to 0.3 m higher than the surrounding sloughs. Large regions of the Everglades can no longer support wading bird rookeries despite having plenty of available food because they are devoid of tree islands. Creating a "better" shallow-depth tree island environment may cause the regional sloughs to dry up more frequently, thus making the situation worse for wading bird foraging. Other areas, such as Shark River Slough (SRS) in ENP, have lost deep water sloughs but retain healthy, functional islands. Creating a deeper slough environment may cause these areas to be less capable of supporting tree islands.

Legacy Impacts

There are many examples of system-wide trends that constrain and severely limit the ability of the Everglades to return to the past because of profound impacts on ecosystem structure, function, or both. The wide-scale flattening of the corrugated landscape that we have described is an excellent example of this. The list also includes (1) a shift to harder, calcium-rich surface waters because of thousands of kilometers of canals cut deep into the underlying limestone; (2) prolonged inflows of relatively high-TP water that have created large areas with high-P soils; (3) strong hydroperiod gradients within the WCAs; (4) a 60% loss of tree islands; (5) a lack of deep-water refugia other than canals; (6) peat soil loss; (7) precipitously small populations of endangered species; (8) dangerously large populations of invasive exotics; (9) increased habitat homogeneity; (10) peat collapse in the coastal Everglades due to hurricanes and accelerating SLR; (11) saltwater intrusion due to SLR; (12) P accumulation in the sediments of lakes and canals; (13) reduced groundwater–surface water interactions; (14) increasing incidents of vegetation die-off; and (15) increasing human dependency on fresh water that is ultimately provided by the Everglades. Although this is a formidable list of roadblocks for our flux-capacitor, it should be noted that by addressing multiple legacies and impacts over a relatively short period of time (i.e., 20 years), the scientific community understands the ecological interactions enough to recognize what remedies are possible today and which fixes can have cumulative benefits. The Everglades scientific community is the ultimate evaluator of each possible future vision via the assessment flow path in Figure 8.2.

Caution: Entering Model Land

The CEPP took advantage of 15 years of CERP modeling and planning. The CERP is the $10 billion restoration plan (underlined to emphasize that a plan is not the same as an allocation of funds) signed by President Bill Clinton in 2000. CEPP is an *incremental adaptive restoration plan* (NRC 2014) that operationalizes key CERP components (which we will describe). It is incremental because its various components will be built sequentially over 10 to 15 years. It is adaptive (per Holling 1978; Walter and Holling 1990) because as this plan is implemented, scientific study will continually address environmental responses, stakeholder concerns, and societal tradeoffs (Sklar et al. 2005). This incremental approach gave stakeholders with conflicting desires the opportunity to evaluate their different visions of the future via workshops where the models were made accessible and stakeholders could use them to essentially drive their time machines into futures of their own making. This participatory modeling approach (Franzen et al. 2011; Van der Belt 2004) allowed the stakeholders to evaluate options while allowing the modelers to optimize their tools. The modelers were able to hone and fine-tune the restoration visions, and the stakeholders gained increased understanding of, and appreciation for, the complexities of the restoration planning process. As a result of this participatory modeling approach, CEPP has four USACE-color-coded, restoration boundary conditions (Fig. 8.3):

1. ***The Redline***: Increased water storage and delivery of clean water across the EAA boundary into WCAs
2. ***The Greenline***: More natural conveyance and distribution of water through WCA-3A and WCA-3B
3. ***The Blueline***: Improved delivery of water to ENP
4. ***The Yellowline***: Management of seepage across the eastern boundary of ENP.

The model that helped determine boundary 1 used a suite of mass balance equations for water and TP to decide on potential water storage designs (SFWMD 2010 and 2011). Boundaries 2 and 3 were combined and put into an inverse model optimization tool (Ali 2009) that sifted through thousands of possibilities. After this process was complete, a hydrodynamic simulation (SFWMD 2005a and 2005b) was used to make a final selection (see "Sheet Flow and Hydrology" section). Boundary 4 used a groundwater–surface water model (SFWMD 2010 and 2011) parameterized using data from a CERP pilot study in which a 50-foot cement subsurface seepage barrier was constructed across a small segment of the eastern boundary of ENP.

Early on in the participatory modeling process, it was critical that all stakeholders, engineers, and scientists agreed on a general hydrologic goal for CEPP. Restoration targets and performance measures (e.g., Table 8.2) were developed in collaboration with researchers from the FCE LTER, U.S. Geological Survey, the Florida Fish and Wildlife Commission, USFWS, and SFWMD. These targets proved that full restoration would not yet be possible, and therefore it was decided that the most important first incremental goal for Everglades rehabilitation was to reduce the intensity and frequency of dry downs—based on historical water levels (McVoy et al. 2011). This practical realization resulted in the hydrologic goals shown in Figure 8.4, where increases of some 300,000 acre-ft of clean (i.e., <10 ppb TP) surface water will be delivered across the Redline, during the dry season, in order to significantly rehydrate SRS.

This maze of models and the associated performance measures (see Table 8.2) took decades to create, but once in place, it became possible to formulate a "restoration design" within 18 months. Constrained by land availability, the mass balance models used for Boundary 1 quickly converged on a single water storage and treatment design north of the Redline that was capable of cleaning and storing an additional 200,000 to 300,000 acre-ft of water. Driven by a need to rehabilitate ridge and slough habitats, increase flows to ENP, and reduce dry-season droughts in SRS and Northern WCA-3A, the models converged on CEPP with a Greenline and Blueline configuration of a new

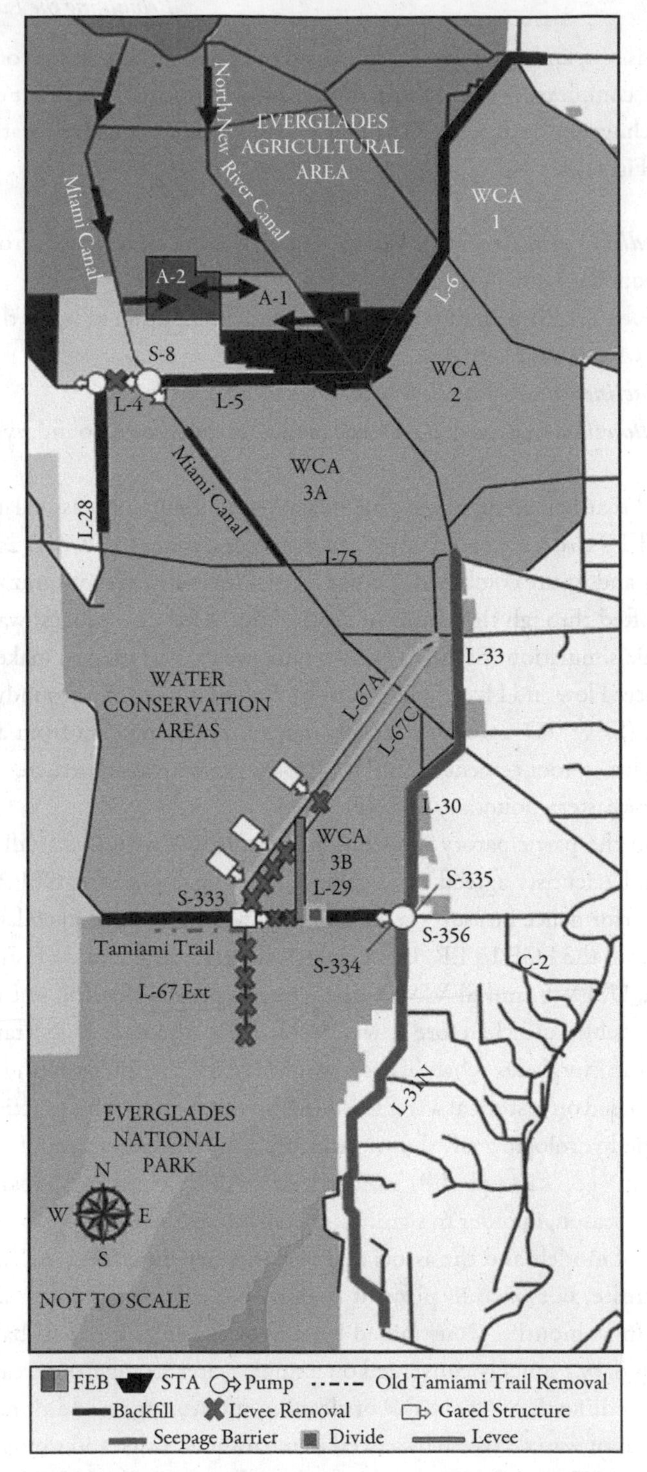

FIGURE 8.3 The CEPP focus was to build a restoration plan that would incrementally add features across critical boundaries (red, green, blue and yellow) and divide structures, such as canals and levees. The final selected plan for CEPP includes a Flow Equalization Basin (FEB), anadditional Stormwater Treatment Area (STA), a spreader canal in NW WCA-3A, backfilling, pumps, levee removal, and a large gated-culvert flowway connecting WCA-3A with the Park. See page 11 of color insert.

TABLE 8.2.

Performance Measure Scores (0 to 100 Scale)

Metric #	Performance Measure Metric	FWO	ALT4
1.1	Inundation Duration in the Ridge and Slough Landscape	68	97
2.1	Sheet Flow in the Ridge and Slough Landscape—Timing	21	32
2.2	Sheet Flow in the Ridge and Slough Landscape—Continuity	35	27
2.3	Sheet Flow in the Ridge and Slough Landscape—Distribution	46	49
3.1	Hydrologic Surrogate for Soil Oxidation—Drought Intensity Index	50	95
4.1	Number and Duration of Dry Events—Number	60	95
4.2	Number and Duration of Dry Events—Duration	26	100
4.3	Number and Duration of Dry Events—Percentage Period of Record	2	100
5.1	Slough Vegetation Suitability—Hydroperiod	53	92
5.2	Slough Vegetation Suitability—Dry Down	69	100
5.3	Slough Vegetation Suitability—Dry-Season Depth	23	65
5.4	Slough Vegetation Suitability—Wet-Season Depth	12	74
	Percentage of Target Habitat Units (HSI × 100)	**44**	**82**

These metrics were developed by the CERP and the Florida scientific community from HSIs for the FWO and CEPP (ALT4).

pump, the filling in of the northern section of the Miami Canal, and a triangular flow-way from WCA-3A to ENP (see Fig. 8.3, right). The hydrodynamic models showed that a FWO scenario had lower water depths in northern ENP, while the CEPP future scenario significantly increased depths and hydroperiods in this region. This CEPP future was also close to desired targets for inundation duration, drought intensity, and slough vegetation suitability (see Table 8.2). The final piece of this restoration design came when the Yellowline models recommended a seepage barrier about 3 km long just south of the Tamiami Trail (see Fig. 8.3, right).

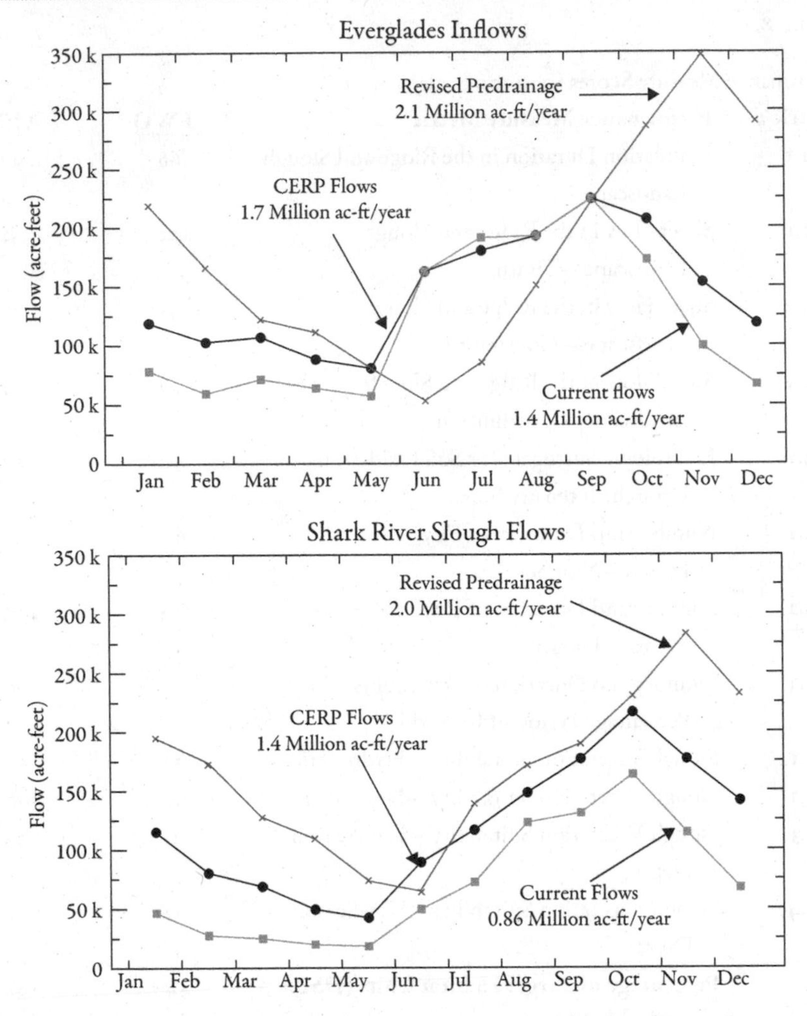

FIGURE 8.4 Top: Historic and current inflows across the northern border of Water Conservation Area 3A (see Redline in Figure 8-3). Bottom: East-west transect across Shark River Slough, the main flowway in Everglades National Park. See text for narrative.

A Future Without Roads?

In *Back to the Future*, Christopher Lloyd ignores the need for a road for takeoff, exclaiming: "Roads? Where we are going, we don't need roads!" Similarly, on returning from the hydrology modeling world in CEPP, Everglades scientists evaluated ecological model output and focused on a future Everglades, exclaiming: "Roads, levees, and canals? Where we need to go, we don't need any roads, levees, or canals!"

All around the world roads, levees, and canals have been engineered to protect human infrastructure and support land development with little regard for the ecosystem services provided by the natural system (Costanza et al. 2008). Now we realize that roads, levees, and canals interfere with the connectivity of landscapes (Forman and Godron 1986; see Chapter 4), landscape structure (Ahern 1991), biodiversity (Debinski and Holt. 2000), and ecosystem productivity (Day et al. 2000). The question becomes: How does one create a vision of restoration that removes damaging elements while balancing the needs of society? This section is focused on topical snapshots of an Everglades future fueled by CEPP, adaptive water management, and co-produced wetland science (see Chapter 9). Each topic should be viewed as a mini-case study of how to use data, hypothesis testing, and prediction to foster a "future without roads" that looks as much like the past as is feasible.

SHEET FLOW AND HYDROLOGY

The primary hydrologic modeling outputs of CEPP were based on outputs from the Glades-LECSA version (RSMGL; SFWMD 2010, 2011) and the regional simulation model (RSM; SFWMD 2005a, 2005b). The RSM is a robust and complex regional-scale model. In the model, it is frequently necessary to implement abstractions of system infrastructure and operations that will, in general, mimic the intent and result of the desired project features without exactly matching the mechanisms by which these results would be obtained in the real world. Additionally, it is sometimes necessary to work within established paradigms and foundations within the model code. Upstream water volumes available for distribution by the RSMGL were calculated from a basin model that tracked the water balance in Lake Okeechobee (SFWMD 2010, 2011) and from a stormwater treatment model that tracked P and water deliveries to the WCAs (Walker and Kadlec 2005; Wang 2012).

The RSM simulation of CEPP was like the DeLorean vehicle in *Back to the Future*, designed to carry the bags of ecological restoration into the future. Obviously, this capacity has limitations. However, according to the stage duration curves used to summarize the hydrologic simulations, this increased capacity is significant (Fig. 8.5). In an environment where an additional 30 cm of water can mean the difference between a landscape with versus without tree island habitats (van der Valk et al. 2012), an additional 12 cm (0.4 ft) of water is a substantial step toward restoration. This is especially true because this additional water will also reduce soil oxidation rates, which is critical to the stability of the corrugated ridge/slough/tree island landscape (McVoy et al. 2011; Sklar and van der Valk 2002). This should also reduce the likelihood of the peat fires that have ravaged the system for the last 100 years (Aich et al. 2014). In the near future, CEPP will do this by preventing SRS from going completely dry. The simulated FWO period between 1965 and 1985 had nine dry periods when there was no water in the sloughs in the northernmost regions of SRS. In CEPP simulations, all of these extremely dry conditions were eliminated without exceeding an upper range of water depths that might be damaging to tree islands (Wu et al. 2002).

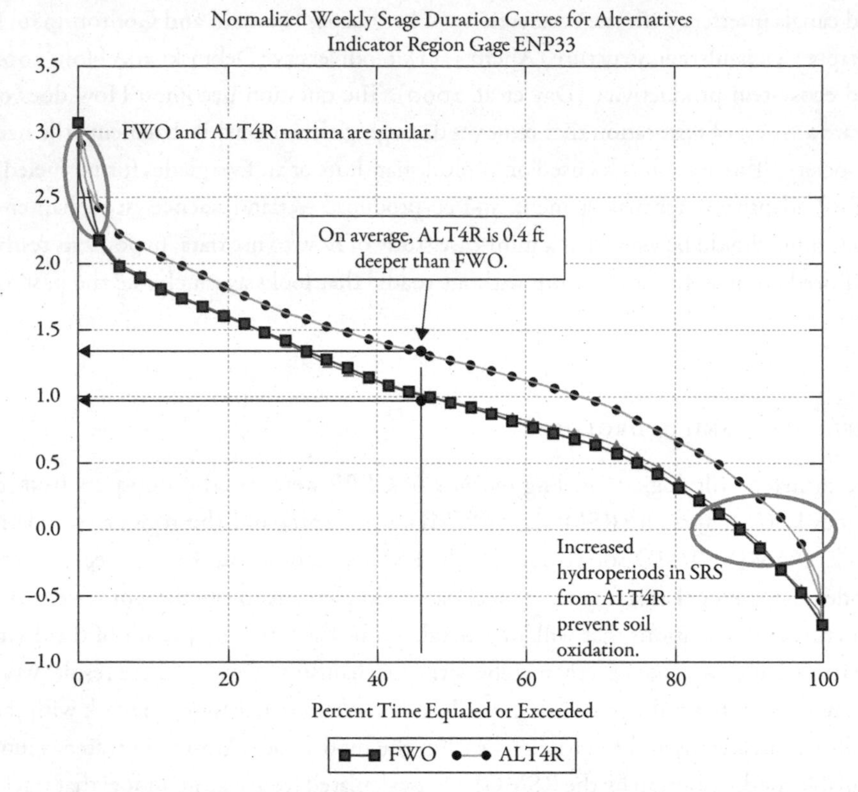

FIGURE 8.5 Shark River Slough (SRS) will significantly improve with Alt4R (the CEPP restoration plan) in comparison to the FWO (future without CEPP) because the recommended plan adds about 0.4 ft. of ponding depths to the entire stage duration curve and because it increases the marsh hydroperiod by about 10%. Indicator Region Gage ENP33 is in the center of SRS, some 30 km south of Tamiami Trail.

FISHING AROUND FOR A BETTER WORLD

Adding water to a parched landscape does not necessarily mean that the fish needed for wading birds and other Everglades wildlife will return when and where they are needed. We have used a fish and invertebrate monitoring program that began in 1996 to calibrate and validate a habitat suitability model that was used to illustrate how fish distributions and population abundance might respond to a future Everglades with slightly altered hydroperiods and depths. The model structure was relatively simple. Aquatic animal abundances were controlled by loss of habitat (including extent of areas inundated), altered hydroperiod, water depth, and frequency of water depths less than 5 cm. The latter is a particularly important phenomenon influencing Everglades aquatic animal communities because as the marsh dries, its surface is exposed, eliminating habitat for aquatic animals and causing high mortality. It takes 3 to 8 years following a drought in a long-hydroperiod marsh (on average >1 year of continuous inundation) for fish and crustacean populations to stabilize (Ruetz et al. 2005; Trexler et al. 2001, 2005). When droughts occurred repeatedly at less than 3- to 8-year intervals, fish and crustacean populations were continually recovering from past droughts and failed to reach densities sufficient to sustain large predators. The model identifies four patterns of population-level responses of wading bird prey species to marsh drying: (1) slow recovery following drought, possibly taking years to regain pre-drought density, typical of bluefin killifish (*Lucania goodie*), least killifish (*Heterandria formosa*), and riverine grass shrimp (*Palaemonetes paludosus*); (2) maximum density attained soon after drying events and lower densities a year or longer after drying, typical of flagfish (*Jordanella floridae*) and marsh killifish (*Fundulus confluentus*); (3) a weak relationship between density and time since drying at a local site, unique in the Everglades to eastern mosquitofish (*Gambusia holbrooki*); and (4) a relationship seen with crayfish that differs from fish and grass shrimp because of their ability to burrow and tolerate moderate amounts of marsh drying.

The differences between a future with and without CEPP were striking (Fig. 8.6). Without CEPP, most fish were in the deep regions of WCA-3A. Population densities of fish tended to be low everywhere else, but especially in SRS. In WCA-1, there was a very slight decrease in fish abundance compared to the FWO scenario, but in WCA-2 this decline was as high as 30%. However, we found substantial increases in ENP and in WCA-3 (see Fig. 8.6). In ENP, northern SRS is currently mediocre wading bird foraging habitat, but with CEPP and its 60% to 90% increase in fish density, it may have the potential to once again be a major foraging and nesting location.

BIRDS AND FOOD AVAILABILITY: WADING IN VAIN?

From 1985 to 2012, systematic reconnaissance flights have been used to document the abundance, flock composition, and spatiotemporal distribution of foraging wading birds across the Greater Everglades system (Bancroft et al. 1994). To develop the wading bird distribution evaluation models (WADEM; Beerens 2014, 2015), systematic reconnaissance flights occurrence data for great egrets (*Ardea alba*), white ibis (*Eudocimus albus*), and wood storks (*Mycteria americana*) were paired with daily hydrologic variables calculated from water depths generated by the Everglades Depth Estimation Network (EDEN; Telis 2006). Four explanatory variables representing hydrologic conditions were used as proxies for prey dynamics: (1) days since dry down and 2) hydroperiod were used as an indicator for long-term prey production (Loftus and Eklund 1994; Trexler 2010); (3) recession rate was used as an indicator for prey concentration dynamics (Beerens et al. 2011; Russell et al. 2002); and (4) daily water depth was used as an indicator of short-term prey availability (Beerens et al. 2011; Gawlik 2002). The interactions among these terms quantified two trends that together supplied fish as food for wading birds: the long-term response of fish populations to deep water (i.e., higher reproduction rates), and a short-term (1–3 weeks) tendency of prey to become concentrated into deeper pools during drying trends. Output from this type of model, averaged over the landscape, served as a surrogate measure of the abundance of high-quality foraging patches.

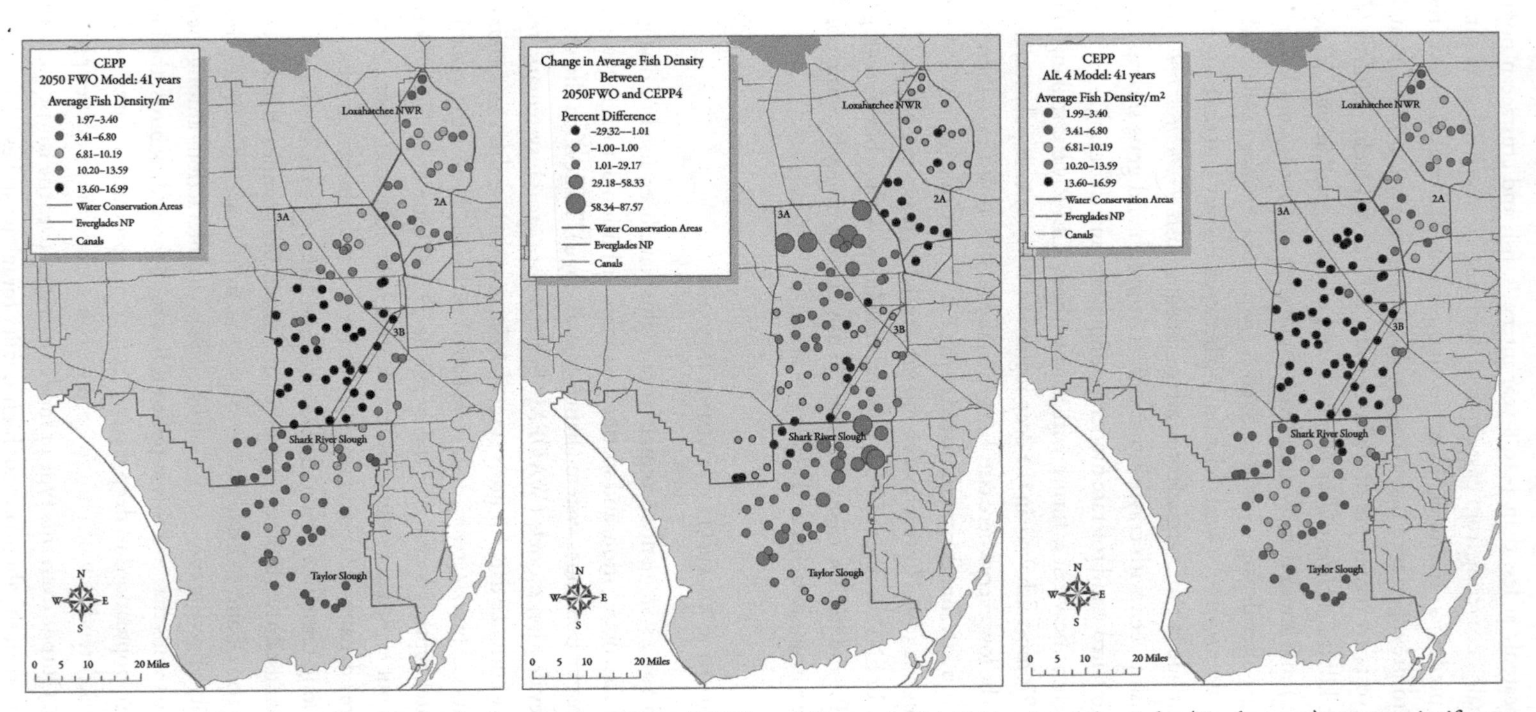

FIGURE 8.6 The difference between Alt4 and the future without CEPP (FWO), according to the fish Habitat Suitability Index (Trexler 2010), suggests significant improvements to the fish populations with restoration. See page 6 in color insert.

This WADEM prediction of bird distribution as a function of hydrology and habitat suitability was one of the more telling stories about the future (Fig. 8.7). Under the FWO scenario, decreased water volume lowered landscape days since dry down, hydroperiods, and resulting prey production, such that prey resources concentrated less when depths were shallow. Not so with the CEPP scenario: the increased volume, flow, and connectivity of the CEPP plan are predicted to increase the foraging response of all wading bird species, but not everywhere. In WCA-1 and WCA-2, foraging conditions declined an average of 10% to 25%, but we saw an improvement of 25% to 100% across vast regions of WCA-3 and ENP (see Fig. 8.7).

IN A WHILE, CROCODILIANS

The Everglades is the only place in the world where it is ecologically correct to say "See you later, alligator; after a while, crocodile," since both species naturally occur there. Both alligators and crocodiles are important players in the coastal Everglades; they are top predators and are indicators of ecological conditions that support a variety of other flora and fauna. Both contribute to food web dynamics, nutrient movements, and creation and maintenance of landscape topography (alligator holes and trails, nests, maintenance of small creeks; see Chapter 4). Alligators occur in freshwater and oligohaline areas, but they can use higher-salinity areas for foraging as long as they have periodic access to fresh water. Historically, some of the highest densities of alligators were in the coastal Everglades, where productivity was high and salinity was relatively low (Craighead 1968; Mazzotti and Brandt 1994). Adult crocodiles occur in the estuaries and coastal areas and can tolerate higher salinities. Hatchlings, however, must have access to fresh or brackish water on a periodic basis to survive and grow. Reduced freshwater flows to the coastal Everglades have led to increased salinities and concurrent changes in where alligators occur, as well as in crocodile growth, survival, and relative abundance. Increasing these freshwater flows is one of the desired outcomes of Everglades restoration, and this is predicted to increase the survival and growth rates of juvenile crocodiles.

The CEPP used a crocodile habitat suitability index (HSI) model to understand how restoration plans might improve conditions for crocodiles and possibly decrease the suitability for alligators (Brandt 2013). The crocodile growth and survival index of this HSI characterized suitable habitat, location of known nest sites, salinity, and prey biomass. This index was calculated for August through December—the period following hatching when hatchlings are most vulnerable to high salinities (Mazzotti 1999, 2007; Moler 1992). We calculated the salinity improvements associated with CEPP as ecological "lift" (=CEPP minus FWO) that ranged from 1% to 12%. All regions along the Florida Bay shoreline and estuarine ecotone improved in the CEPP scenario. However, regions that are currently poor for crocodile growth and survival remained poor. This vision of the future is an improvement but is not enough to ensure that we will meet restoration targets for crocodiles.

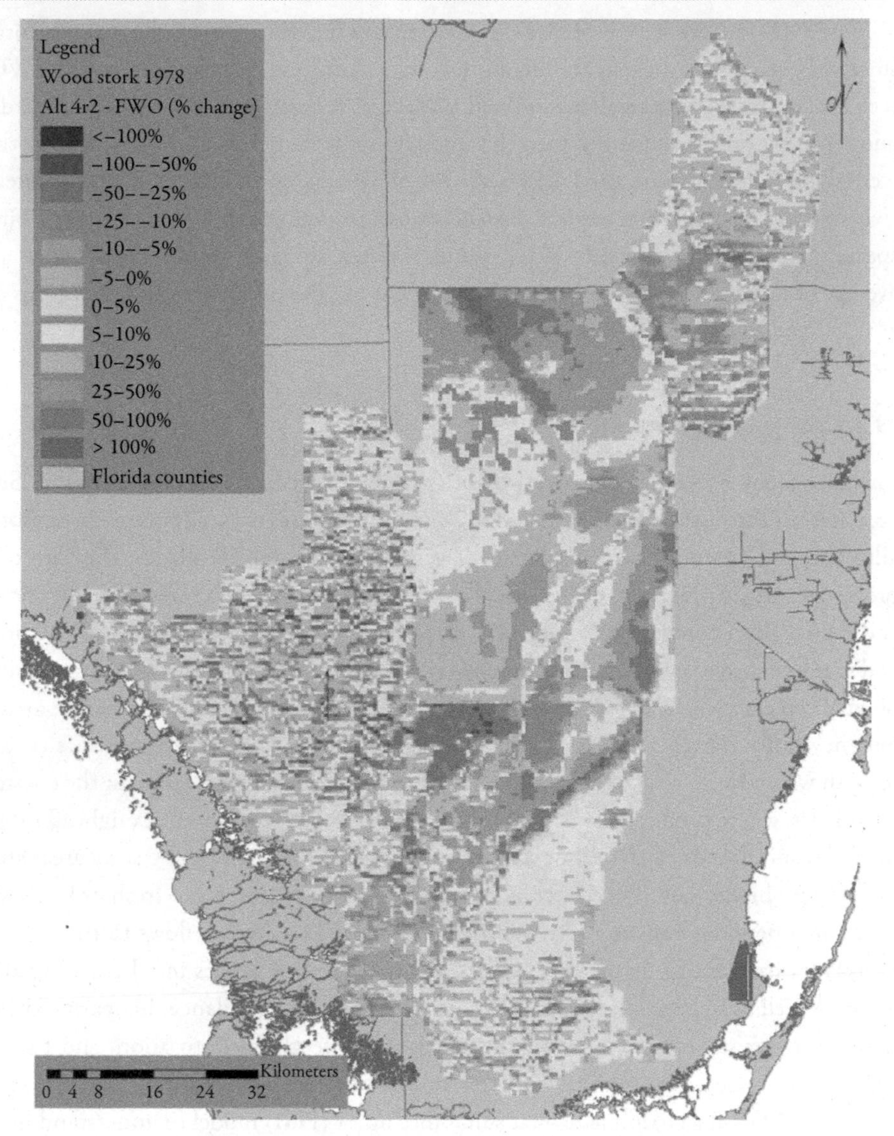

FIGURE 8.7 Mean percent change in Wood Stork spatial foraging conditions in 1978 (an average year) for the CEPP (Alt 4r2) relative to Future WithOut restoration (FWO) according to the WADEM model developed by Beerens (2014). Red demonstrates benefit from restoration, whereas blue represents losses. The large and deep red areas in the northern regions of Water Conservation Area (WCA) 3A and in SRS indicate a 100% improvement compared to a FWO restoration. Areas where the future can get worse by 10-25% are scattered throughout the system but mostly concentrated in WCA-2A and downstream of western Tamiami Trail (see Fig. 8-3). See page 7 of color insert.

A MARL PRAIRIE HOME COMPANION

The southern marl prairies of ENP are a diverse mosaic of tropical hammocks, tree islands, and wet prairie communities (Davis et al. 2005). Endemic to these prairies is the endangered Cape Sable seaside sparrow (*Ammodramus maritimus mirabilis*; CSSS), a small bird with a relatively short lifespan and specific hydrologic requirements for nesting. The CSSS is considered a key indicator species of the marl prairies because of its limited range and sensitivity to hydropattern (Pearlstine et al. 2016). The southern marl prairies that flank SRS to the east and west occur on slightly higher elevations than SRS (Gunderson 1994). A marl prairie hydrologic indicator of successful CSSS breeding and nesting (Pearlstine et al. 2016; SFNRC 2011) includes four metrics: (1) average wet-season water depth; (2) dry-season water depth; (3) hydroperiod; and (4) maximum continuous dry days during the nesting season. Application of this marl prairie hydrologic indicator to the CEPP and FWO scenarios was critical to winning USFWS approval to move forward with CEPP authorization because the CSSS is protected under the Endangered Species Act. In a CEPP future, hydroperiods and water depths in SRS increased, effectively expanding the wetter slough laterally into marl prairie habitats that are preferred by the CSSS. This reduction in habitat extent was counteracted, though, by a conversion of regions that are currently too dry into potential marl prairie habitat that was hydrologically suitable for the sparrow. While the CSSS habitat will move upslope in a CEPP future, its actual extent will not change much. This is an excellent example of how the integration of ecological data and hydrodynamic modeling provides decision makers with an adaptive incremental restoration plan that deals with seemingly conflicting and paradoxical restoration targets.

SAVING THE COASTAL LAKES

The mangrove wetlands and brackish lakes located in the coastal Everglades ecotone have experienced elevated salinities relative to predrainage times, with an estimated 20- to 30-psu increase along the north shore of Florida Bay (McIvor et al. 1994). These increased salinities coincided with declines in submerged aquatic vegetation (SAV) abundance (Craighead 1971; Tabb et al. 1962) and the migratory waterfowl that depend on them as a critical wintertime food source (Ogden 1994). One of the goals of the CEPP is to direct water south to this region to lower salinities in the Florida Bay ecotone and in Florida Bay proper. In this CEPP future, the increased delivery of fresh water will improve water quality (i.e., salinities and the underwater light environment) and thereby enable restoration of historic SAV abundances in these ecotone lakes.

The science of SAV restoration aims to improve our predictions of how SAV communities will respond to restored freshwater flows. This science has long focused on the ecological relationships between SAV communities and water quality (Tabb et al. 1962). This historical decline in cover was originally attributed to the inability of the brackish SAV communities to tolerate elevated salinities. However, recent research has revealed that low light availability caused by persistent phytoplankton blooms fueled by elevated P levels in some ecotone lakes may also be responsible for the SAV decline (Frankovich et al. 2011, 2012). We have found that 5% of the surface light at the sediment surface is a minimum requirement to maintain SAV cover in ecotone lakes and that TP concentrations need to be at or below 2.2 μM to maintain this water clarity. We now understand that both salinities and TP concentrations need to be reduced to restore these SAV communities. Salinity incursions from Florida Bay are largely responsible for supplying P to the ecotone and lakes (Davis et al. 2005), and increasing flows of low-P water from the upstream freshwater Everglades should make this region more oligotrophic (Childers et al. 2006). This CEPP future should reduce phytoplankton blooms, increasing light availability while also reducing salinity. We have seen recent evidence of how these changing conditions will affect SAV cover from the C111 Spreader Canal Western Project. This canal diversion is delivering more fresh water to the central Florida Bay ecotone, with concurrent declines in phytoplankton, nutrients, and salinity and increases in light transparency and SAV coverage (Fig. 8.8). However, initiating SAV establishment and growth in currently unvegetated areas may require even greater water clarity, and thus even lower phytoplankton abundances and lower P concentrations than those needed to maintain existing SAV beds. This is confounded by legacy impacts of P that has already accumulated in lake sediments. This legacy P may fuel elevated phytoplankton production for some time following restoration of greater freshwater flows and lower salinities.

SAVING FLORIDA BAY

The hydrology of Florida Bay is a complex tension between local precipitation, Everglades freshwater inputs, ocean exchange through the Florida Keys tidal passes, and Gulf inputs across the western boundary. Florida Bay hydrology has changed greatly from predrainage conditions (Marshall and Nuttle 2012; Wanless et al. 1994). Currently, only 20% of the fresh water in the Everglades landscape makes it to Florida Bay, as the rest is discharged through canals to the Atlantic Ocean or to the Gulf of Mexico (Light and Dineen 1994). Florida Bay has also become more isolated from oceanic influence during this time with the filling of marine passes between the Keys islands (Fourqurean and Robblee 1999). The result is a bay that has transitioned from an estuary characterized by variable salinities to a more saline marine lagoon. This transition has brought significant ecological change to Florida Bay, including seagrass die-offs, sponge mortality events, phytoplankton blooms, reduced wading bird populations, and diminished fisheries. The seagrass community suffered a catastrophic mortality event in 1987–1989 (Carlson et al. 1990; Robblee et al. 1991) that was partly attributed to salinity stress (Koch et al. 2007). The die-off initially destroyed 4,000 ha of *Thalassia* beds and thinned standing stocks in an additional 23,000 ha (Hall et al. 1999; Robblee et al. 1991). Within the bordering mangrove ecotone, areas of natural sawgrass (*Cladium jamaicense*) and spikerush (*Eleocharis cellulosa*) marsh have died off as the dwarf red mangrove wetlands transgress landward and unvegetated areas with higher salt concentrations, known as the "white zone," are expanding (Egler 1952; Ross et al. 2000).

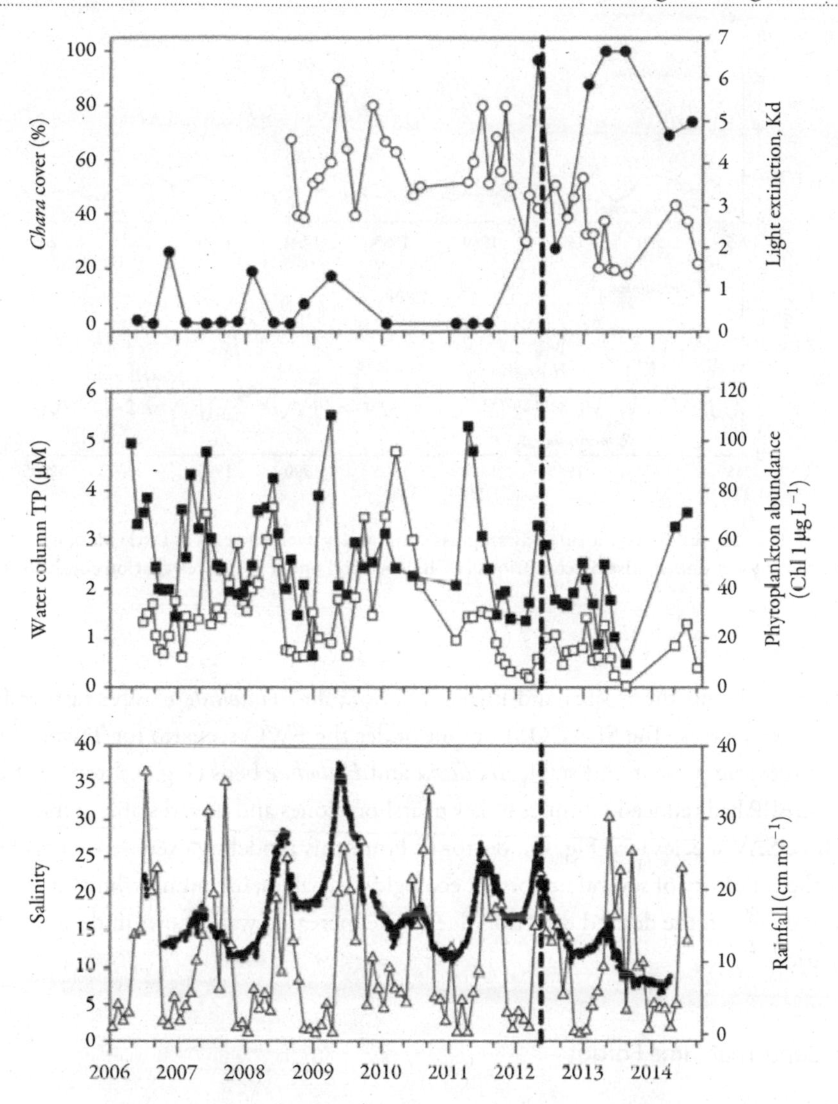

FIGURE 8.8 Top: Time series of *Chara* cover (solid circle), a desirable macro-algae and water column light extinction (K_d; open circle); Middle: Water column total phosphorus (TP) concentration (solid square) and phytoplankton abundance (chl a; open square); Bottom: Salinity (solid circle) and rainfall (open triangle) in West Lake (site 28 in Figure 1 of Frankovich et al. 2011). Rainfall data are monthly sums from NEXRAD data for North Cuthbert Lake (NCL; US National Weather Service). Vertical dashed line indicates the start of water diversion operations in the C111 Spreader Canal Western Project on May 17 2012.

Going back to the future means increasing delivery of fresh water to the southern Everglades and Florida Bay. We have used a dynamic seagrass ecosystem simulation model (SEACOM) to simulate the effects of different restoration scenarios on the SAV communities in Florida Bay (Madden and McDonald 2010). The "desired" future for Florida Bay is a more diverse, mixed SAV community characterized by enhanced *Ruppia*

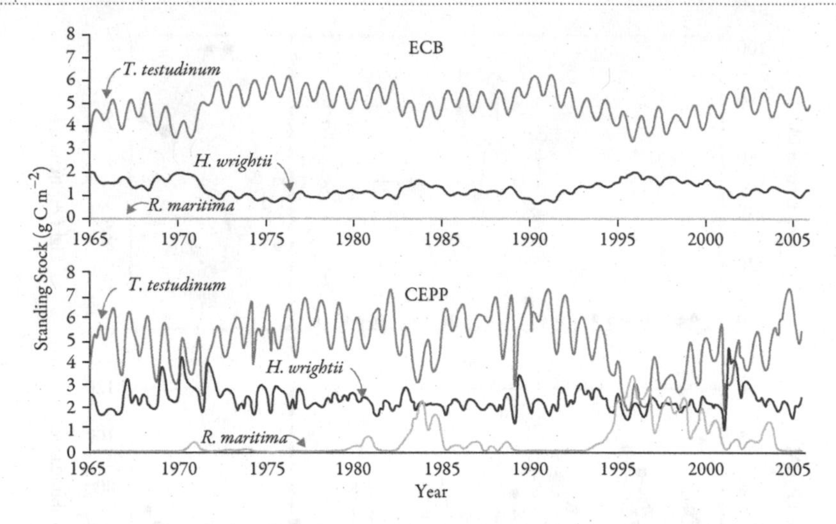

FIGURE 8.9 The predicted Florida Bay seagrass community over 40 years at Little Madeira Bay/Taylor River under existing conditions (ECB; Top) and under CEPP restoration conditions (Bottom).

habitat in the northern sites and mixed *Thalassia* and *Halodule* in sites farther from freshwater sources. The SEACOM output under the FWO scenario for Taylor Slough had no *Ruppia* present and stable *Thalassia* and *Halodule* beds (Fig. 8.9, top). A future with CEPP had reduced salinities in key nearshore zones and mixed stable communities of three SAV species (see Fig. 8.9, bottom). From this modeling exercise we concluded that the trajectory of several important ecological variables, including salinity, nutrients, and SAV, was in the desired direction due to the increased water flows under the CEPP scenario.

The Land That Time Forgot

As is the goal of most national parks, the Everglades is supposed to be a (wet)land left alone, unencumbered by socioeconomic forces, a pristine landscape protected for future generations of people, plants, and wildlife. This is, of course, easier said than done. Some may wish that the Everglades was a land that time forgot, but the encroachment of canals, roads, seas, P, people, and invasive plants and animals has forced society to consider management and restoration strategies that stem these tides of degradation. The CERP is the largest U.S. management program ever proposed to bring back the past—at least ideologically—and the most significant increment of this plan is the CEPP. In this chapter we have highlighted some of the more significant features of a future with CEPP. We demonstrated how long-term datasets were used to quantify the benefits of CEPP, we introduced some of the constraints that make Everglades restoration so complicated, and we gave you a glimpse into the future. What can we surmise, recommend, and applaud from this trip back to the future? The most obvious take-home message is that restoration is not easy—ecologically, hydrologically, socially, politically, or economically. It takes a long time and requires scientific understanding, innovative modeling, and leadership. For a society defined by utilitarian-economic theory, restoration of the Everglades might appear to be too expensive and a burden on society due to a lack of profit margins (Graaf and Batker 2011). But this begs the question: Should we view large landscapes such as south Florida through the lens of a neoclassical macroeconomic market, ignoring nature's goods and services? The expense of CEPP and CERP is insignificant when one considers the value of the natural processes that are critical for sustaining future human well-being and economic growth (Batker et al. 2010; Costanza et al. 1998; Day et al. 2014).

This chapter has shown you the end result of many years of research, modeling, and debate. The results are ecologically significant and socially important, and as we go to press we remain hopeful that the CEPP will soon be funded by Congress ($1.9 billion). We arrived at this scientific and resource management milestone by creating a clear understanding of how to manage the models, the science, and the concerns of society. In short, we have learned the following:

1. Models are essential for creating a vision of the future, but they require:
   - Long-term datasets that capture historical cyclical patterns and trends.
   - An understanding of feedbacks, drivers, and forcing functions.
   - Reasonable levels of calibration and validation.
   - Documentation, review, and approval by the engineering and scientific communities.
2. Ecological uncertainties interfere with creating a clear vision of the future, but they can be addressed by:
   - Focusing the science on ecotones where change is most dynamic, such as the coastal Everglades.
   - Developing research and monitoring to reduce uncertainties associated with changes in water depths, hydroperiods, and flow.
   - Evaluating the impacts of natural and anthropogenic trends into the future.
3. Socioeconomic, legal, and ecological constraints will limit possible futures, but improved environmental conditions and protection can be significant if the following procedures are enacted:

   - Stakeholder participation in scenario development.
   - Adaptive management, particularly for high-risk modifications.
   - Implementation of an incremental process of environmental restoration or rehabilitation.

Acknowledgments

We are particularly grateful to the editors of this book for their time, patience, and insights. We greatly appreciate the edits of Mark Rains and Susan Gray. A special recognition goes to the Jacksonville District of the U.S. Army Corps of Engineers and the South Florida Water Management District for providing the resources for this manuscript and for the planning of the Central Everglades Planning Process. The findings and conclusions in this article are those of the author(s) and do not necessarily represent the views of the U.S. Geological Survey, the U.S. Fish and Wildlife Service, Everglades National Park, or the South Florida Water Management District.

References

Ahern, J. 1991. Planning for an extensive open space system: linking landscape structure and function. *Landscape and Urban Planning* 21(1-2): 131–145.

Aich, S., S.M.L. Ewe, B. Gu, and T.W. Dreschel. 2014. An evaluation of peat loss from an Everglades tree island, Florida, USA. *Mires and Peat* 14: Art. 2.

Ali, A. 2009. Nonlinear multivariate rainfall-stage model for large wetland systems. *Journal of Hydrology* 374(3-4): 338–350.

Bancroft, G.T., A.M. Strong, R.J. Sawicki, W. Hoffman, and S.D. Jewell. 1994. Relationships among wading bird foraging patterns, colony locations, and hydrology in the Everglades. In *Everglades: The Ecosystem and Its Restoration*, edited by S.M. Davis and J.C. Ogden, 615–658. Delray Beach, FL: St. Lucie Press.

Batker, D., I. Torre, R. Costanza, P. Swedeen, and J.W. Day. 2010. Gaining Ground: Wetlands, Hurricanes, and the Economy: The Value of Restoring the Mississippi River Delta. Earth Economics Project Report.

Beerens, J.M. 2014. Development, Evaluation, and Application of Spatio-Temporal Wading Bird Foraging Models to Guide Everglades Restoration. PhD Dissertation, Florida Atlantic University, Boca Raton, FL.

Beerens, J.M., D.E. Gawlik, G. Herring, and M.I. Cook. 2011. Dynamic habitat selection by two wading bird species with divergent foraging strategies in a seasonally fluctuating wetland. *The Auk* 128: 651–662.

Beerens, J.M., E.G. Noonburg, and D.E. Gawlik. 2015. Linking dynamic habitat selection with wading bird foraging distributions across resource gradients. *PLoS ONE* 10(6): e0128182.

Brandt, L. 2013. An Evaluation of Central Everglades Planning Project (CEPP) alternatives using an Index of the crocodile Habitat Suitability Index. U.S. Fish and Wildlife Service, Davie, FL.

Carlson, P., Jr., M.J. Durako, T.R. Barber, L.A. Yarbro, Y. deLama, and B. Hedin. 1990. Catastrophic mortality of the seagrass *Thalassia testudinum* in Florida Bay. Report to Florida Department of Natural Resources.

Carr, R.S. 2002. The archaeology of Everglades tree island. In *Tree Islands of the Everglades*, edited by F. Sklar and A. van der Valk, 187–206. Boston: Kluwer Academic Publishers.

Cattelino, J. 2010. Citizenship and nation in the Everglades. *Anthropology News* 51(1): 11–13.

CEPP (Central Everglades Planning Project) Final Project Implementation Report and Environmental Impact Statement: Appendix F, Recreation. 2014. US Army Corps of Engineers, Jacksonville, FL.

Childers, D.L., J.N. Boyer, S.E. Davis, C.J. Madden, D.T. Rudnick, and F.H. Sklar. 2006. Relating precipitation and water management to nutrient concentrations in the oligotrophic "upside-down" estuaries of the Florida Everglades. *Limnology and Oceanography* 51(1-2): 602–616.

Chimney, M.J., M.K. Nungesser, C. Combs, B. Gu, E. Fogarty-Kellis, J.M. Newman, W. Wagner, and L. Wenkert. 2004. STA Optimization and Advanced Treatment Technologies. In 2004 Everglades Consolidated Report, South Florida Water Management District, West Palm Beach, FL.

Chimney, M.J., Y. Wan, V.V. Matichenkov, and D.C. Calvert. 2007. Minimizing phosphorus release from newly flooded organic soils amended with calcium silicate slag: a pilot study. *Wetlands Ecology and Management* 15: 385–390.

Costanza, R., R. d'Arge, R. de Groot, S. Farber, M. Grasso, B. Hannon, K. Limburg, S. Naeem, R.V. O'Neill, J. Paruelo, R.G. Raskin, P. Sutton, and M. van den Belt. 1998. The value of eco-system services: putting the issues in perspective. *Ecological Economics* 25: 67–72.

Costanza, R., O. Perez-Maqueo, M.L. Martinez, P. Sutton, S.J. Anderson, and K. Mulder. 2008. The value of coastal wetlands for hurricane protection. *AMBIO: A Journal of the Human Environment* 37(4): 241–248.

Craighead, Sr., F.C. 1968. The role of the alligator in shaping plant communities and maintaining wildlife in the southern Everglades. *Florida Naturalist* 41:2-7, 69–74, 94.

Craighead, Sr., F.C. 1971. *The Trees of South Florida, Volume I: The Natural Environments and Their Succession*. Coral Gables, FL: University of Miami Press.

Davis, S.M. 1994. Phosphorus inputs and vegetation sensitivity in the Everglades. In *Everglades: The Ecosystem and Its Restoration*, edited by S.M. Davis and J.C. Ogden, 357–378. Boca Raton, FL: St. Lucie Press.

Davis, S.M., D.L. Childers, J.J. Lorenz, H.R. Wanless, and T.E. Hopkins. 2005. A conceptual model of ecological interactions in the mangrove estuaries of the Florida Everglades. *Wetlands* 25: 832–842.

Day, J.W., L.D. Britsch, S.R. Hawes, G.P. Shaffer, D.J. Reed, and D. Cahoon. 2000. Pattern and process of land loss in the Mississippi Delta: a spatial and temporal analysis of wetland habitat change. *Estuaries* 23: 425–438.

Day, J.W., G.P. Kemp, A.M. Freeman, and D.P. Muth. 2014. *Perspectives on the Restoration of the Mississippi Delta: The Once and Future Delta*. New York: Springer.

Debinski, D.M., and R.D. Holt. 2000. Review: a survey and overview of habitat fragmentation experiments. *Conservation Biology* 14(2): 342–345.

Dengler, M. 2007. Spaces of power for action: governance of the Everglades restudy process (1992–2000). *Political Geography* 26: 423–454.

Doyle, M., and C. Drew (eds.). 2008. *Large-Scale Ecosystem Restoration: Five Case Studies from the United States*. Society for Ecological Restoration International. Washington, DC: Island Press.

Egler, F.E. 1952. Southeast saline Everglades vegetation, Florida and its management. *Vegetatio* 3: 213–265.

Ewing, J. 2013. Florida Agriculture by the Numbers. National Agricultural Statistics Service, Florida Field Office. www.nass.usda.gov/fl.

Faridmarandi, S., and G.M. Naja. 2014. Phosphorus and water budgets in an agricultural basin. *Environmental Science and Technology* 48: 8481–8490.

Forman, R.T.T., and M. Godron. 1986. *Landscape Ecology*. New York: John Wiley and Sons.

Fourqurean, J.W., and M.B. Robblee. 1999. Florida Bay: a history of recent ecological changes. *Estuaries* 22(2B): 345–357.

Frankovich, T.A., J.G. Barr, D. Morrison, and J.W. Fourqurean. 2012. Differing temporal patterns of *Chara hornemannii* cover correlate to alternate regimes of phytoplankton and submerged aquatic-vegetation dominance. *Marine and Freshwater Research* 63: 1005–1014.

Frankovich, T.A., J.W. Fourqurean, and D. Morrison. 2011. Benthic macrophyte distribution and abundance in estuarine mangrove lakes: relationships to water quality. *Estuaries and Coasts* 34: 20–31.

Franzen, F., G. Kinell, J. Walve, R. Elmgren, and T. Soderqvist. 2011. Participatory social-ecological modeling in eutrophication management: the case of Himmerfjarden, Sweden. *Ecology and Society* 16: 4–27.

Gawlik, D.E. 2002. The effects of prey availability on the numerical response of wading birds. *Ecological Monographs* 72: 329–346.

Graaf, J., and D.D. Batker. 2011. *What's the Economy for Anyway?* New York: Bloomsbury Press.

Greer, J.M. 2011. *The Wealth of Nature: Economics as if Survival Mattered*. Gabriopla Island, BC, Canada: New Society Publishers.

Gunderson, L.H. 1994. Vegetation of the Everglades: determinants of community composition. In *Everglades: The Ecosystem and Its Restoration*, edited by S.M. Davis and J.C. Ogden, 323–340. Boca Raton, FL: St. Lucie Press.

Hall, M.O., M.J. Durako, J.W. Fourqurean, and J.C. Zieman. 1999. Decadal changes in seagrass distribution and abundance in Florida Bay. *Estuaries* 22(2B): 445–459.

Holling, C.S. 1978. *Adaptive Environmental Assessment and Management*. London: John Wiley & Sons.

Klein, N. 2014. *This Changes Everything: Capitalism vs. the Climate*. New York: Penguin Books, Limited.

Koch, M.S., S.A. Schopmeyer, O.I. Nielsen, C. Kyhn-Hansen, and C.J. Madden. 2007. Conceptual model of seagrass die-off in Florida Bay: links to biogeochemical processes. *Journal of Experimental Marine Biology and Ecology* 350(1-2): 73–88.

Kolbert, E. 2014. *The Sixth Extinction: An Unnatural History*. New York: Henry Holt and Company.

Light, S.S., and J.W. Dineen. 1994. Water control in the Everglades: a historical perspective. In *Everglades: The Ecosystem and Its Restoration*, edited by S.M. Davis and J.C. Ogden, 47–84. Boca Raton, FL: St. Lucie Press.

Loftus, W.F. and A.M. Eklund. 1994. Long-term dynamics of an Everglades fish community. In *Everglades: The Ecosystem and Its Restoration*, edited by S.M. Davis and J.C. Ogden, 461–483. Delray Beach, FL: St. Lucie Press.

Madden, C.J. and A.A. McDonald. 2010. Seagrass Ecosystem Assessment and Community Organization Model (SEACOM), A Seagrass Model for Florida Bay: Examination of Fresh Water Effects on Seagrass Ecological Processes, Community Dynamics and Seagrass Die-off. South Florida Water Management District, West Palm Beach, FL.

Marshall, F., and W.K. Nuttle. 2012. Hydrologic models predict salinity in Florida Bay. In *Tropical Connections: South Florida's Marine Environments*, edited by W.L. Kruczynski and P.J. Fletcher, 90. Cambridge, MD: University of Maryland Center for Environmental Science, IAN Press.

Mazzotti, F.J. 1999. The American Crocodile in Florida Bay. *Estuaries* 22: 552–561.

Mazzotti, F.J., and L.A. Brandt. 1994. Ecology of the American alligator in a seasonally fluctuating environment. In *Everglades: The Ecosystem and Its Restoration*, edited by S.M. Davis and J.C. Ogden, 485–505. Delray Beach, FL: St. Lucie Press.

Mazzotti, F.J., L.A. Brandt, P.E. Moler, and M.S. Cherkiss. 2007. American Crocodile (*Crocodylus acutus*) in Florida: recommendations for endangered species recovery and ecosystem restoration. *Journal of Herpetology* 41: 122–132.

McCally, D. 2000. *Everglades: An Environmental History*. Gainesville: University Press of Florida.

McIvor, C.C., J.A. Ley, and R.D. Bjork. 1994. Changes in freshwater inflow from the Everglades to Florida Bay including effects on biota and biotic processes: a review. In *Everglades: The Ecosystem and Its Restoration*, edited by S.M. Davis and J.C. Ogden, 117–148. Boca Raton, FL: St. Lucie Press.

McVoy, C.W., W.P. Said, J. Obeysekera, J.A. VanArman, and T.W. Dreschel. 2011. *Landscapes and Hydrology of the Predrainage Everglades*. Gainesville: University Press of Florida.

Miller, J.R., and J.J. Hobbs. 2007. Habitat restoration—do we know what we're doing? *Restoration Ecology* 15: 382–390.

Modeling Department, Everglades Restoration, South Florida Water Management District, West Palm Beach, FL.

Moler, P. 1992. American Crocodile population dynamics. Final Report. Study Number 7532. Bureau of Wildlife Research Florida Game and Fresh Water Fish Commission.

National Research Council (NRC). 2014. *Progress Towards Restoring the Everglades: The Fifth Biennial Review—2014*. Washington, DC: National Academies Press.

Ogden, J.C. 1994. A comparison of wading bird nesting colony dynamics (1931–1946 and 1974–1989) as an indication of ecosystem conditions in the southern Everglades. In *Everglades: The Ecosystem and Its Restoration*, edited by S.M. Davis and J.C. Ogden, 533–570. Boca Raton, FL: St. Lucie Press.

Ogden, J.C., S.M. Davis, K.J. Jacobs, T. Barnes, and H.E. Fling. 2005. The use of conceptual ecological models to guide ecosystem restoration in South Florida. *Wetlands* 25(4): 795–809.

Ogden, L.A. 2011. *Swamplife: People, Gators, and Mangroves Entangled in the Everglades*. Minneapolis: University of Minnesota Press.

Ogden, L., N. Heynen, U. Oslender, P. West, K-A. Kassam, and P. Robbins. 2013. Global assemblages, resilience, and Earth stewardship in the Anthropocene. *Frontiers in Ecology and the Environment* 7(11): 341–347.

Pearlstine, L., A. Lo Galbo, G. Reynolds, J.H. Parsons, T. Dean, M. Alvarado, and K. Suir. 2016. Recurrence intervals of spatially simulated hydrologic metrics for restoration of Cape Sable seaside sparrow (*Ammodramus maritimusmirabilis*) habitat. *Ecological Indicators* 60: 1252–1262.

RECOVER. 2011. Restoration Coordination and Verification Program, US Army Corps of Engineers, Jacksonville District, Jacksonville, Florida, USA; and South Florida Water Management District, West Palm Beach, Florida.

Robblee, M.B., T.R. Barber, P.R. Carlson, M.J. Durako, J.W. Fourqurean, L.K. Muehlstein, D. Porter, L.A. Yarbro, R.T. Zieman, and J.C. Zieman. 1991. Mass mortality of the tropical seagrass *Thalassia testudinum* in Florida Bay (USA). *Marine Ecology Progress Series* 71: 297–299.

Ross, M.S., J.F. Meeder, J.P. Sah, P.L. Ruiz, and G.J. Telesnicki. 2000. The southeast saline Everglades revisited: 50 years of coastal vegetation change. *Journal of Vegetation Science* 11: 101–112.

Ruetz, C.R., III, J.C. Trexler, F. Jordan, W.F. Loftus, and S.A. Perry. 2005. Population dynamics of wetland fishes: spatiotemporal patterns shaped by hydrological disturbance? *Journal of Animal Ecology* 74: 322–332.

Russell, G.J., O.L. Bass, and S.L. Pimm. 2002. The effect of hydrological patterns and breeding-season flooding on the numbers and distribution of wading birds in Everglades National Park. *Animal Conservation* 5: 185–199.

SFNRC. 2011. Marl Prairie Indicator Version 2.0.2. http://www.cloudacus.com/simglades/marlprairie.php.

Simmons, G., and L.A. Ogden. 2010. *Gladesmen: Gator Hunters, Moonshiners, and Skiffers*. Gainesville: University of Florida Press.

Sklar, F.H., M.J. Chimney, S. Newman, P. McCormick, D. Gawlik, S. Miao1, C. McVoy, W. Said, J. Newman, C. Coronado, G. Crozier, M. Korvela, and K. Rutchey. 2005. The scientific and political underpinnings of Everglades restoration. *Frontiers in Ecology and Environment* 3(3): 161–169.

Sklar, F.H., and A. van der Valk (eds.). 2002. *Tree Islands of the Everlades*. Boston: Kluwer Academic Publishers.

South Florida Water Management District. 2005a. Regional Simulation Model—Theory Manual. South Florida Water Management District, West Palm Beach, FL.

South Florida Water Management District. 2005b. Regional Simulation Model—Hydrologic Simulation Engine (HSE) User Manual. South Florida Water Management District, West Palm Beach, FL.

South Florida Water Management District. 2010. Draft Report—Calibration and Validation of the Glades and Lower East Coast Service Area Application of the Regional Simulation Model, Sept. 2010. Hydrologic & Environmental Systems.

South Florida Water Management District. 2011. Model Documentation Report (DRAFT): RSM Glades-LECSA—2015 Future without CERP Baseline V2.1. Hydrologic & Environmental Systems.

Tabb, D.C., D.L. Dubrow, and R.B. Manning. 1962. The ecology of northern Florida Bay and adjacent estuaries. State of Florida Board of Conservation Technical Series No. 39. Miami, Florida.

Telis, P.A. 2006. The Everglades Depth Estimation Network (EDEN) for support of ecological and biological assessments. U.S. Geological Survey Fact Sheet 2006-3087.

Trexler, J.C. 2010. Prey-Based Freshwater Fish Density Performance Measure (Greater Everglades Aquatic Trophic Levels). Draft DECOMP Performance Measure Documentation Sheet. Submitted to the Army Corp of Engineers.

Trexler, J.C., W.F. Loftus, C.F. Jordan, J. Chick, K.L. Kandl, T.C. Ruetz, and O.L. Bass. 2001. Ecological scale and its implications for freshwater fishes in the Florida Everglades. In *The*

Everglades, Florida Bay, and Coral Reefs of the Florida Keys: An Ecosystem Sourcebook, edited by W. Porter and K.G. Porter, 153–181. Boca Raton, FL: CRC Press.

Trexler, J.C., W.F. Loftus, and S. Perry. 2005. Disturbance frequency and community structure in a twenty-five-year intervention study. *Oecologia* 145:140–152.

US Army Corps of Engineers (USACE). 1999. Central and Southern Florida Project Comprehensive Review Study: Final Integrated Feasibility Report and Programmatic Environmental Impact Statement. Jacksonville District, Jacksonville, FL.

US Army Corps of Engineers (USACE). 2014. Central Everglades Planning Project (CEPP): Final Integrated Project Implementation Report and Environmental Impact Statement. Jacksonville District, Jacksonville, FL.

Van der Belt, M. 2004. *Mediated Modeling: A System Dynamics Approach to Environmental Consensus Building*. Washington, DC: Island Press.

Van der Valk, A., B. Bedford, R. Labisky, and J. Volin. 2012. Ecological Effects of Extreme Hydrological Events on the Greater Everglades. Independent Scientific Review Panel Report to RECOVER.

Walker, W., and R. Kadlec 2005. Dynamic Model for Stormwater Treatment Areas, prepared for the U.S. Department of the Interior and the U.S. Army Corps of Engineers. http://www.wwwalker.net/dmsta/

Walters, C.J., and C.S. Holling. 1990. Large-scale management experiments and learning by doing. *Ecology* 71(6): 2060–2068.

Wang, N. 2012. Summary of DMSTA model runs prepared for CEPP. Internal Memorandum to Walter Wilcox, dated December 12, 2012. USACE DASR Project 51—Central Everglades Planning/Analysis Phase/ Plan Formulation/ Baselines/Final_Output_121312/ rsmbn_model_output/ dmsta_model_output/ FWO

Wanless, H.R., R.W. Parkinson, and L.P. Tedesco. 1994. Sea level control on stability of Everglades wetlands. In *Everglades: The Ecosystem and Its Restoration*, edited by S.M. Davis and J.C. Ogden, 199–224. Delray Beach, FL: St. Lucie Press.

Wu, Y., K. Rutchey, W. Guan, L. Vilchek, and F.H. Sklar. 2002. Spatial simulations of tree islands for Everglades restoration. In *Tree Islands of the Everglades*, edited by F. Sklar and A. van der Valk, 469–498. Boston: Kluwer Academic Publishers.

9

Reimagining Ecology Through an Everglades Lens

Evelyn Gaiser, Laura A. Ogden, Daniel L. Childers, and Charles Hopkinson

In a Nutshell

- By exploring the origins of productivity in coastal ecosystems, our research has revealed that:
  - o The karstic and shallow-sloping geology of the south Florida peninsula controls the source, connectivity, and quality of water supplied to the Everglades that determines ecological functions along freshwater to marine gradients. This water source connectivity and quality have been altered by a long history of human modification of the south Florida landscape.
  - o Most phosphorus (P), the limiting nutrient in karstic coastal wetlands, is adsorbed to carbonate bedrock and soil or sequestered by microbial mats in upstream freshwater wetlands, causing extremely low water column concentrations of total P ($<5\ \mu g\ L^{-1}$).
  - o Biogeochemical and productivity gradients are "upside-down," with marine rather than freshwater sources of P predominantly fueling new productivity in coastal forests and estuaries.
  - o Microbial mat communities that are adapted to low-P conditions form the base of a complex food web yet are physically and chemically resistant to grazing, resulting in a truncated biomass pyramid.
  - o P limitation, "upside-down" production gradients, and truncated biomass pyramids appear to be generalized features of wetlands on carbonate platforms.
- Social-ecological studies of the Everglades have promoted a new concept of "wilderness" that embraces the breadth and legacies of human activities in a now-urbanized watershed.

- A transdisciplinary approach has been central to our research; this approach allows us to be informed by, and to effectively influence, resource policy in the Everglades. This approach is a model for protecting threatened coastal resources across the globe.

Introduction

Coastal ecosystem science progressed through two major phases in the last several decades. The first phase exposed estuaries as critical transformers of materials and energy in coastal watersheds, rather than simple conduits for materials from uplands to the sea (Teal 1962). Subsequent studies of energy flow and biogeochemical cycles in estuaries led to new models describing how estuaries transform under changing suites of environmental drivers (Day et al. 1989) and how they regulate regional resource supplies and even global cycles of inorganic nutrients and carbon (Schlesinger and Melack 1981). By the mid-20th century these critical functions were being lost at alarming rates around the globe, threatening coastal resource sustainability. Explorations of the dependency of human well-being on coastal ecosystem services has characterized a second phase of progress in coastal ecology. Such interdisciplinary work has been fostered by social-ecological frameworks that employ long-term data and experiments to understand the function of, and inform the future of, coastal ecosystems (Day et al. 2012a; Hopkinson et al. 2008; Sullivan et al. 2014). This book is replete with examples of this very kind of research.

There are few coastal ecosystems where the intersection of biophysical processes and human activities is more evident than the Florida Everglades. During the early to mid-20th century, development of an extensive drainage system allowed agriculture and people to expand into the Everglades, depleting freshwater supplies to the Everglades and its underlying aquifer (see Chapter 1 for details). Concomitant phosphorus (P) runoff from agricultural lands caused extensive ecosystem alterations in nutrient-poor Everglades marshes (see Chapter 5). A politically embroiled process of Everglades restoration ensued (Sklar et al. 2005; see Chapter 8), beginning with carefully orchestrated, large-scale experiments to establish a water quality criterion for P loading to Everglades wetlands (Gaiser et al. 2005, 2006a). Hydrologic restoration has become dependent on meeting water quality goals while also satisfying the desires of a diverse community of stakeholders in the Everglades landscape, all of which has delayed and reduced the scope of restoration. Held in the balance was the fate of the coastal Everglades—a landscape that is now exceptionally vulnerable to saltwater intrusion caused by reduced freshwater inflows, shallow aquifer pumping for water supply, and accelerating sea level rise (Saha et al. 2012).

The research we have synthesized in this book has focused on the pressing question: How do coupled human–natural drivers of change interact with estuarine and coastal processes to transform coastal landscapes? In this chapter we summarize our contributions to the transformation of ideas in coastal ecology by examining the origins of estuarine productivity, the role of people in coastal landscape transformation, the power of comparative studies to understand ecosystem change, and the importance of knowledge co-production for driving effective ecosystem protection and restoration. We also describe how our work in the coastal Everglades has advanced coastal ecological science and ecology in general. We end with suggested avenues for future interdisciplinary coastal studies.

Origins of Estuarine Productivity

Estuaries are among the most biologically productive ecosystems on the planet, so understanding the factors that regulate their productivity is paramount to predicting change in critical functions such as carbon sequestration, natural resource production, and energy and materials transformations and flows (Day et al. 2012b). In many coastal ecosystems, productivity is governed by the supply of limiting nutrient delivered by the drainage basin. Human activities, such as wetland drainage, agricultural expansion, and urban development, have greatly accelerated these material loads to downstream estuaries. Models based on this understanding of the importance of downstream material transport help predict how processes such as freshwater consumption and diversion; human modification of sediment, nutrient, and contaminant supply; species invasions; regulatory policies; and climate shifts interact to control productivity in the coastal zone (e.g., Cerco and Cole 1993 for Chesapeake Bay; Cloern and Jassby 2012 for San Francisco Bay).

Standard ecological models describing estuarine productivity are difficult to apply to the coastal Everglades, though, because the biogeochemical gradient driving productivity is "upside-down" (Childers et al. 2006). Our early studies showed the freshwater Everglades to be highly oligotrophic (Noe et al. 2001), with the limiting nutrient (P) being supplied to the estuaries by the Gulf of Mexico rather than by freshwater inflows (Chen and Twilley 1999; Childers 2006; Fourqurean et al. 1992). This unusual feature stimulated research to unravel how the changing balance of freshwater and marine sources regulates productivity in the coastal zone (Childers et al. 2006; Fig. 9.1). To do so, we established transects in the two main drainages of the southern Everglades. These two transects have different connections to the Gulf of Mexico, creating a useful comparative context for testing hypotheses about how water source interacts with biologic processes to control estuarine productivity. Along the Shark River Slough transect (SRS; see the preface for details), freshwater and Gulf of Mexico waters mix in a productive mangrove forest. By contrast, the Taylor Slough transect (TS) flows from sawgrass marshes into a low-stature mangrove forest that is separated from the Gulf of Mexico by the expansive seagrass-dominated basins of Florida Bay. A temporal experiment has contributed to this robust spatial design: a large-scale restoration of fresh water is planned for the SRS but not TS, enabling the "upside-down" estuary hypothesis to be tested at contrasting paces

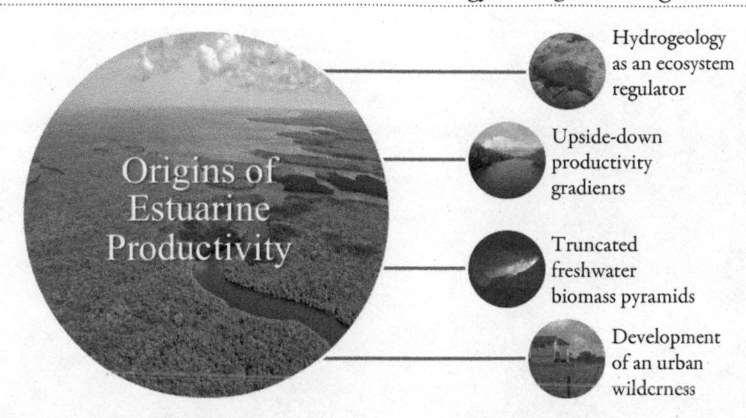

FIGURE 9.1 The FCE program has focused on understanding the origins of productivity in estuaries. Long term studies have underscored the importance of hydrogeology as a regulator of estuarine function. Our research has revealed that the particular hydrogeology of karstic coastal wetlands creates an upside-down pattern of estuarine productivity, where nutrients are supplied from marine rather than freshwater sources. Extreme P limitation in freshwater wetlands causes a truncated biomass pyramid. Understanding ecosystem development in human-dominated coastal watersheds requires a social-ecological perspective to reveal the causes and consequences of coastal transformation. See page 8 of color insert.

and spatial scales, and with more than a decade of data documenting the pre-intervention conditions along both transects (Fig. 9.2).

Ecosystem productivity in the Everglades is a function of the availability and delivery of P, the limiting nutrient, and therefore reflects the balance of its delivery from marine and freshwater sources. This balance is controlled by the underlying geology, climate, and water management. Freshwater inputs to the Everglades include direct rainfall less evapotranspiration, together with water releases from a highly regulated network of canals (Saha et al. 2012; Zapata-Rios and Price 2012; see Chapters 1 and 3). Rainfall is depleted in P, but canals often deliver water enriched in P, mainly from agricultural sources (Gaiser et al. 2006a). Marine water delivery to Everglades estuaries is facilitated by the nearly flat topographic relief that allows marine water to easily intrude landward, both aboveground and belowground (Saha et al. 2011, 2012). This process is driven by accelerating sea level rise (SLR) and is exacerbated by reduced freshwater flows (Gaiser et al. 2006b; Krauss et al. 2011; Ross et al. 2002; Saha et al. 2011). On shorter time scales, daily tides, the annual dry season, and periodic storm surges cause pulses of marine water delivery (Noss 2011; Zhang et al. 2008; see Chapters 3 and 7). Marine water has dramatic effects on ecosystem processes in the coastal zone because P is in relatively greater supply in these source waters (Childers et al. 2006; Price et al. 2006; see Chapter 5) and because salts can also mobilize P desorption from limestone and marl soils (Castañeda-Moya et al. 2010; Smith et al. 2009). This hydrogeologic template thus controls not only the water balance but also biogeochemical fluxes along our two Everglades transects, resulting in the unusual patterns of productivity we have documented. Our understanding of how

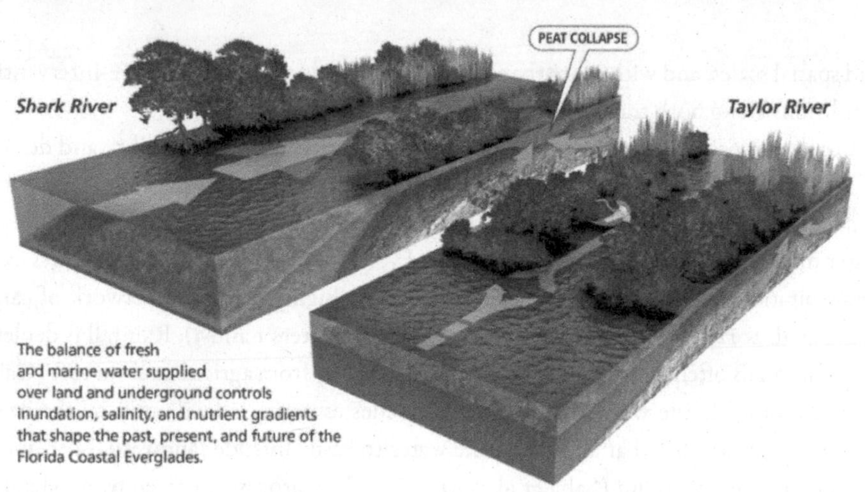

FIGURE 9.2 The upside-down estuary with productivity gradients shaped by the supply of phosphorus (P) from the coast, rather than from upstream marshes. The balance of fresh and marine water supply, therefore, shapes the past, present and future function of the Everglades ecosystem through its effect on P supply, inundation and salinity. Infographic: Hiram Henriquez. See page 1 of color insert.

the natural oligotrophy of the Everglades landscape drives the "upside-down" structure of Everglades estuaries is a unique and important contribution to coastal science (Fig. 9.3). We argue that knowledge of the unique ways in which climate, hydrology, geology, biogeochemistry, ecology, and human systems interact in the coastal Everglades represents new contributions to these disciplines as well.

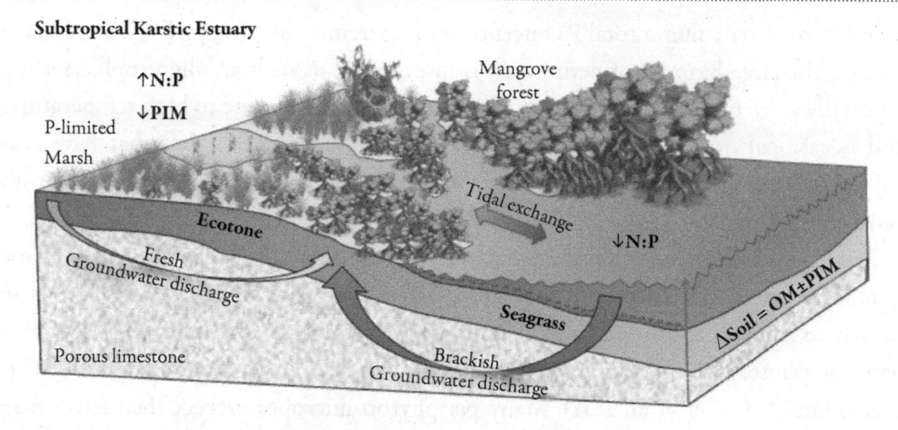

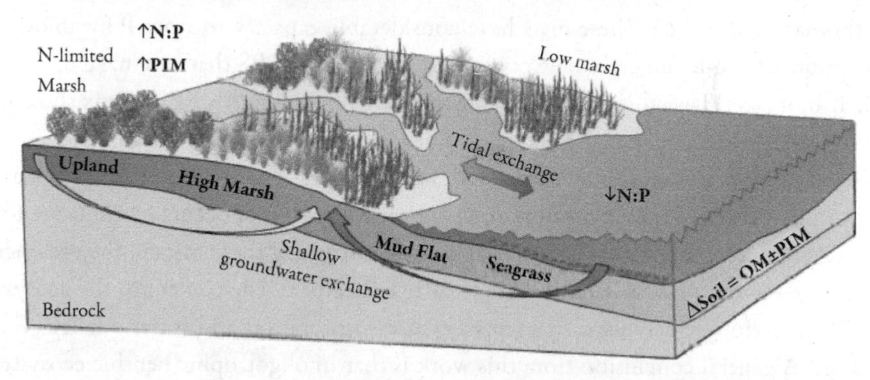

FIGURE 9.3 Depiction of generalized differences between subtropical karst and temperate saltmarsh estuaries. Both types can exhibit decreasing ratios of N:P from freshwater marshes to the estuary, but P concentrations are exceptionally low in karstic systems, and thus productivity is fueled to a greater extent by marine P supplies. Brackish groundwater discharge plays a key role in regulating P supply to karstic marshes, while to a lesser degree in most saltmarsh estuaries. Vertical accretion is driven by organic matter production and precipitation or surge-derived carbonate-rich sediments (Particulate Inorganic Material, PIM) in karstic systems, while upstream sediment supplies dominate the PIM contribution to vertical accretion in temperate estuaries. See page 16 of color insert.

Marsh Oligotrophy and the Paradox of Production

In the shallow waters of the limestone-based, P-limited freshwater Everglades, much of this limiting nutrient is adsorbed to calcium carbonate and is thus relatively invisible to biologic activity. However, a rich microbial community forms thick periphyton mats throughout much of the landscape, attaining standing stocks that are one to two orders of magnitude greater than periphyton found in other benthic aquatic ecosystems (~200 vs. ~50 g ash-free dry biomass m^{-2}, in Everglades vs. other wetlands respectively; Gaiser 2009; Gaiser et al. 2012, 2015). Periphyton mats and microbially active detritus sequester

P and keep water column total P concentrations extremely low (<5 μg L^{-1}; Thomas et al. 2006). The large biomass of periphyton is unexpected in such an oligotrophic setting, especially given other extreme stresses of seasonal drying, exposure to high temperatures, and occasional fires. This production paradox has stimulated studies that have contributed to our understanding of the origins of productivity in oligotrophic systems in general.

Everglades periphyton studies have documented a wide range of adaptations to stressors that are characteristic of the Everglades (*sensu* Gaiser et al. 2011). These adaptations are similar to those of marine stromatolites. In both cases, algae efficiently utilize bicarbonate for photosynthesis, causing calcium carbonate crystals to accumulate and to coprecipitate P (Gaiser et al. 2011). Many periphyton microbes protect themselves from desiccation within this calcium carbonate matrix by forming resistant resting stages or exuding extracellular polymeric substances (EPS) that enable rapid recovery from drying (Thomas et al. 2006). These mats have considerable capacity to store P by abiotic adsorption onto calcium carbonate crystals and within the EPS that can make up 90% of their biomass (Hagerthey et al. 2011). Bacteria living within this EPS matrix show high levels of alkaline phosphatase activity, indicating rapid mobilization and cycling of P (Bellinger et al. 2010). This stimulation of heterotrophic microbial activity by P may result in the widely reported dissolution of calcareous mats that occurs on exposure to excess P (Bellinger et al. 2010; Gaiser et al. 2011). Collapse of these mats in the presence of higher P concentrations is an important early indicator of enrichment in the ecosystem and is therefore widely used to adaptively assess restoration and water quality (Gaiser 2009). A general conclusion from this work is that in oligotrophic benthic ecosystems, water column concentrations of the limiting nutrient are poor metrics of actual nutrient availability (Gaiser et al. 2004). In fact, we have found that in Everglades marshes, a cascade of transformations resulting in hysteretic ecosystem state change is initiated well before water P concentrations are measurably elevated (Gaiser et al. 2005; Hagerthey et al. 2010). A major contribution of our work to the management of oligotrophic ecosystems is thus that water quality impacts should be assessed using nontraditional biogeochemical protocols, including the substitution of microbial indicators for standard water quality measures (Gaiser et al. 2015).

We have identified a second production paradox that may have counterparts in other highly oligotrophic ecosystems. The high production rates of periphyton in the freshwater Everglades ought to support significant standing stocks of primary consumers because these mats form the base of most Everglades food webs. Instead, the Everglades is characterized by surprisingly low standing crops of primary consumers (Fig. 9.4). Our grazing experiments, population manipulations, and surveys of stable isotope distributions have shown that aquatic animal densities are highly regulated by the stress of seasonal drying. But these animals are also limited by a food supply dominated by periphyton mats and periphyton-based detritus that is relatively inedible, unpalatable, and sometimes toxic

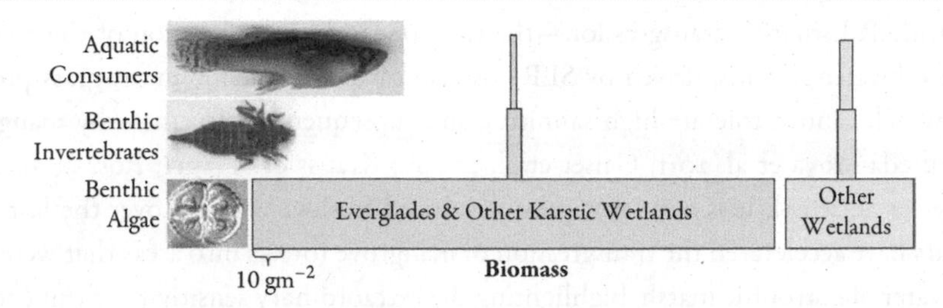

FIGURE 9.4 Despite high periphyton biomass, which forms the base of Everglades food webs, secondary production is very low in the Everglades and other karstic wetlands due to physical and chemical barriers of periphyton to edibility and assimilation by primary consumers (Turner et al. 1999; Gaiser et al. 2015).

because of cyanobacteria and calcium carbonate crystals (Chick et al. 2008; Geddes and Trexler 2003). Our aquatic consumer research has confirmed the long-hypothesized idea that energy in estuarine and wetland ecosystems reaches the food web primarily through less efficient detrital pathways (Williams and Trexler 2006); this reduced efficiency is thus responsible for lower densities and biomass of primary consumers than would be predicted by rates of primary production alone (Gaiser et al. 2012).

Finally, our studies of oligotrophy have also shown that P-limited systems are often characterized by hotspots of higher productivity. Productivity hotspots in the Everglades include the tree islands that cover 3% to 14% of the landscape (Patterson and Finck 1999). Tree islands are hotspots because they accumulate P from the surrounding landscape, with soil P concentrations 3 to 170 times greater than the surrounding marsh (Ross et al. 2006; Wetzel et al. 2005, 2011). Wetzel et al. (2005) proposed that the focused redistribution of P onto tree islands stimulates higher productivity and biomass relative to the surrounding marsh. As tree islands grow, more nutrients are redistributed to the island, creating a positive feedback mechanism that increases the size and vertical stature of the islands. These findings have important implications for self-designing and self-organizing landscapes such as the Everglades (see Chapter 4) and across the world (Eppinga et al. 2009).

The Marsh–Mangrove Ecotone—Production in the Balance

Where sawgrass marshes and spikerush sloughs meet mangrove forests in the oligohaline ecotone, marine supplies of P drive ecosystem production. Along our SRS transect, the mangrove forests are highly productive due to P subsidies from the Gulf of Mexico, while the mangrove forests of our TS transect are dominated by "dwarf" trees that are stunted because P supply is generally limited to brief dry-season groundwater incursions (Ewe et al. 2006; Koch et al. 2012; Price et al. 2010; Rivera-Monroy et al. 2011). Over the last several decades, accelerating SLR has enhanced aboveground and belowground supplies of marine P. Estuarine transgression—the upslope landward migration of the estuaries into freshwater systems, driven by SLR—is also causing a decline in sawgrass production, which cannot tolerate high salinities, and subsequent replacement by mangroves (Castañeda-Moya et al. 2011; Gaiser et al. 2006b; Krauss et al. 2011; Ross et al. 2002; Troxler et al. 2014). It is clear that greatly reduced freshwater flows over the last 50 to 75 years have accelerated the transgression of mangrove forests into areas that were once freshwater oligotrophic marsh, highlighting the extraordinary sensitivity of this ecotone to saltwater encroachment associated with SLR and storm activity. The dynamics of this ecotone movement in the extremely flat coastal Everglades remain uncertain, as the expected restoration of freshwater flow conflicts with accelerating SLR.

While SLR coupled with reduced freshwater flows is accelerating mangrove transgression, recent experiments have demonstrated that this saltwater intrusion also stimulates microbial decomposition in coastal wetland soils, accelerating organic carbon losses (Weston et al. 2011). This phenomenon appears to be contributing to the widely documented peat collapse in coastal marshes across North America (Cahoon et al. 2003; Nyman et al. 2006; Voss et al. 2013). The introduction of salinity to previously freshwater marsh soils appears to push the carbon balance from one of net gain, or stasis, to net loss, as various microbial processes are stimulated by the marine waters (Chambers et al. 2011; Weston et al. 2006) and plant production is stressed (Mendelssohn and Morris 2000). Our ongoing long-term data and experiments will continue to help identify mechanisms by which salinity and saltwater inundation contribute to peat loss in coastal wetlands and inform management of coastal landscapes threatened by accelerating SLR.

Further down the estuary, in our full-stature mangrove forests, soil accretion rates (Smith et al. 2009; Smoak et al. 2013) match or exceed the average rate of SLR reported for Key West, Florida—for now (Zhang et al. 2009, 2011). Our research has shown that these mangrove forests sequester globally relevant quantities of CO_2, and that these high carbon uptake rates are sensitive to the press of SLR and to pulsed storm disturbances (Barr et al. 2011; Bouillon et al. 2008;). We have also found that storm surges actually enhance estuarine production even while the hurricanes themselves can be extremely damaging to mangrove forests (see Chapter 7 for our story about the effects of Hurricane Wilma in October 2005). At the same time, our mangrove forest eddy flux tower data have shown a suppression of carbon loss from the soils during tidal inundation (Barr et al. 2010; Schedlbauer et al. 2010). There are also large uncertainties about the tidal advection of organic and inorganic carbon into adjacent estuarine waters, about downstream mineralization, and about air–water CO_2 fluxes (Bouillon et al. 2007, 2008; Duarte and Prairie 2005; Miyajima et al. 2009). While data from our mangrove forest flux tower—the first of its kind in the world—have informed the importance of tropical estuarine wetlands to global carbon budgets, there are still gaps in our knowledge that we will continue to fill.

Seagrass Meadows and the Future of Shallow Blue Carbon

The Florida Bay estuary and its expansive, shallow seagrass meadows is bounded by the Florida Keys, the Gulf of Mexico, and the dwarf mangrove forests of the southern Everglades. Ecosystem productivity in these seagrass meadows is regulated by P supply and salinity (Herbert and Fourqurean 2008), which are functions of freshwater supply, tides, storms, and SLR. New results from comparative Long Term Ecological Research Program (LTER) research suggest that carbon storage in seagrass sediments rivals that of tropical forests and that Florida Bay sediments are carbon-rich compared to seagrass systems worldwide (Fourqurean et al. 2012). This recently identified massive storage of "blue" carbon should be valued for its potential role in mitigating climate change. However, an alarming loss of seagrass ecosystems occurred during the 20th century. If present rates of loss continue, up to 299 Tg of carbon could be lost to the atmosphere (roughly 10% of all emissions attributed to land use change). The future status of this seagrass carbon sink in Florida Bay will be determined by the extent to which freshwater restoration balances the continued influence of SLR and, ultimately, how the extent of Florida Bay changes as the estuarine complex inevitably transgresses upslope and landward.

New Perspectives on Coastal Ecology from Comparative International Studies

The unusual productivity dynamics we have observed in the coastal Everglades have prompted research in other coastal systems to determine how these iconic features may inform general theory about coastal ecosystem transformations. We are also interested in using the Everglades as a sentinel of change and a model for resource management in other similar ecosystems. Much of this comparative research has been conducted through partnerships fostered through the International LTER Network (see Gosz 1996) and includes collaborations with scientists from the Caribbean to Australia (Gaiser et al. 2015). These systematic comparisons with wetlands in similar climates and geologic conditions caused us to rethink the notion that the Everglades is novel and unique. While it is certainly iconic in its expanse, notoriety, and research (see Chapter 2), we have found seemingly novel Everglades characteristics, such as the "upside-down estuary" and "productivity paradox," in other coastal karstic systems around the world.

Our international explorations to find Everglades analogs began with studies of other estuaries located on carbonate platforms. We pursued the "upside-down" model on the Yucatán Peninsula of Mexico and in Shark Bay, Western Australia (Kendrick et al. 2012; Price et al. 2012). As it turned out, these other karst estuaries functioned similarly, and all had under-recognized marine surface and groundwater sources of their limiting nutrient (Lagomasino et al. 2015; Price et al. 2012; Stalker et al. 2014). They all showed P-limited, oligotrophic characteristics and had P supplied primarily from marine sources

(Burkholder et al 2013; Fourqurean et al. 2012). This broader international work has shown that changes in surface water and groundwater flows, land–sea interactions, and previously unknown or "hidden" sources of the limiting nutrient are important to the productivity dynamics of coastal ecosystems. In short, the broader contributions of our "upside-down" estuary model of the coastal Everglades has been validated by research in similar ecosystems and coastal landscapes around the world (Fig. 9.5).

Our international research in other karstic ecosystems has also focused on understanding food webs under the productivity paradox conditions we have already described. Research in the freshwater karstic landscapes of the Sian Ka'an Biosphere Reserve (Mexico), New River Lagoon (Belize), Black River Morass (Jamaica), and the dolomite limestone wetlands of the Great Lakes (called "alvars"; Gaiser et al. 2012; La Hée and Gaiser 2012) showed that these wetlands were surprisingly similar to the Everglades (Gaiser et al. 2015). Water column total P concentrations were low, except in areas with a long history of enrichment. Common algal taxa dominated the periphyton mats across regions (La Hée and Gaiser 2012). Periphyton in these diverse karstic

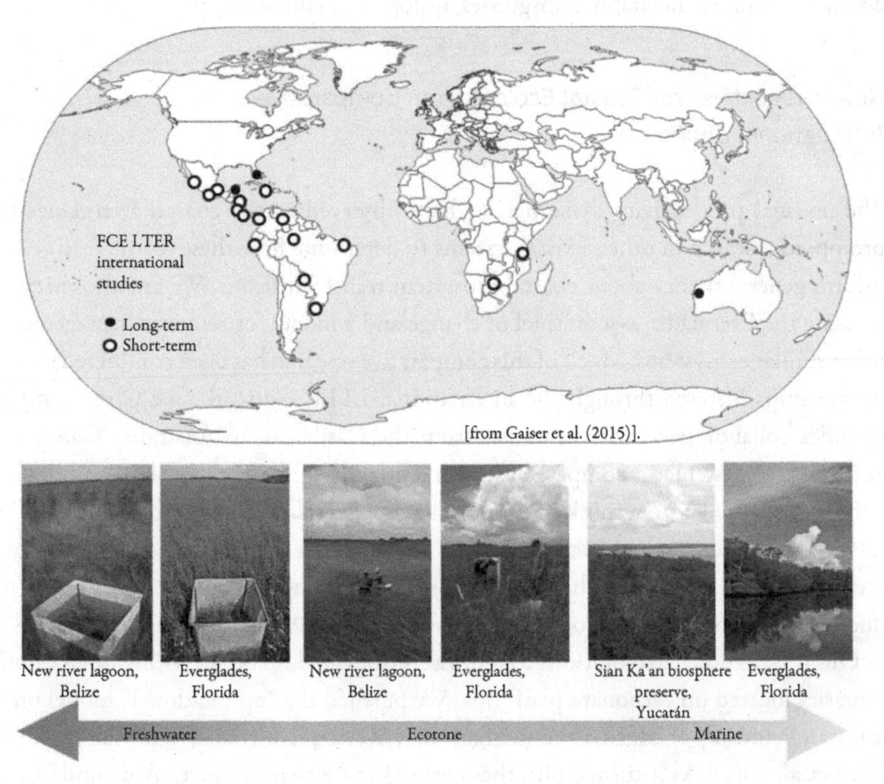

FIGURE 9.5 Sites of long- and short-term comparative studies led by FCE researchers to test hypotheses about the origins of productivity and approaches to wetland restoration in other comparable coastal and inland wetlands around the globe (from Gaiser et al. 2015). Photos depict ecosystem types arrayed along the coastal gradient, showing similarities in Everglades features to those of karstic ecosystems elsewhere.

systems are productive throughout the year, and they are thus important contributors to the carbon cycle through their impact on inorganic and organic transformations (Troxler et al. 2013). Further up the food chains, we found that primary consumers in these other karstic wetlands also showed much lower densities than would be predicted by the rates of primary production, supporting our Everglades findings that periphyton mats are a low-quality food resource for consumers (Trexler et al. 2015). Interestingly, in Shark Bay, Australia, we found that mangrove productivity provided little direct subsidy to the refractory food web carbon (Heithaus et al. 2011). This finding seems to directly challenge the "outwelling hypothesis" of estuarine wetland productivity (Childers et al. 2000; Odum 1957), with the caveat that this theoretical construct and empirical evidence supporting it never explicitly included karstic coastal systems.

Global comparative studies have also been important for understanding the origins of mangrove forest productivity in the coastal Everglades. High production typifies mangrove forests of the Caribbean region and appears to be promoted by biogeochemical and hydrological forces unique to these subtropical karstic estuaries (Castañeda-Moya et al. 2010; Jardel et al. 2013; Rivera-Monroy et al. 2008; Stalker et al. 2014). Climate disturbance is a critical modulator of these biogeochemical and hydrologic controls, over long-time frames and across broad spatial scales (Calderon-Aguilera et al. 2012; Farfán et al. 2014). Expanding our study domain to the tropics has provided fertile ground to develop hypotheses addressing how a wide range of drivers (e.g., deforestation, eutrophication, urban development) affect the sustainability of global mangrove resources (e.g., water quality, habitat, fisheries; Rivera-Monroy et al. 2004). New collaborations with scientists throughout South America are focused on quantifying mangrove carbon storage and productivity using remote sensing tools that can identify both biophysical and social drivers of change (Simard et al. 2008; Suarez-Abelenda et al. 2014; Uchida et al. 2014). We are using an integrated social-ecological framework for this work because mangrove restoration programs in the neotropics often fail due to a lack of flexible scenarios for mangrove vulnerability that can guide effective conservation policy (Lewis 2009; Twilley and Rivera 2005).

Reframing Ecology of Wildernesses in Urban Landscapes

The evolution and expansion of our social-ecological framing has transformed concepts of human interactions with "wilderness" in urban landscapes locally, and in ways that have permeated collaborations within the LTER Network and internationally. For reasons we have described in this chapter and in Chapter 2, the Everglades has long served as an icon of isolated wilderness (see also Ogden 2011). In fact, the Department of Interior has designated 98% of Everglades National Park (ENP)—much of what we consider "Everglades" today—as "wilderness," a legal designation for areas considered "primeval" and "untrammeled by man" (McCloskey 1966). The "trouble with wilderness", as

William Cronon (1995) has so eloquently argued, is that it is a paradigm that positions people as profoundly alien and intrusive to landscapes that have been valued primarily for their non-human life and attributes. In wilderness, the "human" is generally positioned as a negative agent of change ("impacts"). In the case of the Everglades, humans have been conceptualized as external to the landscape itself ("adjacent communities," "stakeholders," or "visitors"). Conservation, protection, and restoration are socially significant goals associated with wilderness. At the same time, we have challenged our research to rethink wilderness in ways that enable more holistic and inclusive approaches to science and stewardship (Chapin et al. 2011). In short, challenging the wilderness paradigm entailed expanding our conceptual and geographic boundaries of research to include the messiness of human history, politics, and trajectories of social difference (and indifference).

Except for the urban LTER programs in Phoenix and Baltimore, and the experimental agricultural LTER program at the Kellogg Biological Station in Michigan, the majority of the LTER sites have been established in protected areas that no longer have human communities living within their boundaries. Our coastal Everglades LTER program is no exception. Needless to say, this lack of people (aside from natural resource staff, recreationalists, and scientists) shaped the research that these LTER programs have conducted and the types of scientists involved in their programs. They have a shared mission to collect and maintain a long-term record of similar biophysical datasets, which has enabled us to ask comparative questions about environmental change at broad scales. While this programmatic commitment is one of the most significant aspects of LTER science, it has also inherently constrained the involvement of social scientists in LTER research while also minimizing direct societal influences of LTER knowledge.

Of course, all LTER sites have complex human histories and are a part of much broader social-ecological contexts (political, economic, cultural). More than a decade ago, LTER scientists recognized the artificiality of these conceptual, geographic, and disciplinary boundaries and the limitations these boundaries imposed on their ability to understand change in environmental patterns and processes (*sensu* Collins et al. 2011). This recognition—that even LTER sites that seem to epitomize *wilderness* in its purest incarnation are actually *social-ecological systems*—mirrors advances in ecological theory and the contributions of LTER scientists to this theory (Grimm et al. 2000, 2008; Liu et al. 2007; Polsky et al. 2014).

In the 1990s, scientists and resource agency staff involved in Everglades restoration efforts began to call for a broader approach to understanding the drastic decline in Everglades ecosystems and biota. Marrying the conceptual framing of sustainability science with ecosystem ecology, Everglades restoration advocates called for a "socio-ecological" approach to science and emphasized the need for regional planning (Ogden 2008). Though restoration efforts have taken much longer than anyone anticipated, this initial social-ecological and regional framing continues to fundamentally shape restoration implementation activities (see Chapter 8).

As we discuss in the section "The Co-production of Knowledge, Science, and Management," many of us came to our Everglades science within this restoration framework, and the FCE program has always had strong ties to the agencies leading the restoration effort. Perhaps this culture of collaborative science helped us become an "early adopter" of the efforts to regionalize or expand the spatial scale and disciplinary boundaries of many LTER research programs. In 2006, we regionalized by expanding the boundaries of our research program to include the urban, suburban, and agricultural communities adjacent to ENP (Fig. 9.6). As this synthesis demonstrates, this expansion allowed us to ask questions about (1) the role of environmental politics in the restoration process; (2) the social, political, and hydrologic aspects of regional water management

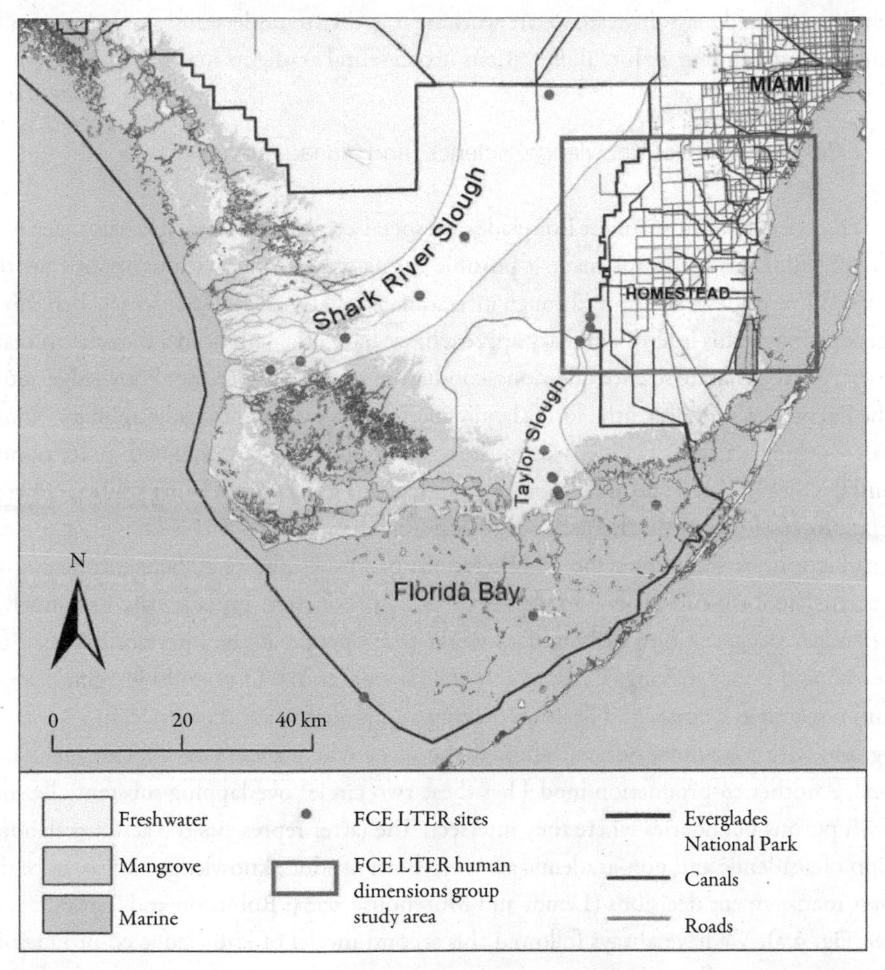

FIGURE 9.6 The study domain of the FCE LTER program showing our 2006 conceptual and programmatic expansion into urban Miami-Dade County. This allowed us to examine social-ecological connections among the human-dominated urban and immediately adjacent wildlands of South Florida.

practices; and (3) the causes and consequences of agricultural land conversion within our study area.

For us, regionalization began as a strategy to facilitate the involvement of social scientists in the program and as a practical way to study the Everglades as a social-ecological system. Along the way, this became a process of *reconceptualization* as well. Instead of thinking of the Everglades from strictly the wilderness context, we were able to reconceptualize the Florida Coastal Everglades (FCE) as also an urban LTER site. In doing so, we have been able to interpret the boundaries of natural versus human and urban versus rural more broadly. Social theorists have long understood that urban–rural boundaries are not porous and that the city and the hinterlands are mutually constituted forms of life (Williams 1973). We no longer talk about the Everglades as a "natural system" that excludes the human. Nor do we talk about "the human," as if people were an undifferentiated mass. Instead, we are working together to understand the trajectories of human and non-human life, along various urban–rural gradients in south Florida.

The Co-production of Knowledge, Science, and Management

The reconceptualization of the Everglades as a social-ecological system in a heterogeneous urban/wildlands landscape made it possible for us to address questions about how the coastal Everglades transforms through interconnected human and biophysical behaviors. In addition to this interdisciplinary approach, we have also embraced a transdisciplinary structure to generate science questions, conduct research, and produce knowledge about the Everglades and the south Florida landscape. Here, we define "transdisciplinary" as not just working across disciplines within the academy, but also working with practitioners and decision makers who are responsible for managing the system being studied (Hirsch Hadorn et al. 2008). In the parlance of sustainability science, this is known as the co-production, or sometimes the co-generation, of knowledge (e.g., Schellnhuber et al. 2005). There are two general models for this co-production process. The first involves boundary organizations, or boundary work, that operate at the interface between research and policy making (Clark et al. 2011; Guston 2001). One could imagine science and research as a circle, and decision making and policy as another circle, in a Venn diagram, with a boundary organization as the place where these two circles touch (Fig. 9.7). Another co-production model has these two circles overlapping substantially, and with porous boundaries where they intersect. The latter represents the active collaboration of academic and non-academic actors to both produce knowledge and to make the best management decisions (Lemos and Morehouse 2005; Robinson and Tansey 2006; see Fig. 9.7). We have always followed this second model of knowledge co-production through active collaboration.

Arguably, the two most important regional players in Everglades management and restoration are ENP (and thus the U.S. Department of Interior) and the South Florida

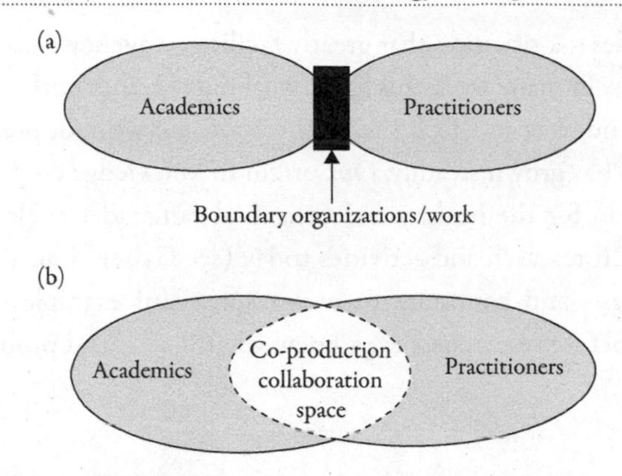

FIGURE 9.7 (A) Venn-type diagram of the boundary organization or boundary work model for knowledge co-production; (B) Venn-type diagram of the collaboration interface model for knowledge co-production. Note that in (B) the boundaries where the two circles overlap is porous. On the left side this signifies the free movement of academic scientists into and out of direct collaboration with agency scientists, depending on respective agendas. On the right side this signifies the movement of FCE-based science knowledge from agency scientists/collaborators to decision-makers and policy-makers in their respective agencies. Modified from Figure 1 in Pohl et al. (2010).

Water Management District (SFWMD). Our strongest and longest-lived relationships are with these two critical agencies. Both have always included a strong in-house science staff as an important component of their management repertoire, with the (presumed) logic being that these staff scientists will provide timely, sage knowledge to agency managers and decision makers on the ecohydrologic condition of the Everglades. This strategy of supporting in-house scientists greatly facilitated our initial co-production of knowledge and science with ENP and the SFWMD, although through time we have grown our co-production collaborations to include a number of other relevant institutions. As noted before, scientists from ENP and the SFWMD have been active collaborators in our science since before the FCE program was funded in 2000, and they have been responsible for moving knowledge across the porous boundary of their respective agencies' "co-production circle" (see Fig. 9.7). At the same time, FCE scientists have always felt free to move in and out of this co-production "collaboration space" (see Fig. 9.7) by working with agency scientists when it makes the most sense. The sustainability science literature is rich with examples of how such co-production of knowledge and science leads to the best, most timely, and most strategic decisions about resource management, sustainable development, and a host of other challenges.

It is worth reflecting a bit and asking the question: How did we come to be so successful at knowledge co-production? It almost goes without saying that in the mid- to late 1990s, when the ideas that ultimately became the FCE program were first gelling, researchers knew little about the concept of co-production, or even the term. By working together, in the "co-production collaboration space" shown in Figure 9.7, many of us already had a history of cooperating on research. Some collaborators were, in fact, friends as well as colleagues—a situation that greatly facilitated agency–academia interactions and collaborations. In many ways, this made working together both rewarding and fun! Since FCE was funded as an LTER site in 2000, this network of professional and personal interactions has grown steadily. Our origin in knowledge co-production is probably a major reason for the healthy and congenial camaraderie felt by everyone who participates in FCE research and activities today (see Gaiser et al. 2016). In short, the FCE program was—and continues to be—a successful example of knowledge co-production long before researchers even knew what it was, and probably before it was even cool!

Future Directions in Coastal Ecosystem Science

The time is ripe to expand the boundaries of knowledge co-production to help ensure the sustainability of social-ecological coastal systems through the 21st century. Over half of the world's human population now lives in the coastal zone (i.e., within 70 km of an ocean). They are vulnerable to rising seas and they face ever-increasing costs (in energy and dollars) of engineered solutions that attempt to replace natural ecosystem services (Day et al. 2016). Given this conundrum, we need to expand our LTER conversations and audiences in order to find timely solutions to some of the most socially relevant issues of the day. One promising direction is to formalize the comparative approach we have already taken through our international collaborations and expand cross-site research within the existing network of coastal LTER sites as well as through partnerships with other coastal zone agencies (e.g., the U.S. Environmental Protection Agency's Coastal Bays Program, the National Oceanic and Atmospheric Administration's National Estuarine Research Reserves). The coastal LTER sites span a full range of coastal biogeographic provinces of the United States, representing a broad array of ecosystem drivers including temperature; salinity; watershed influence; tidal range; primary production resource base; and extent of human influence, manipulation, or control. Experiments across sites making use of these natural gradients will provide critical new data that will help us to understand the response of coastal systems to myriad perturbations in the 21st century. By fully embracing the social-ecological framework that has been so effective across the LTER network (Collins et al. 2011), our project shows that integrating human dimensions themes into long-term science can, in fact, both inform and transform knowledge and management practices. Our explorations of the human agencies of change have reconceptualized the Everglades as an urban wilderness, which has led to a new generation of conceptual models of coastal ecosystem structure and function. An expanded, highly collaborative network of coastal LTER programs would powerfully advance both coastal ecosystem ecology and the decisions necessary to sustain them, while enhancing the services provided to people at a time of unprecedented change.

References

Barr, J.G., V.C. Engel, J.D. Fuentes, J.C. Zieman, T.L. O'Halloran, T.J. Smith III, and G.H. Anderson. 2010. Controls on mangrove forest–atmosphere carbon dioxide exchanges in western Everglades National Park. *Journal of Geophysical Research Biogeosciences* 115: G02020.

Barr, J.G., V. Engel, T.J. Smith, and J.D. Fuentes. 2011. Hurricane disturbance and recovery of energy balance, CO_2 fluxes and canopy structure in a mangrove forest of the Florida Everglades. *Agricultural and Forest Meteorology* 153: 54–66.

Bellinger, B.J., M.R. Gretz, D.S. Domozych, S.N. Kiemle, and S.E. Hagerthey. 2010. Composition of extracellular polymeric substances from periphyton assemblages in the Florida Everglades. *Journal of Phycology* 46: 484–496.

Bouillon, S., A. Borges, E. Castañeda-Moya, K. Diele, T. Dittmar, N.C. Duke, E. Kristensen, S.Y. Lee, C. Marchand, J.J. Middelburg, V.H. Rivera-Monroy, T.J. Smith III, and R.R. Twilley. 2008. Mangrove production and carbon sinks: A revision of global budget estimates. *Global Biogeochemical Cycles* 22: 1–12.

Bouillon, S., F. Dehairs, B. Velimirov, G. Abril, and A.V. Borges. 2007. Dynamics of organic and inorganic carbon across contiguous mangrove and seagrass systems (Gazi Bay, Kenya). *Journal of Geophysical Research* 112: G02018.

Burkholder, D.A., J.W. Fourqurean, and M.R. Heithaus. 2013. Spatial pattern in stoichiometry indicates both N-limited and P-limited regions of an iconic P-limited subtropical bay. *Marine Ecology Progress Series* 472: 101–115.

Cahoon, D.R., P. Hensel, J. Rybczyk, K.L. McKee, C.E. Proffitt, and B.C. Perez. 2003. Mass tree morality leads to mangrove peat collapse at Bay Islands, Honduras after Hurricane Mitch. *Journal of Ecology* 91(6): 1093–1105.

Calderon-Aguilera, L.E., V.H. Rivera-Monroy, L. Porter-Bolland, A. Martínez-Yrízar, L. Ladah, M. Martínez-Ramos, J. Alcocer, A.L. Santiago-Pérez, H.A Hernandez Arana, V.M. Reyes-Gómez, D.R. Pérez-Salicrup, V. Diaz-Nuñez, J. Sosa-Ramirez, J. Herrera-Silveira, and A. Búrquez. 2012. An assessment of natural and human disturbance effects on Mexican ecosystems: current trends and research gaps. *Biological Conservation* 21: 589–617.

Castañeda-Moya, E., R.R. Twilley, V.H. Rivera-Monroy, B. Marx, C. Coronado-Molina, and S.E. Ewe. 2011. Patterns of root dynamics in mangrove forests along environmental gradients in the Florida Coastal Everglades, USA. *Ecosystems* 14: 1178–1195.

Castañeda-Moya, E., R.R. Twilley, V.H. Rivera-Monroy, K.Q. Zhang, S.E. Davis, and M. Ross. 2010. Sediment and nutrient deposition associated with Hurricane Wilma in mangroves of the Florida Coastal Everglades. *Estuaries and Coasts* 33: 45–58.

Cerco, C., and T. Cole. 1993. Three-dimensional eutrophication model of Chesapeake Bay. *Journal of Environmental Engineering* 119: 1006–1025.

Chambers, L.G., K.R. Reddy, and T.Z. Osborne. 2011. Short-term response of carbon cycling to salinity pulses in a freshwater wetland. *Soil Science Society of America Journal* 75(5): 2000–2007.

Chapin, F. S. III, M.E. Power, S.T.A. Pickett, A. Freitag, J.A. Reynolds, R.B. Jackson, D.M. Lodge, C. Duke, S.L. Collins, A.G. Power, and A. Bartuska. 2011. Earth stewardship: science for action to sustain the human-earth system. *Ecosphere* 2: art89.

Chen, R.R., and R.R. Twilley. 1999. Patterns of mangrove forest structure and soil nutrient dynamics along the Shark River estuary, Florida. *Estuaries* 22: 955–970.

Chick, J. H., P. Geddes, and J. C. Trexler. 2008. Periphyton mat structure mediates trophic interactions in a subtropical wetland. *Wetlands* 28: 378–389.

Childers, D.L. 2006. A synthesis of long-term research by the Florida Coastal Everglades LTER Program. *Hydrobiologia 569*(1): 531–544.

Childers, D.L., J.N. Boyer, S.E. Davis, C.J. Madden, D.T. Rudnick, and F.H. Sklar. 2006. Relating precipitation and water management to nutrient concentration patterns in the oligotrophic "upside down" estuaries of the Florida Everglades. *Limnology and Oceanography* 51: 602–616.

Childers, D.L., J.W. Day, Jr., and H.N. McKellar, Jr. 2000. Twenty more years of marsh and estuarine flux studies: revisiting Nixon (1980). In *Concepts and Controversies in Tidal Marsh Ecology*, edited by M.P. Weinstein and D.Q. Kreeger, 385–414. Dordrecht: Springer Netherlands.

Clark, W.C., T.P. Tomich, M. van Noordwijk, D. Guston, D. Catacutan, N.M. Dickson, and E. McNie. 2011. Boundary work for sustainable development: Natural resource management at the Consultative Group on International Agricultural Research (CGIAR). *Proceedings of the National Academy of Sciences USA* 113: 4615–4622.

Cloern, J.E., and A.D. Jassby. 2012. Drivers of change in estuarine-coastal ecosystems: Discoveries from four decades of study in San Francisco Bay. *Reviews of Geophysics* 50(4): RG4001.

Collins, S.L., S.R. Carpenter, S.M. Swinton, D.E. Orenstein, D.L. Childers, T.L. Gragson, N.B. Grimm, J.M. Grove, S.L. Harlan, J.P. Kaye, A.K. Knapp, G.P. Kofinas, J.J. Magnuson, W.H. McDowell, J.M. Melack, L.A. Ogden, G.P. Robertson, M.D. Smith, and A.C. Whitmer. 2010. An integrated conceptual framework for long-term social-ecological research. *Frontiers in Ecology and the Environment* 9: 351–357.

Cronon, W. 1995. The trouble with wilderness; or, getting back to the wrong nature. In *Uncommon Ground: Rethinking the Human Place in Nature*, edited by W. Cronon, 69–90. New York, NY: W. W. Norton & Co.

Day, J.W., J. Agboola, Z. Chen, C. D'Elia, D.L. Forbes, L. Giosan, P. Kemp, C. Kuenzer, R.R. Lane, R. Ramachandran, J. Syvitski, and A. Yañez-Arancibia. 2016. Approaches to defining deltaic sustainability in the 21st century. *Estuarine, Coastal and Shelf Science* 183(B): 275–291.

Day, J.W., B.C. Crump, W.M. Kemp, and A. Yáñez-Arancibia (eds.). 2012b. *Estuarine Ecology* (2nd ed.). Hoboken, NJ: John Wiley and Sons, Inc.

Day, J.W., C.A.S. Hall, W. Kemp, and A. Yáñez-Arancibia. 1989. *Estuarine Ecology*. Hoboken, NJ: John Wiley and Sons, Inc.

Day, J.W., A. Yáñez-Arancibia, and W.M. Kemp. 2012a. Human impact and management of coastal and estuarine ecosystems. In *Estuarine Ecology* (2nd ed.), edited by J.W. Day, B.C. Crump, W.M. Kemp, and A. Yáñez-Arancibia, 483–495. Hoboken, NJ: John Wiley and Sons, Inc.

Duarte, C.M., and Y.T. Prairie. 2005. Prevalence of heterotrophy and atmospheric CO_2 emissions from aquatic ecosystems. *Ecosystems* 8(7): 862–870.

Eppinga, M.B., P.C. de Ruiter, M.J. Wassen, and M. Rietkerk. 2009. Nutrients and hydrology indicate the driving mechanisms of peatland surface patterning. *American Naturalist* 173: 803–818.

Ewe, S., E. Gaiser, D. Childers, V. Rivera-Monroy, D. Iwaniec, J. Fourqurean, and R. Twilley. 2006. Spatial and temporal patterns of aboveground net primary productivity (ANPP) in the Florida Coastal Everglades LTER (2001–2004). *Hydrobiologia* 569: 459–474.

Farfán, L.M., E. D'Sa, K. Liu, and V.H. Rivera-Monroy. 2014. Tropical cyclone impacts on coastal regions: the case of the Yucatán and the Baja California Peninsulas, Mexico. *Estuaries and Coasts* 37: 1388–1402.

Fourqurean, J.W., G.A. Kendrick, L.S. Collins, R.M. Chambers, and M.A. Vanderklift. 2012. Carbon and nutrient storage in subtropical seagrass meadows: examples from Florida Bay and Shark Bay. *Marine and Freshwater Research* 63: 967–983.

Fourqurean, J.W., J.C. Zieman, and G.V.N. Powell. 1992. Phosphorus limitation of primary production in Florida Bay: evidence from the C:N:P ratios of the dominant seagrass *Thalassia testudinum*. *Limnology and Oceanography* 37: 162–171.

Gaiser, E. 2009. Periphyton as an indicator of restoration in the Everglades. *Ecological Indicators* 9: S37–S45.

Gaiser, E. 2016. The benefits of long-term environmental research, friendships, and boiled peanuts. In *Long-Term Environmental Research: Changing the Nature of Scientists*, edited by M. Willig and L. Walker, 177–184. New York: Oxford University Press.

Gaiser, E.E., E.P. Anderson, E. Castañeda-Moya, L. Collado-Vides, J.W. Fourqurean, M.R. Heithaus, R. Jaffe, D. Lagomasino, N. Oehm, R.M. Price, V.H. Rivera-Monroy, R. Roy Chowdhury, and T. Troxler. 2015. New perspectives on an iconic landscape from comparative international long-term ecological research. *Ecosphere* 6(10): 1–18.

Gaiser, E., A. Gottlieb, S. Lee, and J. Trexler. 2015. The importance of species-based microbial assessment of water quality in freshwater Everglades wetlands. In *Microbiology of the Everglades Ecosystem*, edited by J. Entry, K. Jayachandrahan, A. Gottlieb, and A. Ogram, 115–130. Boca Raton, FL: Science Publishers.

Gaiser, E., P. McCormick, and S. Hagerthey. 2011. Landscape patterns of periphyton in the Florida Everglades. *Critical Reviews in Environmental Science and Technology* 41(S1): 92–120.

Gaiser, E., J. Richards, J. Trexler, R. Jones, and D. Childers. 2006a. Periphyton responses to eutrophication in the Florida Everglades: cross-system patterns of structural and compositional change. *Limnology and Oceanography* 51: 617–630.

Gaiser, E., L. Scinto, J. Richards, K. Jayachandran, D. Childers, J. Trexler, and R. Jones. 2004. Phosphorus in periphyton mats provides best metric for detecting low-level P enrichment in an oligotrophic wetland. *Water Research* 38: 507–516.

Gaiser, E., J. Trexler, J. Richards, D. Childers, D. Lee, A. Edwards, L. Scinto, K. Jayachandran, G. Noe, and R. Jones. 2005. Cascading ecological effects of low-level phosphorus enrichment in the Florida Everglades. *Journal of Environmental Quality* 34: 717–723.

Gaiser, E., J. Trexler, and P. Wetzel. 2012. The Everglades. In *Wetland Habitats of North America: Ecology and Conservation Concerns*, edited by D. Batzer and A. Baldwin, 231–252. Berkeley: University of California Press.

Gaiser, E., A. Zafiris, P. Ruiz, F. Tobias, and M. Ross. 2006b. Tracking rates of ecotone migration due to salt-water encroachment using fossil mollusks in coastal south Florida. *Hydrobiologia* 569: 237–257.

Geddes, P., and J.C. Trexler. 2003. Uncoupling of omnivore-mediated positive and negative effects on periphyton mats. *Oecologia* 136:585–595.

Gosz, J.R. 1996. International long-term ecological research: priorities and opportunities. *Trends in Ecology and Evolution* 11(10): 444.

Grimm, N.B., S.H. Faeth, N.E. Golubiekski, C.H. Redman, J. Wu, X. Bai, and J.M. Briggs. 2008. Global change and the ecology of cities. *Science* 319(5864): 756–760.

Grimm, N.B., J.M. Grove, S.T.A. Pickett, and C.L. Redman. 2000. Integrated approaches to long-term studies of urban ecological systems. *BioScience* 50: 571–584.

Guston, D. H. 2001. Boundary organizations in environmental policy and science: an introduction. *Science, Technology, & Human Values* 26: 399–408.

Hagerthey, S., B. Bellinger, K. Wheeler, M. Gantar, and E. Gaiser. 2011. Everglades periphyton: a biogeochemical perspective. *Critical Reviews in Environmental Science and Technology* 41(S1): 309–343.

Hagerthey, S.E., J.J. Cole, and D. Kilbane. 2010. Aquatic metabolism in the Everglades: dominance of water column hterotrophy. *Limnology and Oceanography* 55(2): 653–666.

Heithaus, E.R., P.A. Heithaus, M.R. Heithaus, D. Burkholder, and C.A. Layman. 2011. Trophic dynamics in a relatively pristine subtropical fringing mangrove community. *Marine Ecology Progress Series* 428: 49–61.

Herbert, D.A., and J.W. Fourqurean. 2008. Ecosystem structure and function still altered two decades after short-term fertilization of a seagrass meadow. *Ecosystems* 11: 688–700.

Hirsch Hadorn, G., H. Hoffmann-Riem, S. Biber-Klemm, W. Grossenbacher-Mansuy, D. Joye, C. Pohl, U. Wiesmann, and E. Zemp 2008. *Handbook of Transdisciplinary Research*. Dordrecht: Springer.

Hopkinson, C.S., and J.W. Day. 1977. A model of the Barataria Bay salt marsh ecosystem. In *Ecosystem Modeling in Theory and Practice: An Introduction with Case Histories*, edited by C.A.S. Hall and J.W. Day, 236–265. New York: Wiley Interscience.

Hopkinson, C.S., A. E. Lugo, M. Alber, A. Covich, and S. van Bloem. 2008. Understanding and forecasting the effects of sea level rise and intense windstorms on coastal and upland ecosystems: the need for a continental-scale network of observatories. Frontiers in Ecology and the Environment. 6: 255–263.

Jardel, P., M. Maass, and V.H. Rivera-Monroy (eds.). 2013. *La Investigación Ecológica de Largo Plazo en México*. Mexico: Editorial Universitaria- Universidad de Guadalajara.

Kendrick, G.A., J.W. Fourqurean, M.W. Fraser, M.R. Heithaus, G. Jackson, K. Friedman, and D. Hallac. 2012. Science behind management of Shark Bay and Florida Bay, two P-limited subtropical systems with different climatology and human pressures. *Marine and Freshwater Research* 63: 941–951.

Koch, G., D.L. Childers, E. Gaiser, and R. Price. 2012. Hydrological conditions control P loading and aquatic metabolism in an oligotrophic, subtropical estuary. *Estuaries and Coasts* 35: 292–307.

Krauss, K.W., A.S. From, T.W. Doyle, T.J. Doyle, and M.J. Barry. 2011. Sea-level rise and landscape change influence mangrove encroachment onto marsh in the Ten Thousand Islands region of Florida, USA. *Journal of Coastal Conservation* 15: 629–638.

Lagomasino, D., R.M. Price, J. Herrera-Silveira, F. Miralles-Wilhelm, G. Merediz-Alonso, and Y. Gomez-Hernandez. 2015. Connecting groundwater and surface water sources in groundwater dependent coastal wetlands and estuaries: Sian Ka'an Biosphere Reserve, Quintana Roo, Mexico. *Estuaries and Coasts* 38(5): 1744–1763.

La Hée, J., and E. Gaiser. 2012. Benthic diatom assemblages as indicators of water quality in the Everglades and three tropical karstic wetlands. *Freshwater Science* 31: 205–221.

Lemos, M.C., and B.J. Morehouse 2005. The co-production of science and policy in integrated assessments. *Global Environmental Change* 15: 57–68.

Lewis, R.R. 2009. Knowledge overload, wisdom underload. *Ecological Engineering* 35: 341–342.

Liu, J., T. Dietz, S.R. Carpenter, M. Alberti, C. Folke, E. Moran, A. Pell, P. Deadman, T. Kratz, J. Lubchenco, E. Orstrom, Z. Ouyang, W. Provencher, C. Redman, S. Schneider, and W. Taylor. 2007. Complexity of coupled human and natural systems. *Science* 317(5844): 1513–1516.

McCloskey, M. 1966. The Wilderness Act of 1964: its background and meaning. *Oregon Law Review* 45(4): 288–321.

Mendelssohn, I.A., and J.T. Morris. 2000. Ecophysiological controls on the growth of *Spartina alterniflora*. In *Concepts and Controversies in Tidal Marsh Ecology*, edited by N.P. Weinstein and D.A. Kreeger, 59–80. Dordrecht: Kluwer.

Miyajima, T., Y. Tsuboi, Y. Tanaka, and I. Koike. 2009. Export of inorganic carbon from two Southeast Asian mangrove forests to adjacent estuaries as estimated by the stable isotope composition of dissolved inorganic carbon. *Journal of Geophysical Research* 114: G01024.

Noe, G., D.L. Childers, and R.D. Jones. 2001. Phosphorus biogeochemistry and the impacts of phosphorus enrichment: why are the Everglades so unique. *Ecosystems* 4: 603–624.

Noss, R.F. 2011. Between the devil and the deep blue sea: Florida's unenviable position with respect to sea level rise. *Climate Change* 107: 1–16.

Nyman, J.A., R.J. Walters, R.D., DeLaune, and W.H. Patrick. 2006. Marsh vertical accretion via vegetative growth. *Estuarine, Coastal and Shelf Science* 69: 370–380.

Odum, H.T. 1957. Trophic structure and productivity of Silver Springs, Florida. *Ecological Monographs* 27: 55–112.

Ogden, L. 2008. The Everglades ecosystem and the politics of nature. *American Anthropologist* 110(1): 21–32.

Ogden, L. 2011. *Swamplife: People, Gators, and Mangroves Entangled in the Everglades*. Minneapolis: Minnesota Press.

Patterson, K., and R. Finck. 1999. Tree islands of the WCA3 aerial photointerpretation and trend analysis project summary report. SFWMD Report, St. Petersburg, FL.

Pohl, C., S. Rist, A. Zimmermann, P. Fry, G.S. Gurung, F. Schneider, C.I. Speranza, B. Kiteme, S. Boillat, E. Serrano, G.H. Hadorn, and U. Wiesmann. 2010. Researchers' roles in knowledge co-production: experience from sustainability research in Kenya, Switzerland, Bolivia and Nepal. *Science and Public Policy* 37(4): 267–281.

Polsky, C., J.M. Grove, C. Knudson, P.M. Groffman, N. Bettez, J. Cavender-Bares, S.J. Hall, J. B. Heffernan, S.E. Hobbie, K.L. Larson, J.L. Morse, C. Neill, K.C. Nelson, L.A. Ogden, J. O'Neil-Dunne, D.E. Pataki, R. Roy Chowdhury, and M.K. Steele. 2014. Assessing the homogenization of urban land management with an application to US residential lawn care. *Proceedings of the National Academy of Sciences USA* 111(12): 4432–4437.

Price, R.M., M.R. Savabi, J.L. Jolicoeur, and S. Roy. 2010. Adsorption and desorption of phosphate on limestone in experiments simulating seawater intrusion. *Applied Geochemistry* 25: 1085–1091.

Price, R.M., G. Skrzypek, P.F. Grierson, P.K. Swart, and J.W. Fourqurean. 2012. The use of stable isotopes of oxygen and hydrogen in identifying water exchange of in two hypersaline estuaries with different hydrologic regimes. *Marine and Freshwater Research* 63: 952–966.

Price, R.M., P.K. Swart, and J.W. Fourqurean. 2006. Coastal groundwater discharge—an additional source of phosphorus for the oligotrophic wetlands of the Everglades. *Hydrobiologia* 569: 23–36.

Rivera-Monroy, V.H., J.A. Benitez, J. Euan, H. Gonzalez, J. Herrera, M. Perez, V. Reyes, E. Rodriguez, and D. Valdes. 2008. Ecohidrologia y demanda de agua en Mexico. *Ciencia y Desarrollo* 34: 25–29.

Rivera-Monroy, V.H., R. Twilley, D. Bone, D.L. Childers, C. Coronado-Molina, I.C. Feller, J. Herrera-Silveira, R. Jaffe, E. Mancera, E. Rejmankova, J.E. Salisbury and E. Weil. 2004. A conceptual framework to develop long term ecological research and management objectives in the tropical coastal settings of the wider Caribbean region. *BioScience* 54: 843–856.

Rivera-Monroy, V.H., R.R. Twilley, S.E. Davis, D.L. Childers, M. Simard, R. Chambers, R. Jaffe, J.N. Boyer, D. Rudnick, E. Castañeda-Moya, S. Ewe, R.M. Price, C. Coronado-C. Molina, M. Ross, T.J. Smith, B. Michot, W. Nuttle, T. Troxler, and G.B. Noe. 2011. The role of the Everglades Mangrove Ecotone Region (EMER) in regulating nutrient cycling and wetland productivity in South Florida. *Critical Reviews in Environmental Science and Technology* 41: 633–669.

Robinson, J., and J. Tansey. 2006. Co-production, emergent properties and strong interactive social research: the Georgia Basin Futures Project. *Science and Public Policy* 33: 151–160.

Ross, M.S., E.E. Gaiser, J.F. Meeder, and M.T. Lewin. 2002. Multi-taxon analysis of the "White Zone", a common ecotonal feature of South Florida coastal wetlands. In *The Everglades, Florida Bay, and Coral Reefs of the Florida Keys: An Ecosystem Sourcebook*, edited by J.W. Porter and K.G. Porter, 205–238. Boca Raton, FL: CRC Press.

Ross, M.S., S. Mitchell–Bruker, J. Sah, S. Stothoff, P. Ruiz, D. Reed, K. Jayachandran, and C. L. Coultas. 2006. Interaction of hydrology and nutrient limitation in the ridge and slough landscape of the southern Everglades. *Hydrobiologia* 569: 37–59.

Saha, A.K., C.S. Moses, R.M. Price, V. Engel, T.J. Smith, and G. Anderson. 2012. A hydrological budget (2002–2008) for a large subtropical wetland ecosystem indicates marine groundwater discharge accompanies diminished freshwater flow. *Estuaries and Coasts* 35(2): 459–474.

Saha, A.K., S. Saha, J. Sadle, J. Jiang, M.S. Ross, R.M. Price, L. Sternberg, and K.S. Wendelberger. 2011. Sea level rise and South Florida coastal forests. *Climate Change* 107: 81–108.

Schedlbauer, J., S. Oberbauer, G. Starr, and K.L. Jimenez. 2010. Seasonal differences in the CO_2 exchange of a short-hydroperiod Florida Everglades marsh. *Agricultural and Forest Meteorology* 150: 994–1006.

Schellnhuber, H.J., P.J. Crutzen, W.C. Clark, and J. Hunt. 2005. Earth systems analysis for sustainability. *Environment* 47: 10.

Schlesinger, W.H., and J.M. Melack. 1981. Transport of organic carbon in the world's rivers. *Tellus* 33:172–187.

Simard, M., V.H. Rivera-Monroy, J.E. Mancera-Pineda, E. Castañeda-Moya, and R.R. Twilley. 2008. A systematic method for 3D mapping of mangrove forests based on shuttle radar topography mission elevation data, ICEsat/GLAS waveforms and field data: application to Cienaga Grande de Santa Marta, Colombia. *Remote Sensing of Environment* 112: 2131–2144.

Sklar, F.H., M.J. Chimney, S. Newman, P. McCormick, D.W. Gawlik, S. Miao, C. McVoy, W. Said, J. Newman, C. Coronado, G. Crozier, M. Korvela, and K. Rutchey. 2005. The ecological-societal underpinnings of Everglades restoration. *Frontiers in Ecology and the Environment* 3: 161–169.

Smith, T.J., G. Anderson, K. Balentine, G. Tiling, G.A. Ward, and K. Whelan. 2009. Cumulative impacts of hurricanes on Florida mangrove ecosystems: sediment deposition, storm surges and vegetation. *Wetlands* 29: 24–34.

Smoak, J.M., J.L. Breithaupt, T.J. Smith III, and C.J. Sanders. 2013. Sediment accretion and organic carbon burial relative to sea-level rise and storm events in two mangrove forests in Everglades National Park. *Catena* 104: 58–66.

Stalker, J.C., R.M. Price, V.H. Rivera-Monroy, J. Herrera-Silveira, S. Morales, J.A. Benitez, and D. Alonzo-Parra. 2014. Hydrologic dynamics of a subtropical estuary using geochemical tracers, Celestún, Yucatán, Mexico. *Estuaries and Coasts* 37(6): 1376–1387.

Suarez-Abelenda, M., T.O. Ferreira, M. Camps-Arbestain, V.H. Rivera-Monroy, F. Macias, G.N. Nobrega, and X.L. Otero. 2014. The effect of nutrient-rich effluents from shrimp farming on mangrove soil carbon storage and geochemistry under semi-arid climate conditions in northern Brazil. *Geoderma* 213: 551–559.

Sullivan, P., E. Gaiser, D. Surrat, D. Rudnick, S. Davis, and F. Sklar. 2014. Wetland ecosystem response to hydrologic restoration and management: The Everglades and its urban-agricultural boundary (FL, USA). *Wetlands* 34: S1–S8.

Teal, J. 1962. Energy flow in the salt marsh ecosystem of Georgia. *Ecology* 43: 614–624.

Thomas, S., E. Gaiser, M. Gantar, and L. Scinto. 2006. Quantifying the responses of calcareous periphyton crusts to rehydration: a microcosm study (Florida Everglades). *Aquatic Botany* 84: 317–323.

Trexler, J., E. Gaiser, and J. Kominoski. 2015. Edibility and periphyton food webs, specific indicators. In *Microbiology of the Everglades Ecosystem*, edited by J. Entry, K. Jayachandrahan, A. Gottlieb, and A. Ogram, 155–179. Boca Raton, FL: Science Publishers.

Troxler, T., E. Gaiser, J. Barr, J. Fuentes, R. Jaffe, D. Childers, L. Collado-Vides, V. Rivera-Monroy, E. Castañeda-Moya, W. Anderson, R. Chambers, M. Chen, C. Coronado-Molina, S. Davis, V. Engel, C. Fitz, J. Fourqurean, T. Frankovich, J. Kominoski, C. Madden, S. Malone, S. Oberbauer, P. Olivas, J. Richards, C. Saunders, J. Schedlbauer, F. Sklar, T. Smith, J. Smoak, G. Starr, R. Twilley, and K. Whelan. 2013. Integrated carbon budget models for the Everglades terrestrial-coastal-oceanic gradient: current status and needs for inter-site comparisons. *Oceanography* 26: 98–107.

Turner, A.M., J.C. Trexler, C.F. Jordan, S.J. Slack, P. Geddes, J.H. Chick, and W.F. Loftus. 1999. Targeting ecosystem features for conservation: standing crops in the Florida Everglades. *Conservation Biology* 13(4): 898–911.

Troxler, T., D.L. Childers, and C.J. Madden. 2014. Drivers of decadal-scale change in southern Everglades wetland macrophyte communities of the coastal ecotone. *Wetlands* 34(Suppl 1): 81–90.

Twilley, R.R., and V.H. Rivera-Monroy. 2005. Developing performance measures of mangrove wetlands using simulation models of hydrology, nutrient biogeochemistry, and community dynamics. *Journal of Coastal Research* 40: 79–93.

Uchida, E., V.H. Rivera-Monroy, A. Gold, and H. Uchida. 2014. US-Tanzania: Building a Research Collaboration on the Dynamics of Mangrove Ecosystem Services and Poverty Traps. Final Report to the NSF Catalyzing International Collaborations program. National Science Foundation, Rhode Island, USA.

Voss, C.M., R.R. Christian, and J.T. Morris. 2013. Marsh macrophyte responses to inundation anticipate impacts of sea-level rise and indicate ongoing drowning of North Carolina marshes. *Marine Biology* 160: 181–194.

Weston, N.B., R.E. Dixon, and S.B Joye. 2006. Ramifications of increased salinity in tidal freshwater sediments: geochemistry and microbial pathways of organic matter mineralization. *Journal of Geophysical Research: Biogeosciences* 111(G1):14.

Weston, N.B., M.A. Vile, S.C. Neubauer, and D.J. Velinsky. 2011. Accelerated microbial organic matter mineralization following salt-water intrusion into tidal freshwater marsh soils. *Biogeochemistry* 102: 135–151.

Wetzel, P.R., F.H. Sklar, C.A. Coronado, T.G. Troxler, S.L. Krupa, P.L. Sullivan, S. Ewe, R. Prize, S. Newman, and W.H. Orem. 2011. Biogeochemical processes on tree islands in the Greater Everglades: initiating a new paradigm. *Critical Reviews in Environmental Science and Technology* 41(S1): 670–701.

Wetzel, P. R., A.G. van der Valk, S. Newman, D.E. Gawlik, T.T. Gann, C.A. Coronado-Molina, D.L. Childers, and F.H. Sklar. 2005. Maintaining tree islands in the Florida Everglades: nutrient redistribution is the key. *Frontiers in Ecology and the Environment* 3: 370–376.

Williams, A.J., and J.C. Trexler. 2006. A preliminary analysis of the correlation of food-web characteristics with hydrology and nutrient gradients in the southern Everglades. *Hydrobiologia* 569(1): 493–504.

Williams, R. 1973. *The Country and the City*. New York: Oxford University Press.

Zapata-Rios, X., and R.M. Price. 2012. Estimates of groundwater discharge to a coastal wetland using multiple techniques: Taylor Slough, Everglades National Park. *Hydrogeology Journal* 20: 1651–1668.

Zhang, K., J. Dittmar, M. Ross, and C. Bergh. 2011. Assessment of sea level rise impacts on human population and real property in the Florida Keys. *Climate Change* 107: 129–146.

Zhang, K., M. Simard, M.S. Ross, V.H. Rivera-Monroy, P. Houle, P.L. Ruiz and R.R. Twilley, and K. Whelan. 2008. Airborne laser scanning quantification of disturbances from hurricanes and lightning strikes to mangrove forests in Everglades National Park, USA. *Sensors* 8: 2262–2292.

Acknowledgments

The Florida Coastal Everglades Long Term Ecological Research Program (FCE LTER) has been funded by the National Science Foundation since its inception in 2000 (Grants No. DEB-1832229, DEB-1237517, DBI-0620409, and DEB-9910514). Additional funding for FCE-related research and activities has been generously provided by Everglades National Park, the South Florida Water Management District, the U.S. Geological Survey, the National Oceanic and Atmospheric Administration, the U.S. Department of Energy, Florida Seagrant, and others.

Commonly Used Acronyms and Abbreviations

Al	Aluminum
ALT4R	Central Everglades Planning Process restoration plan
AMO	Atlantic Multi-decadal Oscillation
BMP	Best management practices
BOD	Biological oxygen demand
C	Carbon
C&SF	Central & Southern Florida Flood Control Project
Ca	Calcium
CDOM	Chromophoric dissolved organic matter
CEPP	Central Everglades Planning Process
CERP	Comprehensive Everglades Restoration Plan
CISRERP	Committee of Independent Scientific Review of Everglades Restoration Progress
CSSS	Cape Sable seaside sparrow (*Ammodramus maritimus mirabilis*)
DCI	Directional connectivity index
DIC	Dissolved inorganic carbon
DIP	Dissolved inorganic phosphorus
DMSTA	Stormwater treatment model that tracked phosphorus and water deliveries to the water conservation areas
DOC	Dissolved organic carbon

DOM	Dissolved organic matter
DON	Dissolved organic nitrogen
DOP	Dissolved organic phosphorus
EAA	Everglades Agricultural Area
EDEN	Everglades Depth Estimation Network
EEM	Excitation emission matrix
EEM-PARAFAC	Excitation emission matrix (EEM) fluorescence coupled with parallel factor analysis (PARAFAC)
EFA	Everglades Forever Act
ENP	Everglades National Park
ENSO	El Niño-Southern Oscillation
EPA	Environmental Protection Agency
ER	Ecosystem respiration
ERTP 2012	Everglades Restoration Transition Plan
ET	Evapotranspiration
FAS	Florida Aquifer System
FCE	Florida Coastal Everglades
FCE LTER	Florida Coastal Everglades Long Term Ecological Research Program
FDEP	Florida Department of Environmental Protection
Fe	Iron
Floc	Flocculent organic particles
FWO	Future Without Restoration
GEE	Gross ecosystem exchange
GPP	Gross primary productivity
HSI	Habitat suitability index
IP	Inorganic phosphorus
K	Gas transfer velocity
LEC	Lower East Coast
LECSA	Lower East Coast Service Area
LNWR	Arthur R. Marshall Loxahatchee National Wildlife Refuge
LORS 2008	Lake Okeechobee Regulation Schedule
LTER	Long Term Ecological Research
LUE	Light-use efficiency
N	Nitrogen
NAO	North Atlantic Oscillation
NECB	Net ecosystem C balance
NEE	Net ecosystem exchange
NEP	Net ecosystem production
NPCA	National Parks and Conservation Association

NPP	Net primary productivity
P	Phosphorus
PARAFAC	Parallel factor analysis
PDO	Pacific Decadal Oscillation
PIP	Particulate inorganic phosphorus
POC	Particulate organic carbon
POP	Particulate organic phosphorus
PP	Particulate phosphorus
R_E	Ecosystem respiration rates
RECOVER	REstoration, COordination and VERification
RSM	Regional simulation model
RSMBN	Basin model that tracked the water balance in Lake Okeechobee
RSMGL	Regional Simulation Model Glades-LECSA version
S-12A, S-12B, S-12C, S-12D	Water control structures located along the northern boundary of Everglades National Park
SAV	Submerged aquatic vegetation
SEACOM	A dynamic seagrass ecosystem simulation model
SFWMD	South Florida Water Management District
SFWMM	South Florida Water Management Model
SRP	Soluble reactive phosphorus
SRS	Shark River Slough
STA	Stormwater treatment areas
TN	Total nitrogen
TP	Total phosphorus
TS/PH	Taylor Slough-Panhandle
UNESCO	United Nations Educational, Scientific and Cultural Organization
USACE	U.S. Army Corps of Engineers
USFWS	U.S. Fish & Wildlife Service
USSC	U.S. Sugar Corporation
WADEM	Wading bird distribution evaluation models
WCA	Water Conservation Area
WCA-3A	Water Conservation Area 3A
WCA-3B	Water Conservation Area 3B

Index

For the benefit of digital users, indexed terms that span two pages (e.g., 52–53) may, on occasion, appear on only one of those pages.

Page numbers followed by *t*, *f*, or *b* denote tables, figures, and boxes respectively.